AF694501

AN INTRODUCTION TO LIVING CELLS:

THE BUILDING BLOCKS OF LIFE

AN INTRODUCTION TO LIVING CELLS:
THE BUILDING BLOCKS OF LIFE

Prof. Ramesh Chandra and Dr. Neha Gopal

Delve PUBLISHING

www.delvepublishing.com

An Introduction to Living Cells: The Building Blocks of Life

Prof. Ramesh Chandra and Dr. Neha Gopal

Delve Publishing
224 Shoreacres Road
Burlington, ON L7L 2H2
Canada
www.delvepublishing.com
Email: orders@arclereducation.com

© 2024 Delve Publishing

ISBN: 978-1-77469-932-4 (Hardcover)

This book contains information obtained from highly regarded resources. Reprinted material sources are indicated and copyright remains with the original owners. Copyright for images and other graphics remains with the original owners as indicated. A Wide variety of references are listed. Reasonable efforts have been made to publish reliable data. Authors or Editors or Publishers are not responsible for the accuracy of the information in the published chapters or consequences of their use. The publisher assumes no responsibility for any damage or grievance to the persons or property arising out of the use of any materials, instructions, methods or thoughts in the book. The authors or editors and the publisher have attempted to trace the copyright holders of all material reproduced in this publication and apologize to copyright holders if permission has not been obtained. If any copyright holder has not been acknowledged, please write to us so we may rectify.

Notice: Registered trademark of products or corporate names are used only for explanation and identification without intent of infringement.

Delve Publishing publishes wide variety of books and eBooks. For more information about Delve Publishing and its products, visit our website at www.delvepublishing.com.

ABOUT THE AUTHORS

Prof. Ramesh Chandra, Fellow of the Royal Society of Chemistry, London is a distinguished scientist and a pioneer leader in the field of Biomedical Sciences. He is known for his appetite for academic progress and vision for establishing state-of-the-art Institutions. He has received global recognition for his work in academia and research in Biomedical Sciences and Medicinal Chemistry. Apart from academia, he has been actively involved in other areas of development too. He is official resource person to the Minister and Department of Science and Technology in the kingdom of Lesotho, South Africa as well as President of the Indian Chemical Society (2004-2006), Member, Planning Commission, Government of U.P. India; Secretary - Zaheer Science Foundation, New Delhi and Non-Official Director, Rashtriya Ispat Nigam Ltd (RINL), Government of India. On 08th March, 2022, Prof. Ramesh Chandra assumed the Vice-Chancellorship of Maharaja Surajmal Brij University, Bharatpur, Rajasthan. Evidently, under his charge the University witnessed inception of new administrative structures and research facilities to the introduction of innovative management concepts, interdisciplinary courses, and global academic initiatives. Prof. Chandra has been Vice-Chancellor, Bundelkhand University, Jhansi for six years (1999-2005). He established 36 institutes and that, too, with minimal financial support of the State Government. These institutes are in the fields Pure and Applied Radiation Biology, Pharmacy, Criminology and Forensic Sciences, Mass Communication & Journalism, Law, Ayurveda and Alternate Medicine, Planning and Architecture, Computer Science, Engineering, Biotechnology, Hotel Management and others. In 1991, Prof. Chandra, established Dr. B. R. Ambedkar Center for Biomedical Research (ACBR) – pioneering Biomedical Education and Research Institute at Delhi University, which has acquired an international status in the field of Biomedical Research through its original and creative research. As an advisor, he has been instrumental in various academic developments of Indian Institute of Information Technology (IIIT - Allahabad). Prof. Chandra started his research career at the University of Delhi; thereafter he went to pursue his research at The New York Hospital-Cornell University Medical College, New York jointly at The Rockefeller University, New York; State University of New York at Stonybrook as Assistant Research Professor. He conducted advanced research at the Harvard University Medical School-Massachusetts General Hospital (MGH) jointly at MIT, Cambridge, USA. Prof. Chandra was awarded Capital Foundation Dr A.P.J. Abdul Kalam Award for outstanding Scientific Research and Administrative Qualities (2023) in recognition of his duties for more than three decades. An outstanding Scientist and revered teacher, Prof. Chandra has been the recipient of several professional national/ international recognitions too: J. William Fulbright Scholarship (1993); The Rockefeller Foundation USA-Biotechnology Career Award (1993); Award of the Highest Honor of Soka University, Tokyo, Japan (2000); Millennium Plaques of Honor (Life Time Achievement Award for contribution in Science & Technology) by the Indian Science Congress Association (ISCA) {2017-2018}, Lifetime Achievement Award of the Indian Chemical Society (ICS) (2003); UGC Research Scientist Award (1988); UGC Career Award (1993); Rajib Goyal Award for Young Scientists (2002); Fellowship of: Indian Academy of Biomedical Sciences (IABS); International Academy of Physical Sciences; Institution of Chemists, India; Indian Chemical Society and others.

Dr. Neha Gopal is a Biotechnologist and a nanotechnology researcher. Currently, she is working as an Assistant Professor in Maharaja Surajmal Brij University, Bharatpur, Rajasthan. She is also involved with the administration by working as Deputy Dean, Self Financing Scheme (SFS) Board and Deputy Dean, Women's Cell. Dr. Neha did her Bachlor in Science (B.Sc.) from Deen Dayal Upadhyay Gorakhpur University, Gorakhpur, Uttar Pradesh with subjects Industrial Microbiology, Botany and Zoology and Master of Science at Banasthali University, Jaipur, Rajasthan, India in Applied Microbiology and Biotechnology. Dr. Neha holds her Doctorate Degree from Amity University, Noida, India on the topic entitled "Development of Electrochemical Biosensor for diagnosis of Liver Cancer", awarded in the year 2022. After submission of her Ph.D. thesis, she joined Department of Chemistry, University of Delhi as Senior Project Associate and worked on the project entitled "Electrochemical ultra-sensing platform integrated with biomarker coated on graphene detecting neonatal sepsis" under the supervision of Prof. Ramesh Chandra, funded by Department of Biotechnology (DBT), Government of India. Dr. Neha also has six months training experience at Fortis Escort Heart Institute, Okhla, New Delhi, India under guide Dr. Anita Arora on the title "Microbiological profile of lower respiratory tract samples" with studies on occurrence and sensitivity patterns of various microorganisms in 2012. Dr. Neha holds expertise in fabricating nanomaterial and polymer based thin films, development of composites, bioelectrodes and biosensing platforms for the diagnosis of diseases. She has developed electrochemical biosensors for the diagnosis of diseases such as diabetes, hepatocellular carcinoma and neonatal sepsis. Her current research is focused on the development of nanomaterial based electrochemical biosensors for the detection of analytes and diagnosis of diseases. She has experience in performing nano-technological scientific research, using various characterization methods including DIY enzyme-linked immunosorbent assay

(ELISA), Fourier transform infrared (FTIR) spectroscopy, Scanning electron microscopy (SEM), Transmission electron microscopy (TEM), Atomic Force Microscopy (AFM), Thermogravimetric analysis (TGA), Raman spectroscopy, Cyclic Voltammetry (CV), Electrochemical Impedance Spectroscopy (EIS) and other electrochemical measurements. She has experience in maintaining bacterial culture, performing biochemical tests, staining, and inferring results of microbiological experiments. She has knowledge of molecular biology techniques such as gel electrophoresis, Polymerase Chain reaction (PCR), designing primers for amplification of nucleic acids, maintaining impeccable standards in managing laboratory protocols and work ethics, safely disposing of waste and accurately calibrating, handling and maintaining laboratory instruments. Dr. Neha is the author of multiple research and review papers published by high impact national and international journals such as Artificial Cells, Nanomedicine and Biotechnology (Taylor & Francis), Analytical Methods, Royal (Royal Society of Chemistry), Fullerenes, Nanotubes and Carbon Nanostructure (Taylor & Francis), Asian Journal of Chemistry and Advances in Polymer Sciences and Technology. She is co-inventor of a patent entitled "Flexible CNT polymer nanocomposite as a universal electrochemical platform for enzyme based biosensor" filed in 2020 with applicants Amity University, Noida, Uttar Pradesh and Ministry of Science and Technology, Department of Biotechnology.

TABLE OF CONTENTS

LIST OF FIGURES

PREFACE

Welcome to "Introduction to Living Cells: The Building Blocks of Life." In the pages of this book, we delve into the captivating world of living cells, the fundamental units of life that serve as the foundation for all organisms. From the simplest unicellular organisms to the complex multicellular organisms that make up our diverse ecosystems, these tiny entities hold the key to understanding life's essence. Throughout our exploration, we aim to foster a deeper understanding of cellular biology, highlighting its historical discoveries, contemporary insights, and future implications. We begin with a reflection on the history of cell biology, acknowledging the contributions of early pioneers like Robert Hooke, Theodor Schwann, Matthias Schleiden, and Rudolf Virchow, whose work laid the groundwork for the cell theory. The development of microscopy and its advancements in cell observation also form an essential part of our journey.

Cellular diversity plays a pivotal role in shaping life's intricacies. We examine both unicellular and multicellular organisms, contrasting prokaryotic and eukaryotic cells and delving into the functions of membrane-bound organelles that define the latter. Cell structure and function come into focus as we explore the roles of various organelles, such as the plasma membrane, nucleus, endoplasmic reticulum, Golgi apparatus, mitochondria, and chloroplasts. Understanding the interplay of these components in cellular processes is vital in comprehending life's underlying mechanisms.

Transport mechanisms form another crucial aspect of cell biology. We unravel the mysteries of passive and active transport, endocytosis, exocytosis, cell communication, and cell adhesion, which are vital for maintaining cellular integrity and communication within the organism. Cell division and genetics form the bedrock of life's continuity. We explore the intricacies of both mitosis and meiosis and delve into the structure of DNA, replication, transcription, translation, and Mendel's laws of inheritance. Cellular metabolism provides the energy necessary for life-sustaining processes. We delve into cellular respiration, photosynthesis, metabolic pathways, and the role of enzymes as catalysts for cellular reactions.

As we examine cell diversity and specialization, we encounter stem cells, muscle cells, nerve cells, and specialized plant cells. Understanding these diverse cell types is essential in comprehending the functioning of multicellular organisms. Cell reproduction and development are vital processes in the maintenance and growth of living organisms. We explore fertilization, embryonic development, stem cells, regeneration, aging, and cancer biology, as they all contribute to the intricate fabric of life.

This book is designed to cater to a wide range of audiences. Undergraduate students pursuing degrees in biology, life sciences, biochemistry, or related fields will find it to be an indispensable companion. Practitioners in various medical and scientific disciplines will benefit from its comprehensive insights. Policy makers and government officials concerned with health, biotechnology, and environmental issues will find it valuable in making informed decisions. Moreover, the book aims to foster scientific literacy among the general public, enhancing their appreciation for the wonders of living cells.

In conclusion, we express our heartfelt wish that readers embark on an exciting journey through the pages of "Introduction to Living Cells: The Building Blocks of Life." We hope that the exploration of cell biology, encompassing historical milestones, contemporary insights, and future implications, will leave readers with a newfound appreciation for the remarkable complexity of life at the cellular level. Whether you are an undergraduate student, a practitioner, a policy maker, or a curious individual, we believe that this book will spark your curiosity and fascination for the intricate world of living cells. Enjoy the exploration!

Chapter

1 THE CELL THEORY AND HISTORICAL DISCOVERIES

CONTENTS

Welcome to the first chapter of "Introduction to Living Cells: The Building Blocks of Life." In this chapter, we will explore the cell theory and historical discoveries that have shaped our understanding of the fundamental unit of life – the cell. We will delve into the contributions of visionaries such as Robert Hooke, Matthias Schleiden, Theodor Schwann, and Rudolf Virchow, who laid the foundation of cell theory. Their work led to the proposition that all living organisms consist of one or more cells and that cells are the basic units of life. Furthermore, we will examine the pivotal role of microscopes in the advancement of cell biology. These powerful instruments allowed scientists to observe cellular structures in greater detail, uncovering the mysteries of life at the cellular level. The evolution of microscopy, from simple magnifying lenses to sophisticated electron microscopes, has had a profound impact on our understanding of cellular processes.

In addition, we will explore the diversity of cells between unicellular and multicellular organisms. Unicellular organisms, such as bacteria and protozoa, consist of a single cell that performs all necessary functions for survival.

On the other hand, multicellular organisms, including plants, animals, and fungi, are composed of numerous specialized cells that collaborate to form complex tissues and organ systems. Understanding this distinction is crucial in comprehending the complexities of life forms. Another aspect we will delve into is prokaryotic cells. These ancient organisms lack a defined nucleus and membrane-bound organelles, setting them apart from eukaryotic cells. We will explore their structure and functions and highlight their essential roles in various ecological processes, such as nutrient cycling and decomposition.

Finally, we will examine eukaryotic cells, the basis of all complex life forms. These cells have a well-defined nucleus and various membrane-bound organelles, each contributing to specific cellular processes. Understanding the structure and functions of eukaryotic cells is fundamental to comprehending the intricacies of life and the interconnectedness of cellular processes. By exploring these key themes in the first chapter, we aim to provide readers with a solid foundation in cell biology. The historical context, advancements in microscopy, understanding of cell diversity, and distinctions between prokaryotic and eukaryotic cells will pave the way for a comprehensive exploration of living cells – the building blocks of life – in the subsequent chapters of this book.

1.1 THE FOUNDATION OF CELL THEORY: CONTRIBUTIONS OF HOOKE, SCHLEIDEN, SCHWANN, AND VIRCHOW

The foundation of cell theory represents a crucial milestone in the history of biology, shaping our understanding of life at its most fundamental level (Hummert et al., 2021; Husnik et al., 2021; Hudson & Wieland, 2023). This theory proposes that all living organisms are composed of one or more cells and that cells are the basic units of life. The formulation of this theory was a result of the groundbreaking contributions of four visionary scientists – Robert Hooke, Matthias Schleiden, Theodor Schwann, and Rudolf Virchow (see figure 1). The foundation of cell theory, indeed, marks a critical juncture in the history of biology, effectively revolutionizing our comprehension of life at its most rudimentary form. To elucidate this essential milestone, a thorough exploration of the discovery and gradual development of cell theory provides an appropriate case study. Instrumental to shaping our initial knowledge of cells were the pioneering works of scientists like Robert Hooke and Antonie van Leeuwenhoek in the 17th century.

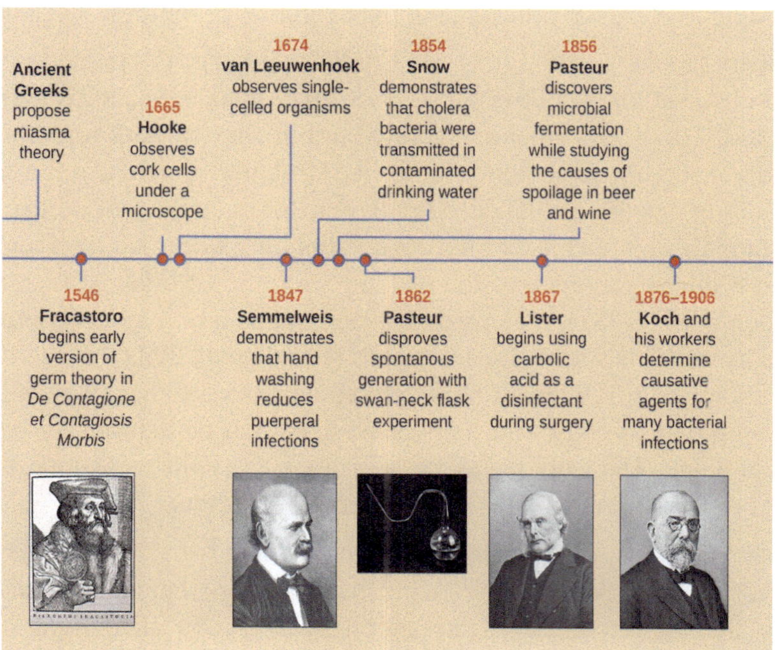

Figure 1. Foundations of modern cell theory.

Source: Lumen Learning (https://ecampusontario.pressbooks.pub/microbio/chapter/foundations-of-modern-cell-theory/)

Leeuwenhoek's development of an advanced microscope and Hooke's consequent observation of cork cells illuminated the role of cells as life's primary building blocks, planting the seeds of subsequent discoveries (Critcher & Huang, 2022; Cui et al., 2022; Cristiani et al., 2023). However, it was only in the 19th century, with the works of Theodor Schwann and Mathias Schleiden, that the true essence of cellular theory gained prominence. Schleiden's momentous declaration that plants constitute units of individual cells marked the beginning of a new era in biology, a sentiment echoed and expanded by Schwann, who avowed that animals were also made of cells. This understanding revolutionized biology by shifting the field's focus from organism-level study to cell-level analysis. Consequently, this shift facilitated an unprecedented investigation into the minute intricacies of life and the discovery of intriguing concepts such as cellular division, multiplication, specialization, and death, culminating in an extensive and nuanced knowledge of the principles that govern life. In retrospect, through augmenting our comprehension of life's basic unit, cell theory has fashioned the way for the emergence of new disciplines in biology such as cellular biology, genetic engineering, and molecular biology, thereby substantiating its standing as a crucial milestone in the evolution of biology.

Robert Hooke (1635-1703), an English scientist, is credited with the earliest observation of cells (Panieri et al., 2022; Park et al., 2022; Panstruga et al., 2023). In 1665, Hooke examined a thin slice of cork under a crude microscope and noticed a honeycomb-like structure, which he referred to as "cells" due to their resemblance to small rooms. Although Hooke did not fully understand the significance of his observation, his work laid the foundation for future investigations into the cellular structure of organisms. Matthias Schleiden (1804-1881), a German botanist, made a pivotal contribution to cell theory with his study of plant tissues (Nieto et al., 2022; Nygaard et al., 2021; Nyberg et al., 2023). In 1838, Schleiden proposed that all plant tissues were composed of cells, and he hypothesized that cells were the building blocks of plants. His work led to the realization that cells were not exclusive to cork but were a fundamental component of all plants.

Robert Hooke, an eminent English scientist born in the year 1635 and died in 1703, holds the distinguished honor of being recognized for making the earliest known observation of cells. To shed light on this significant scientific discovery, a comprehensive examination of Hooke's groundbreaking experiment conducted in the mid-17th century serves to validate this claim. Well-established for his multifaceted interests spanning physics, architecture, and biology, Hooke is primarily celebrated for his scientific endeavor documented in "Micrographia," a pioneer study viewed as a seminal scientific text in early microscopy. It was in this work that Hooke detailed his examination of thin slices of cork through the lens of a rudimentary compound microscope, a device that was an exciting novelty of his era. His meticulous observations led him towards a critical revelation; he perceived through the lens tiny, empty spaces contained by walls, resembling a honeycomb structure. Hooke, drawing from the architectural terminology of his time, referred to these spaces as 'cells,' which interestingly, in this context, bore a connotation of 'small rooms'. Though Hooke wrongly assumed these cells to be nonliving components due to the cork's nature as a plant tissue long devoid of life, his identification of cells as fundamental structural units was a pioneering insight that laid the groundwork for the development of cell theory. These observations by Hooke paved the way for future scientists like Anton Van Leeuwenhoek, who, following Hooke's trail, went a step further to observe living cells in the natural world. Albeit his misconception about the nature of cells, Hooke's initial observations were the beacon that illuminated a path towards a granular understanding of living organisms, thereby establishing Hooke as a significant force in the genesis of cell biology.

Theodor Schwann (1810-1882), a German physiologist, extended Schleiden's observations to animals (Ostrop et al., 2021; Ovciarikova et al., 2022; Ouyang et al., 2023). In 1839, Schwann proposed that all animal tissues were also composed of cells, and he further emphasized the parallel between plant and animal cells. Schwann's insights marked a significant step in unifying plant and animal biology under the common framework of cell theory. Theodor Schwann, a German physiologist born in the year 1810 and died in 1882, earned his renowned place in the annals of biological sciences by broadening Schleiden's findings onto the domain of animals. Known as a significant player in biological microscopy, Schwann's exploration began with his interests in examining different animal tissues. His inquiries found the impetus they required through his association and consultations with his fellow German scientist, Mathias Jakob Schleiden. Schleiden, a botanist, had earlier stipulated the fundamental role of cells in the constitution of plants, arguing, on the basis of observational evidence, that each plant consists of individual cells. Schwann, intrigued by Schleiden's insights, felt the need to determine if such observations might apply equally well to animals. Hence, he undertook rigorous experiments and meticulous examinations of animal tissues. His observations, interestingly, echoed Schleiden's claims, but with respect to animals. Schwann found that animal tissues were also, in fact, composed of cells. Thus, his findings, known as Schwann's cell doctrine, imparted credibility to the view that all living organisms, be they plants or animals, consisted of cells. This fundamental understanding not only substantiated the universality of cell theory but also marked a transformational moment in the annals of biological sciences. Schwann's extension of Schleiden's observations to the animal kingdom formed the bedrock of present cellular biology, reinforcing the cell as the basic unit of life, and therefore, underscoring the vital role both scientists played in the formation and establishment of cell theory.

Rudolf Virchow (1821-1902), a German pathologist, provided the final crucial contribution to cell theory in 1855 (Wolin, 2021; Wong et al., 2021; Wood et al., 2021; Woods & Gladfelter, 2021). Virchow's famous aphorism "omniscellula e cellula," meaning "every cell arises from a pre-existing cell," emphasized the continuity of life through cell division. This concept challenged the prevailing idea of spontaneous generation and solidified the notion that cells were the basic units of life and that they could only arise from pre-existing cells. Rudolf Virchow, a German pathologist whose lifetime spanned from 1821 to 1902, is esteemed for having rendered the final crucial contribution to cell theory in 1855. A central figure in the medical reformation in the 19th century, Virchow not only contributed across diverse disciplines - from medical pathology to anthropology - but also left a lasting impact on the foundation of cell biology. His profound influence rests primarily on his articulation of an epoch-making observation in 1855; his dictum "Omnis cellula e cellula," translated as "every cell originates from another existing cell," provided the cornerstone of what is now recognized as the tenet of modern cell theory. Virchow's assertion, which challenged the prevailing belief in spontaneous generation, was a result of his rigorous studies on cellular pathology. His studies of diseased tissues led him to conclude that diseases do not strike at a whole organ or an individual at large, rather they begin at the cellular level. In cells undergoing division, Virchow observed that every new cell was the product of the division of a pre-existing cell. This powerful idea, thus, ended the notion that cells can arise out of nothing, thereby providing the third and crucial postulate to cell theory. The affirmation of this premise revolutionized the biological sciences, the understanding of diseases, and the approach to medical treatment, making it a critical milestone in the annals of medical history and biology. By affirming the concept of cellular pathology, Virchow greatly influenced medical science, leading to significant advancements in the designing of treatments aimed at cellular management.

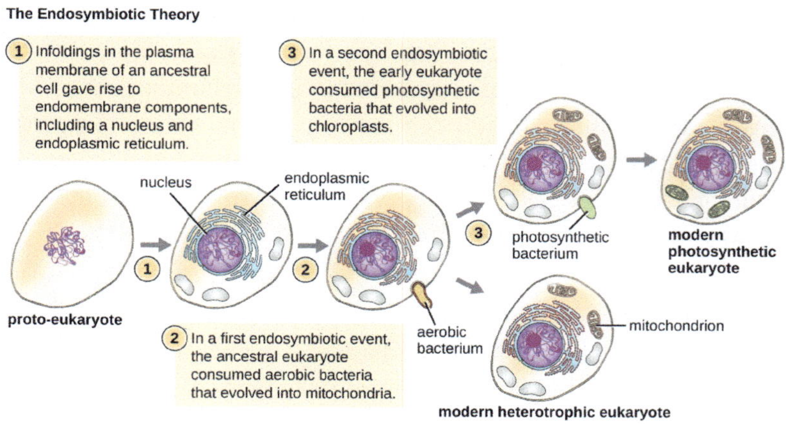

Figure 2. The endosymbiotic theory.

Source: Lumen Learning https://courses.lumenlearning.com/suny-microbiology/chapter/foundations-of-modern-cell-theory/

One of the outcomes was the development and application of the endosymbiotic theory (see figure 2). The development and application of the endosymbiotic theory, also known as endosymbiosis theory, qualifies as a significant evolutionary mechanism and encompasses a fascinating chapter in the field of biology (Lanz et al., 2021; Larson & McFarlane, 2021; Le Goff et al., 2022). This theory, pioneered by American biologist Lynn Margulis in the 1960s, hypothesizes that several key organelles of eukaryotes originated as symbioses between separate single-celled organisms. Margulis's conception of endosymbiosis drew upon earlier ideas posited by Russian botanist Konstantin Mereschkowski in the early twentieth century but brought forth substantial observational and experimental evidence that elevated the hypothesis to a widely accepted theory. The crux of the endosymbiotic theory revolves around the idea that mitochondria and plastids such as chloroplasts, two integral organelles of present-day eukaryotic cells, were once independent bacteria that were engulfed by another, larger prokaryotic cell. Over time, as Margulis argued, these entities ceased to be separate and became part and parcel of the host cell, eventually evolving into organelles. This theory is substantiated by several compelling lines of evidence such as the self-replication of mitochondria and chloroplasts, their possession of a unique DNA separate from the nucleus, and their resemblance to extant bacterial species in terms of size and structure. The application of the endosymbiotic theory has shaped the understanding of eukaryotic cell evolution, offering a convincing explanation for the complexity of these cells. Furthermore, its implications extend to understanding genetic transfer mechanisms, the origins of multicellularity, and the adaptation capacity of organisms. By providing new perspectives on evolutionary biology and cell advancement, the endosymbiotic theory affirmatively underscores the power of symbiotic relationships in the genesis and development of complex life forms.

The combined work of Hooke, Schleiden, Schwann, and Virchow established the cell theory as a fundamental principle in biology. Their contributions revolutionized the way scientists perceived living organisms, shifting the focus from the macroscopic view of whole organisms to the microscopic level of cells. The cell theory provided a unifying framework that connected diverse organisms and laid the groundwork for modern cell biology. Moreover, the cell theory had far-reaching implications beyond biology. It influenced various scientific fields and even impacted medical practices (Donato et al., 2021; Don & Kalebic, 2022; Domcke & Shendure, 2023). Understanding the cellular basis of life has been instrumental in the development of medical treatments, disease diagnosis, and advancements in biotechnology. Furthermore, the cell theory has fostered the exploration of cells' intricate structure and functions, leading to numerous discoveries in cell biology (Bonilha, 2022; Boukhatmi, 2021; Boumard & Bardin, 2021). Advancements in microscopy and other scientific techniques have allowed scientists to peer into the inner workings of cells, uncovering the complexities of cellular processes. In conclusion, the foundation of cell theory owes its existence to the remarkable contributions of Robert Hooke, Matthias Schleiden, Theodor Schwann, and Rudolf Virchow. Their pioneering work transformed our understanding of life, introducing the concept of cells as the building blocks of living organisms. The cell theory remains a cornerstone of modern biology and continues to be a source of inspiration for scientists, as they unravel the mysteries of life at the cellular level. The second section of this chapter will consider advancements in cell biology as far as they pertain to microscopes and cell observations.

1.2 MICROSCOPES AND CELL OBSERVATION: ADVANCEMENTS IN CELL BIOLOGY

Microscopes and cell observation have been the driving force behind the remarkable progress in the field of cell biology (Efstathiou & Draviam, 2021;

Duncan, 2022, 2023; Dunkley et al., 2022). Advancements in microscopy have allowed scientists to explore the intricacies of living cells, revealing a world of complexity and wonder that was once hidden from human eyes. The development of microscopes and their continuous improvement over the years have revolutionized our understanding of cellular structures, functions, and processes. The inception and progressive advancements in microscopy and cell observation have undeniably served as the impetus behind the awe-inspiring strides made in the field of cell biology. Tracing the historical framework, the origins of cell biology can be intricately linked to the invention and enhancement of the microscope reciprocally amplifying the progress made in both fields. Emblematic of this is the pioneering work of Robert Hooke in the 17th century, who, with the aid of a rudimentary compound microscope, made the monumental discovery of cells in the structure of cork.

Hooke's observations, although limited by the constraints of prevailing microscope designs, marked the preliminary understanding of cell existence. Following Hooke, Anton Van Leeuwenhoek, equipped with more advanced optical microscopes, managed to observe living cells, thereby taking the exploration a step further.

The advancements in microscopes continued symbiotically with the intricacies unraveled in cell biology, with each frontier breached in observation, further details emerged in understanding cells. The development of electron microscopes in the 20th century, for instance, paved the way for viewing subcellular structures, augmenting the resolution and detailing of cell anatomy immeasurably.

This allowed cell biologists to delve deeper into cellular complexities (see figure 3), opening doors to discoveries like cellular organelles and the double helical structure of DNA. Further advancements in microscopy techniques in recent years, including confocal and fluorescent microscopy, have enabled scientists to view cells in their natural state, track the dynamics within living cells, and explore cellular processes in their innate context, intensifying our understanding of the life cycle, division, and functioning of cells notably. Therefore, without contention, microscopes and cell observation have acted as the primary driving force behind the growth seen in the field of cell biology, substantiating the interdependence between technological innovation and scientific discovery.

The early history of microscopy dates back to the 17th century when pioneers like Antonie van Leeuwenhoek first observed microorganisms through simple, single-lens microscopes (Kim et al., 2023; Klingemann, 2023; Kroll et al., 2022; Laiman et al., 2023).

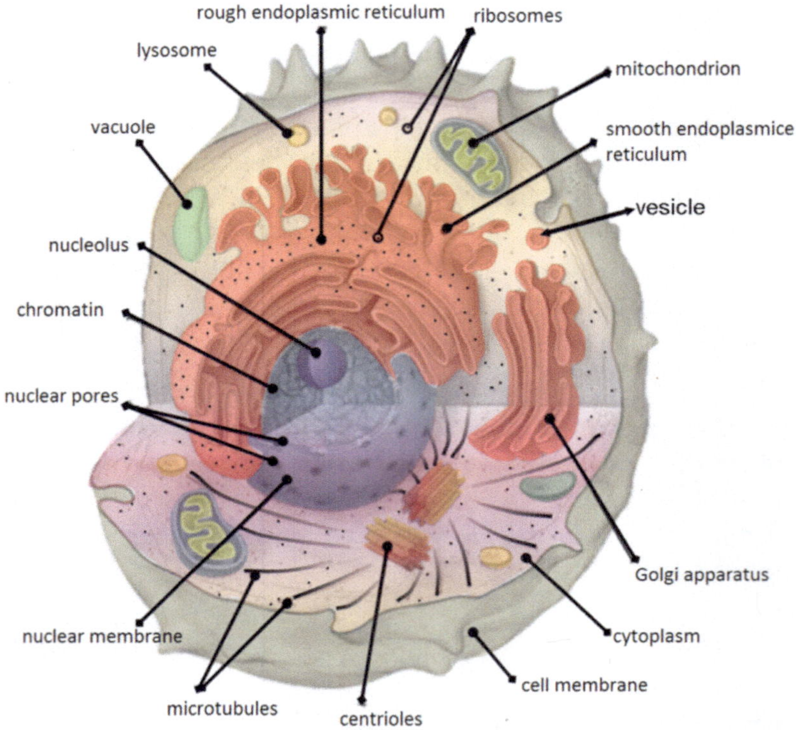

Figure 3. The minute details of an animal cell.

Source: By Bingbongboing - https://centoshelp.org/mario/, CC BY-SA 4.0, https://commons.wikimedia.org/w/index.php?curid=54735767

Leeuwenhoek's discovery of bacteria and other microscopic organisms laid the foundation for the field of microbiology. These early microscopes provided the first glimpses of the microscopic world, but they were limited in magnification and resolution. The early history of microscopy, while intriguing, is firmly entrenched in the 17th century, when notable pioneers such as Antonie van Leeuwenhoek made significant strides by observing organisms through simple, single-lens microscopes. A closer analysis of this key period through the lens of van Leeuwenhoek's work, primarily his empirical observations, offers an illuminating context. Born in Delft, the Netherlands, in 1632, Antonie van Leeuwenhoek can be characterized as a self-taught scientist who brought forth an unprecedented level of detail in his observations, primarily concerning small creatures or "animalcules", as he termed them. Utilizing his remarkable lens-grinding skills, he designed simple microscopes, contrary to the compound microscopes of his era, characterized by a single high-quality lens that offered unparalleled magnification, reportedly able to magnify objects up to 200 times their actual size. What distinguished van Leeuwenhoek in the realm of microscopy was not merely his technical proficiency but the meticulousness of his observations. Through his microscopes, he made several groundbreaking discoveries such as the first visual accounts of bacteria, protists, spermatozoa, erythrocytes, and the banded pattern of muscular fibers. His observations, enriched by detailed pictorial documentation and comprehensive descriptions, ushered in an era of microscopic exploration and significantly broadened scientific knowledge about the varieties of minute life forms that remained unnoticed to the naked eye. His letters to the Royal Society of London provided a significant catalyst for future scientific investigations in the microscopic world. It is, therefore, beyond doubt that Antonie van Leeuwenhoek, through his innovative approach to simple microscopy and investigative curiosity, left an indelible imprint in the early history of microscopy, paving the way for later advancements in viewing the world too small for the human eye to perceive.

In the 19th century, advancements in optics and manufacturing techniques led to the development of compound microscopes with multiple lenses, providing greater magnification and improved image quality (Efstathiou & Draviam, 2021; Duncan, 2022, 2023; Dunkley et al., 2022). With the compound microscope, scientists could now observe more intricate cellular structures, such as the nucleus, cell organelles, and even details of cell division. In the 19th century, remarkable strides in the spheres of optics and manufacturing techniques culminated in the development of compound microscopes furnished with multiple lenses. These sophisticated microscopes offered significant enhancements in both magnification and image quality. The case of Carl Zeiss Optics, a German manufacturer of optical systems, illustrates this turning point with utmost clarity. Founded in 1846 by Carl Zeiss, a skilled mechanic and business strategist, the company embarked on a mission to improve the existing microscope designs of the era. Zeiss engaged the expertise of physicist Ernst Abbe, who reimagined the principles of microscope construction. Abbe, equipped with his understanding of the wave nature of light, proposed a formula to calculate the maximum resolution achievable by a microscope: the Abbe diffraction limit. This fundamental insight led to the development of apochromatic lenses offering superior magnification, reduced chromatic aberration, and improved image quality. The manufacturing procedures also integrated cutting-edge practices like the incorporation of new types of glass, precise assembly techniques, and exhaustive testing methods. Through these innovations, Zeiss's microscopes brought unseen details of the microscopic world to light, enabling observers to discern structures as tiny as half a micrometer. This period, consequently, was a turning point in biological research, inviting transformative discoveries like the structure of cells, microorganisms, and tissue architecture. Hence, it is clear how the development of compound microscopes with multiple lenses in the 19th century was a monumental leap forward, one that has had an enduring impact on the fields of biology, medicine, and beyond. Figure 4 is an illustration of some of the early observational tools that supported cell microbiology.

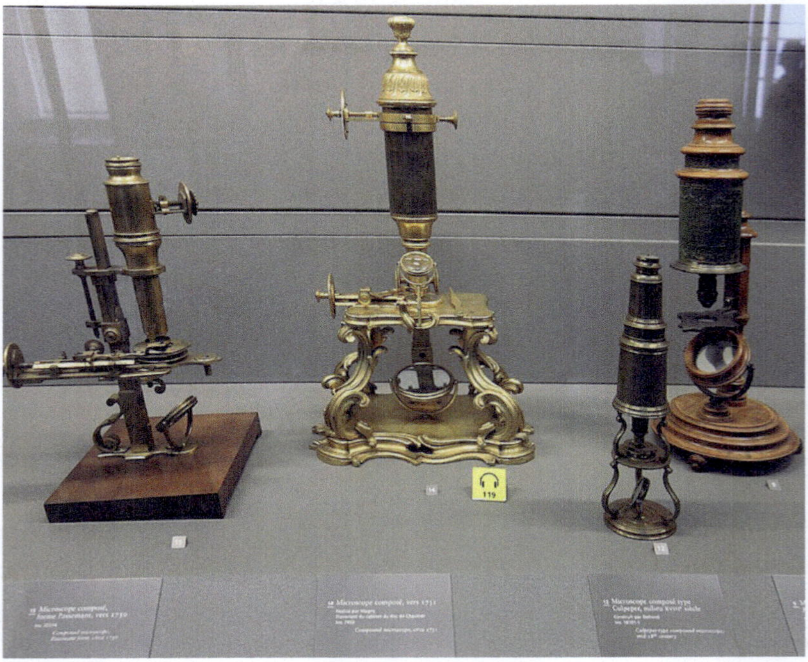

Figure 4. Illustration of early observational tools that support cell microbiology.

Source: By Edal Anton Lefterov - Own work, CC BY-SA 3.0, https://commons.wikimedia.org/w/index.php?curid=11622769

The breakthrough moment came in the late 19th and early 20th centuries with the invention of electron microscopy (Scott et al., 2022; Scott-Solomon & Hsu, 2022; Sezin et al., 2022). Ernst Ruska and Max Knoll built the first electron microscope in 1931, using electron beams to achieve much higher magnification and resolution than traditional light microscopes. This innovation allowed scientists to visualize cellular structures at an unprecedented level of detail, including ultrastructural components like ribosomes, mitochondria, and the endoplasmic reticulum. Ernst Ruska and Max Knoll, both German electrical engineers, remedied a longstanding inadequacy in the realm of microscopy by constructing the first electron microscope in 1931. This instrument, particularly notable for its application of electron beams, offered significantly enhanced magnification and resolution compared to conventional light microscopes. To gain a clearer understanding of this groundbreaking innovation, a profound inspection of the contributions of Ruska and Knoll is integral. The challenge they confronted was bypassing the resolution limit posed by light microscopes due to the diffraction of light, a limitation illuminated by Ernst Abbe in the 19th century.

Realizing that the short wavelengths of electrons could overcome this limitation, Ruska and Knoll embarked on a mission to substitute light waves in microscopes with electron beams. Their earliest model, though primarily intended to visualize electronic components, functioned essentially as an electron microscope. The principle, in simple terms, involves the transmission of high-speed electrons through a minuscule specimen to form an image with higher resolution compared to those formed by light microscopes. The path-breaking reinforcement of the electron microscope's ability to magnify objects up to a million times (compared to the maximum 2000 times magnification offered by light microscopes) revolutionized the arena of microscopy, paving the way for researchers to observe minute details at the molecular level. The electron microscope, hence, became a formidable tool, extensively employed in both biological and material sciences, revealing unprecedentedly minute details such as virus structures, intracellular components, metallurgical crystalline arrangements and atomic particles. Thus, Ruska and Knoll's construction of the first electron microscope not only established a new era in microscopy but also expanded the realms of human knowledge about the minuscule world.

Transmission electron microscopy (TEM) and scanning electron microscopy (SEM) became essential tools in cell biology research (Liao et al., 2021; Liedtke et al., 2022; K. Lim et al., 2023). TEM enables scientists to study the internal structures of cells, while SEM provides a three-dimensional view of the cell's surface.

The combination of light microscopy and electron microscopy has enabled researchers to explore cells and their functions at various scales, from macroscopic tissues to the nanometer level of molecules. The introduction and subsequent developments in TEM and SEM have been pivotal to the progress and various breakthroughs in cell biology research. A tangible paradigm that validates the indispensability of these tools transpires with the elucidation of the inner architecture of the cell. TEM, through its ability to visualize unstained and ultra-thin sliced specimens by transmitting electrons, specializes in providing high-resolution images and has enabled the revelation of minute cellular structures and the functional regions of organelles. An exemplary application of TEM is depicted in the seminal work of George Palade, commonly recognized as the founding father of modern cell biology. Palade employed TEM to decode the structure and function of cellular organelles such as ribosomes, the endoplasmic reticulum, and mitochondria, advancing our understanding of the sophistication synonymous with the interiors of a cell.

SEM, on the other hand, while offering a slightly lower resolution than TEM, specializes in producing high-definition three-dimensional images of the surface of specimens by scanning them with a focused beam of electrons. SEM excels at generating detailed topographical surveys of cell surfaces, enabling scientists to explore surface-dependent phenomena such as cell adhesion, mobility, and interaction. For instance, SEM has proved instrumental in the study of morphologies of various cells and pathogens, cell-to-cell interactions, and the process of infection. Hence, the conjunction of TEM and SEM has endowed cell biology with a diversified toolkit that can probe into the microenvironment of the cell, appreciating the intricate details of life at a sub-cellular level. Through their distinct but complementing roles, TEM and SEM have become essential companions in cell biology research, providing captivating insights into the realm of the infinitesimally small.

The advent of fluorescence microscopy further expanded the capabilities of cell observation (X. C. Wang et al., 2021; X. Wang & X. W. Chen, 2023; Y. F. Wang & H. R. Chen, 2023). This technique uses fluorescent molecules to label specific cellular components, making them visible under specific wavelengths of light. Fluorescence microscopy allows researchers to study live cells in real-time, observe dynamic processes such as cell division, and investigate interactions between molecules within cells. The advent of fluorescence microscopy has indisputably broadened the horizons in the realm of cell observation, endowing scientists with unparalleled depth of knowledge and tools in visualizing cellular structures and processes. This cutting-edge technology employs the phenomena of fluorescence and phosphorescence, revolutionizing the conventional approach to microscopy through an enhanced depiction of complex cellular dynamics. To illustrate this argument in a more explicit context, let us explore the case of neuronal structure analysis, a pivotal aspect in the field of neuroscience. As neuronal structures exhibit intricate characteristics, traditional microscopy techniques have been vastly constrained in rendering comprehensive visualization and understanding of these structures. However, with the introduction of fluorescence microscopy, neuroscientists have been granted an expanded capability to investigate not just the neuron's anatomy but also its functional aspects. By using fluorescent stains or proteins, this advanced microscope highlights the neuron's dendrites, axons and synapses, allowing scientists to unravel the complexities of the neural block with augmented precision. Therefore, the introduction of this technique has undeniably made it possible to observe and study the neuron's interaction with other neurons, its membrane potential and other essential aspects in real-time. The potential of fluorescence microscopy can, thus, be verified through the monumental progress it has generated in the field of neuroscience, allowing the perception of neurons to transcend beyond their mere structure to understand their functional relevance in the nervous system. The advent of fluorescence microscopy, therefore, undoubtedly marks a significant advancement in the field of cell observation.

In recent years, super-resolution microscopy has pushed the limits of what was previously thought possible in optical microscopy (Tolic & Pavin, 2021; Tang et al., 2021; Tanaka et al., 2022; Thakur et al., 2022). Techniques like stimulated emission depletion (STED) microscopy and single-molecule localization microscopy (SMLM) have broken the diffraction barrier, enabling scientists to visualize cellular structures with molecular precision. In recent years, there has been a remarkable transformation in the domain of cell biology with the advent of super-resolution microscopy, which has surpassed the conceived boundaries of traditional optical microscopy. Super-resolution microscopy, by transcending the diffraction limit which had previously constricted the resolution of light microscopy to around 200 nanometers, has empowered researchers to view biological structures of much smaller sizes with exceptional clarity. An illustrative case study demonstrating the potential of this novel tool is the advancement in understanding the architecture of nuclear pore complexes (NPCs). NPCs, critical structures involved in the regulation of substances that cross the nuclear envelope, have dimensions that lie just beyond the optical resolution limit. Therefore, their detailed visualization using conventional microscopy techniques remained a daunting challenge for cell biologists. However, the advent of super-resolution microscopy, in particular Stochastic Optical Reconstruction Microscopy (STORM) technique, has led to significant strides in elucidating the intricate details of NPCs. STORM facilitated the detection of distinct subunits and their organization within NPCs, insights that were inconceivable with previous microscopy tools. Additionally, with the ability of super-resolution techniques to provide dynamic information, the mechanism by which NPCs mediate cargo transport has been meticulously elaborated, shedding light on nuclear-cytoplasmic transport's regulatory layers. Consequently, it is evident that super-resolution microscopy's introduction has revolutionized cell biology by pushing the limits of optical microscopy, enabling the detailed examination of cellular structures and processes that were previously unobservable.

The integration of microscopy with other cutting-edge technologies, such as genomics, proteomics, and bioinformatics, has opened up new avenues of research in cell biology (Chen et al., 2021; Chen et al., 2022; Chen et al., 2023). High-throughput imaging systems and automated analysis tools have accelerated the study of cellular processes on a large scale, contributing to the development of systems biology. The integration of microscopy with other state-of-the-art techniques, like genomics, proteomics, and bioinformatics, has indeed unfurled new trajectories of research in the arena of cell biology. Through an amalgamation of these disciplines, we can now analyze cells at the molecular level with meticulous detail. To substantiate this claim, we can consider the instance of "Cellular Genomics", an innovative interdisciplinary approach that is reshaping the landscape of genomic studies. Genomics, by elucidating the blueprint of life, has been a cornerstone of molecular biology research. However, its use on the individual cellular level had been limited due to the "population averaging" problem of traditional sequencing technologies. The advent of Single-cell Sequencing (scSeq), where genomic sequencing is performed at an individual cellular level, has surmounted this problem, allowing scientists to dissect the cellular heterogeneity at the DNA level. However, scSeq solely cannot provide spatial information on genetic arrangements. Therefore, the integration of microscopy with scSeq is leveraged, through platforms like 'spatial transcriptomics' and 'fluorescence in situ hybridization (FISH) based assays', to provide contextual information about where the gene expressions are taking place within cells, tissues and organ systems. This amalgamation illuminates the correlation between gene expression patterns with morphological cellular phenotypes and their position in a tissue or organ system, unlocking a new realm of insights in cell biology, developmental biology, neurobiology, and cancer research. In essence, the union of microscopy with cutting-edge technologies, such as genomics, is providing unprecedented opportunities and insights into cellular and molecular biology.

In summary, advancements in microscopy and cell observation have been integral to the progress of cell biology and the understanding of cellular processes. From the humble beginnings of single-lens microscopes to the sophisticated techniques of electron microscopy and super-resolution imaging, these innovations have allowed scientists to peer into the inner workings of cells, revealing a dynamic and complex world that continues to be a source of fascination and discovery. As technology continues to evolve, we can expect even more exciting breakthroughs in the field of cell microbiology, offering deeper insights into the building blocks of life – the living cells. The next section will compare unicellular with multicellular organisms to further extend our understanding of cell diversity.

1.3 UNICELLULAR VS. MULTICELLULAR ORGANISMS: UNDERSTANDING CELL DIVERSITY

The distinction between unicellular and multicellular organisms lies at the heart of understanding cell diversity in the field of cell microbiology (Ricci & Srivastava, 2021; Reiche et al., 2022; Reuschl et al., 2022). These two categories represent fundamentally different ways of organizing life, each with unique characteristics, advantages, and challenges. The juxtaposition of unicellular and multicellular organisms is pivotal to the comprehensive understanding of cell diversity within the discipline of cell microbiology. To elucidate this, consider the case of "microbial population studies", an essential aspect in grasping the complexity and adaptability of microorganisms within diverse environmental settings. Unicellular organisms, such as bacteria or yeast, lack the intricate organization that is quintessential to multicellular organisms, yet they exhibit robust adaptability to an array of varying environmental stresses. For instance, Escherichia coli, a unicellular organism, displays remarkable adaptability by altering its gene expressions according to environmental stressors. On the other hand, multicellular organisms possess a more complex, hierarchical biological system, with each cell type performing specialized functions contributing to the overall organism's well-being. A compelling example of this is observed in the human body wherein heart cells, liver cells, and brain cells, although originating from the same basic cell type in the embryonic stage, modify and specialize over the course of development to perform diverse roles. In the case of pathogenic infections, the body's diverse cell types coordinate their response through the immune system to fend off the infection. Thus, the dichotomy between unicellular and multicellular organisms reflects not just on the variations in their physical complexity but extends towards the adaptability, survivability and coordination mechanisms. This distinction, hence, lies central to understanding cell diversity in the field of cell biology allowing researchers to decipher mechanisms of adaptation, cellular differentiation, and cell communication.

Unicellular organisms, such as bacteria, archaea, and various protists, consist of a single cell that performs all the necessary functions for survival (Stancliffe & Patti, 2022; Stratman et al., 2022; Stowers, 2023). Despite their simplicity in structure, these microorganisms exhibit remarkable diversity in their metabolism, physiology, and ecological roles. Their unicellular nature provides advantages in rapidly adapting to changing environments and exploiting a wide range of habitats. Some unicellular organisms, like bacteria, reproduce rapidly through binary fission, allowing them to multiply and colonize new environments quickly. Unicellular organisms, encompassing bacteria, archaea, and various protists, indeed consist of a singular cell, yet they harbor all the necessary machinery to ensure their survival, representing a tour de force of biological efficiency and adaptability. This is powerfully evidenced in the case study of Paramecium caudatum, a protist that exemplifies the ability of a single cell to accommodate multiple functions essential for survival. Paramecium caudatum is an exemplar of a protist that leverages cilia for motility, which permits it to navigate its environment efficiently in search of food and to evade potential threats. In addition, it employs two types of nuclei - a larger macronucleus for metabolic control and a smaller micronucleus for sexual reproduction, thereby striking a balance between daily metabolic activities and continuity of its lineage. Moreover, it possesses a complex feeding mechanism, encapsulating food vacuoles that fuse with lysosomes to digest prey, and even a rudimentary defense mechanism in the form of trichocysts that allow it to ward off predators or competitors. Therefore, examining unicellular organisms such as Paramecium caudatum presents an intriguing perspective on how a single cell can ingeniously carve out multiple functions necessary for survival, thereby driving home the point that complexity in form does not always equate to complexity in function.

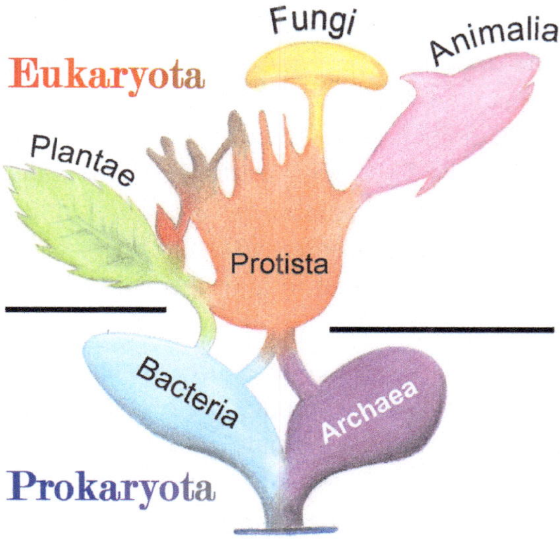

Figure 5. Graphical illustration of unicellular and multicellular organisms.

Source: By Maulucioni y Doridí - Own work, CC BY-SA 3.0, https://commons.wikimedia.org/w/index.php?curid=25888693

Multicellular organisms, on the other hand, are composed of specialized cells that collaborate to form complex tissues and organ systems (see figure 5). This division of labor allows multicellular organisms to achieve greater complexity and functionality compared to unicellular organisms (Pandey et al., 2021; Panicker et al., 2021; Paliulis et al., 2023). Multicellular organisms indeed offer an encapsulation of sheer complexity and coordination; their existence relies on a harmonious collaboration between specialized cells that engage in intricate processes resulting in the formation of various tissues and organ systems. An illustrative example is the human body, a complex labyrinth composed of trillions of cells that are relentlessly orchestrated into a myriad of tissues and organ systems, each with distinct functionalities. Consider, for instance, the formation and functioning of the human brain, arguably the most complex structure known to mankind. The brain is composed of about 100 billion neurons, the fundamental units of the nervous system. These neurons, via synapses, create an expansive network of nearly 100 trillion interconnections. Though each neuron is unique in size, structure, and function, they all share the primary responsibility of transmitting nerve impulses, resulting in the execution of various functions such as sensing, muscle control, and complex cognitive processes. Glia, another type of cell in the brain, nourish neurons, form the myelin insulating sheath around the neurons, and help maintain homeostasis. Thus, the vivacity and intricacy of the human brain, formed by the collaboration of specialized cells, draw a vivid picture of the complexity inherent in multicellular organisms. This complexity fundamentally diversifies the multitude of capabilities and adaptabilities of multicellular organisms when compared to their unicellular counterparts, and offers us a glimpse into the unfathomable layers of coordination and specialization embedded within every multicellular organism.

Cell specialization in multicellular organisms is a key aspect of understanding cell diversity (Riley et al., 2021; Robaszkiewicz et al., 2021; Roeder et al., 2022). Cells differentiate into various types, each with a unique structure and function tailored to its role in the organism. Examples include muscle cells for movement, nerve cells for communication, and epithelial cells for lining tissues. The formation of tissues, where similar specialized cells group together, allows for even more complex functions to be performed. Cell specialization in multicellular organisms is indeed a key aspect of comprehending cell diversity. In a multicellular organism, while each cell shares a common genetic code, the manifestation of this code differs considerably, a fact embodied by specialized cells that diversify to perform specific tasks.

This distinctive phenomenon illuminates the different characteristics and functions of cells, and unequivocally highlights the inherent diversity within a solitary organism. To illustrate this concept, an incisive case study can be derived from the human body. Composed of approximately thirty-seven trillion cells, the human body entails myriad cell types, each having its own specialized role. Consider cardiomyocytes, the specialized cells that make up the bulk of our heart muscle – the myocardium. These cells are fundamental for the heart's primary function: pumping blood throughout the body. Cardiomyocytes exhibit specific adaptations, such as densely packed mitochondria and specific proteins like contractile proteins, that enable them to continually contract and relax, a critical process for the circulation of blood. This is markedly different from neurons, the principal components of the nervous system, which are specialized for transmitting nerve impulses, and thus, they feature long projections and synapses for electrical communication. Therefore, the specialization of cells as seen in cardiomyocytes and neurons offers a compelling understanding of the diversity of cells within a multicellular organism. Overall, cell specialization unlocks a deeper layer of understanding about cell diversity and the intricate orchestration of life processes in multicellular organisms.

Unicellular organisms are considered more ancient, representing the earliest forms of life on Earth (Featherston & Cotruvo, 2021; Ferrari et al., 2022; Ferrari et al., 2023). Multicellularity, on the other hand, evolved relatively later in the history of life. The transition from unicellular to multicellular forms involved significant changes in genetic regulation and cell communication, ultimately leading to the development of complex organisms with differentiated cell types. Unicellular organisms, encompassing bacteria, archaea, and various protists, are deemed more ancient and represent the earliest forms of life on Earth, a statement authenticated by substantial scientific evidence. A noteworthy case comes to light through the study of stromatolites, which are sedimentary structures resulting from the metabolic activities of cyanobacteria, a unicellular organism.

The oldest known stromatolites, located in Western Greenland, date back about 3.7 billion years, providing solid fossil evidence of these ancient unicellular organisms. Furthermore, research based on molecular clocks, which allow the estimation of divergence times between different species based on genetic sequences, also corroborates the antiquity of unicellular life forms. This technique estimates the divergence of bacteria and archaea, both unicellular, to about 3.9 billion years, essentially aligning with the Earth's geological evidence. Such findings thus demonstrate the longevity of unicellular organisms, situating them as the pioneering bearers of life on Earth. This early emergence notwithstanding, the resilience and adaptability of unicellular organisms should not be underestimated, as they continue to thrive in environments ranging from the deep sea to the human gut, indeed testifying to their indomitable survival strategies honed over billions of years of evolution. Hence, unicellular organisms, through their early introduction and continued prevalence, represent an essential milestone in the comprehension of life's chronicle on Earth. Figure 6 highlights the minute components of a cell under a microscope.

Cell diversity between unicellular and multicellular organisms has significant ecological implications (Patriarca et al., 2021; Pauletto et al., 2021; Passanha et al., 2022). Unicellular organisms play essential roles in various ecological processes, such as nutrient cycling, decomposition, and symbiotic relationships with other organisms. They are crucial for maintaining ecological balance and supporting higher trophic levels in food webs. The diversity between unicellular and multicellular organisms indeed carries substantial ecological implications, a topic of great interest in modern ecological research.

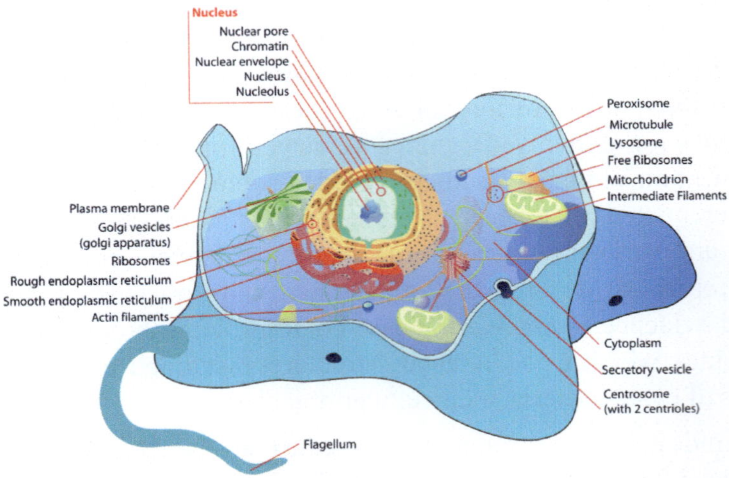

Figure 6. The components of an animal cell under a microscope.

Source: By LadyofHats (Mariana Ruiz) - Own work using Adobe Illustrator. Image renamed from Image:Animal cell structure.svg, Public Domain, https://commons.wikimedia.org/w/index.php?curid=4266142

To illustrate this, consider the biological phenomenon known as 'primary productivity', the process through which light energy is converted into organic matter via photosynthesis. This process is mainly spearheaded by unicellular organisms like cyanobacteria and single-celled algae in aquatic environments, and by multicellular plants on land. Phytoplankton, unicellular photosynthetic organisms in the ocean, account for nearly half of the global primary productivity, thus forming the foundational basis of marine food webs. They sequester carbon dioxide, thereby playing a crucial role in the global carbon cycle and mitigating climate change. On the other hand, multicellular plants, through the process of photosynthesis, form the basis of terrestrial ecosystems, turning sunlight into energy-dense organic molecules that serve as the food source for herbivores. Furthermore, through transpiration, they play a significant role in the water cycle, influencing precipitation patterns. Additionally, multicellular organisms such as fungi perform a critical ecological role by decomposing organic matter and recycling nutrients within ecosystems. Therefore, the contrast and diversity present between unicellular and multicellular organisms are not merely academic curiosities; instead, they drive the primary ecological processes that sustain the life and dynamics of our planet's ecosystems.

Multicellular organisms have a more significant impact on ecosystem structure and functioning (Urmi et al., 2021; Usmani et al., 2021; Turner et al., 2022). Their complex interactions, niche specialization, and large biomass contribute to the stability and diversity of ecosystems. Multicellular organisms form the foundation of various ecological communities and ecosystems, influencing the distribution and abundance of other organisms. The assertion that multicellular organisms have a more significant impact on ecosystem structure and functioning is a pertinent one, rooted in the extensive ecological roles these organisms perform in various biomes. To delve into this subject, let us refer to the integral role of trees, a type of multicellular organism, in forest ecosystems. Trees, through their extensive biomass, contribute enormously to carbon storage - sequestering carbon dioxide from the atmosphere and thus playing a pivotal role in mitigating climate change. Furthermore, trees also form the basis of forest food chains, with their leaf fall providing nutrients for a host of fungi and invertebrates, and their fruits and seeds serving as food for a multitude of animals. Moreover, trees confer habitat to a diverse array of organisms including birds, mammals, insects and epiphytic plants, thereby promoting biodiversity. In the context of hydrology, trees balance the water cycle by reducing runoff and erosion, maintaining watershed health, and influencing localized precipitation patterns through transpiration. In addition, trees interconnect underground through a network of mycorrhizal fungi, creating a complex system known as the "Wood Wide Web", where resources like nitrogen, phosphorus and water, and even signals of danger, are exchanged. Thus, through the collective actions of trees, one can comprehend how multicellular organisms significantly mold ecosystem structure and regulate its functions. Being an integral node connecting several ecological processes, multicellular organisms like trees underscore the interconnectedness and diversity of life on our planet.

In conclusion, understanding cell diversity between unicellular and multicellular organisms is fundamental to the study of cell microbiology. Unicellular organisms showcase the simplicity and adaptability of single-celled life forms, while multicellular organisms demonstrate the complexity and specialization that arise from the collaboration of diverse cell types. This knowledge not only enhances our understanding of basic cellular biology but also provides insights into evolutionary processes, ecological interactions, and the marvels of life's organization and diversity. The next section will consider the structure and functions of prokaryotic cells.

1.4 PROKARYOTIC CELLS: STRUCTURE AND FUNCTIONS

Prokaryotic cells represent one of the two main categories of cells, the other being eukaryotic cells (Gan et al., 2022; Garcia-Gutierrez & Garcia-Dominguez, 2022; Geers et al., 2022). These single-celled microorganisms are among the most abundant and diverse forms of life on Earth, thriving in a wide range of environments, from the depths of the ocean to extreme temperatures and acidic conditions. Understanding the structure and functions of prokaryotic cells is crucial in grasping the fundamentals of cell microbiology. Prokaryotic cells indeed epitomize one of the two fundamental categories of cells, the other being eukaryotic cells, delineating a profound bifurcation in the domain of cellular biology. This segregation, founded on structural and functional distinctions, underpins the cellular heterogeneity in the biological world.

Prokaryotic cells, a term derived from Greek words meaning "before the nucleus", primarily encompass bacteria and archaea. These cells are typified by their relative simplicity, characterized by their lack of a true membrane-bound nucleus and other organelles, save for ribosomes, and typically a small size. Pivoting to a case study for a grounded understanding, let us consider Escherichia coli, a well-studied, gram-negative bacterium residing in the human gut. In E. coli, like other prokaryotes, DNA is arranged in a singular circular chromosome situated in a region called the nucleoid, contrasting the multiple linear chromosomes enclosed within a nuclear envelope in eukaryotic cells. Furthermore, E. coli, possesses a plasma membrane and a cell wall, with an additional outer membrane, distinctive to gram-negative bacteria. Metabolic processes such as respiration occur in the plasma membrane in prokaryotes, unlike eukaryotes where they occur in specialized organelles like mitochondria. These distinctions between E. coli and eukaryotes elucidate the fundamental differences between prokaryotic and eukaryotic cells. Thus, the classification into prokaryotic and eukaryotic cells demarcates the broadest organizational diversity exhibited by cells, providing the basic framework for understanding life's myriad forms.

Prokaryotic cells are characterized by their simplicity in structure compared to eukaryotic cells (Riley et al., 2021; Robaszkiewicz et al., 2021; Roeder et al., 2022). Unlike eukaryotic cells, prokaryotic cells lack a true nucleus and membrane-bound organelles. Instead, the genetic material, a single circular DNA molecule, is found in a region called the nucleoid, which is not enclosed by a membrane. Surrounding the nucleoid is the cytoplasm, a fluid-like substance where various cellular processes occur. Prokaryotic cells are indeed characterized by their simplicity in structure, particularly when juxtaposed against the intricacy of eukaryotic cells. This distinction has been amply confirmed by numerous microbiological studies through decades of exhaustive research. One of the quintessential examples of a prokaryotic cell is the bacterium Escherichia coli. Known for its prevalent occurrence in the human gut, E. coli exemplifies the simplicity of prokaryotic cells. In E. coli, genetic material, consisting of a singular circular chromosome, cohabits the cytoplasm freely as prokaryotic cells lack a nucleus, a prominent trait distinguishing them from eukaryotic cells. Additionally, prokaryotic cells like E. coli, devoid of organelles such as mitochondria, chloroplasts or a complex internal membrane system, must execute cellular processes in more basic ways. For instance, the electron transport chain, an integral part of cellular respiration in eukaryotic cells and localized to the mitochondria, takes place in the plasma membrane in E. coli and other prokaryotes. Yet, despite their simplicity, prokaryotic cells efficiently execute critical biochemical processes vital for survival, underscoring the principle that biological complexity is not a prerequisite for the efficient conduct of life's essential activities. Thus, the simplicity of prokaryotic cells (see figure 7), as epitomized by E. coli, underlines the impressive adaptability of life in its numerous manifestations.

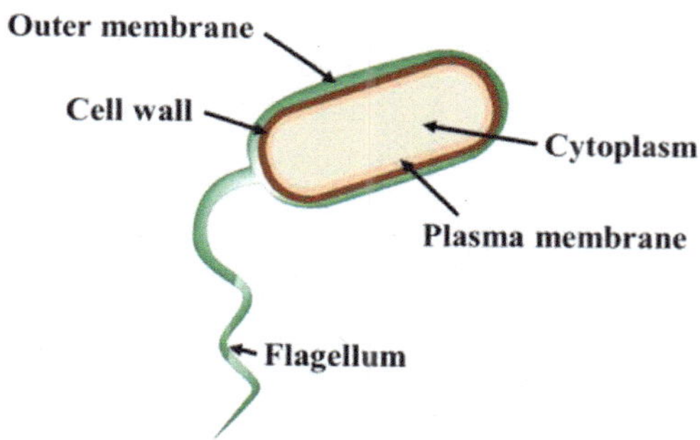

Figure 7. Prokaryotic cell structure, characteristics, and function (E. coli).

Source: L Yahav G, Pawar S, Lipovsky A, Gupta A, Gedanken A, Duadi H, Fixler D. Probing Polarity and pH Sensitivity of Carbon Dots in Escherichia coli through Time-Resolved Fluorescence Analyses. Nanomaterials. 2023; 13(14):2068. https://doi.org/10.3390/nano13142068

The cell membrane, or plasma membrane, encloses the cytoplasm, separating the cell's interior from its external environment (Wolin, 2021; Wong et al., 2021; Wood et al., 2021; Woods & Gladfelter, 2021). This selectively permeable membrane plays a vital role in regulating the passage of molecules in and out of the cell. The cell membrane, or plasma membrane, indeed functions as the crucial barrier separating the cellular interior, known as the cytoplasm, from the external environment. This characteristic, universal to all cells, enables the preservation and regulation of the cell's internal milieu, a concept known as homeostasis, essential for the cell's survival and functioning. An insightful case study elucidating this idea can be drawn from the human red blood cell (RBC). The primary function of RBCs, namely the transport of oxygen to tissues and removal of carbon dioxide, is contingent on its plasma membrane. This selectively permeable membrane allows the passive diffusion of oxygen and carbon dioxide across it, enabling the exchange of these gasses between the RBC and its surrounding plasma. Moreover, this membrane performs a critical regulatory role by housing the protein complexes like Sodium-Potassium Pump and Anion-Exchanger, involved in maintaining the ionic balance and cell volume respectively, ensuring the optimal functioning and structural integrity of the RBC. The plasma membrane, hence, encloses and protects the cell's vital components, facilitates communication with the environment, and regulates the transport of substances across it, emphasizing its immense importance in cellular life. Therefore, the cell membrane, through its essential role in demarcating and regulating the cellular environment, stands as an indispensable element in the understanding of cellular physiology and survival.

Many prokaryotic cells possess an additional protective layer called the cell wall (Magalhaes et al., 2021; Mahaldashtian et al., 2023; Maib & Murray, 2022). The cell wall provides structural support and protection against osmotic stress. The composition of the cell wall varies among different groups of prokaryotes. For example, bacteria typically have a peptidoglycan cell wall, while archaea may have a cell wall made of different materials, such as pseudomurein. Indeed, many prokaryotic cells are endowed with a supplementary protective layer, known as the cell wall, which imparts rigidity, strength, and resilience against environmental adversities. This trait serves as a determinant of cell shape and offers overall structural integrity. For illustration, consider the remarkable case of Bacillus subtilis, a gram-positive, rod-shaped bacterium commonly found in soil. The cell wall of B. subtilis, as in other gram-positive bacteria, primarily consists of peptidoglycan, a sturdy lattice-like structure made up of sugars and amino acids. The high proportion of peptidoglycan gives B. subtilis its characteristic violet coloring in a gram-stain lab technique, lending the category of 'gram-positive'. This robust wall not only provides protection against physical disturbances but also defends the bacterium from osmotic lysis – rupture due to an imbalance in inside vs. outside solute concentrations. Additionally, the cell wall of B. subtilis acts as a shield against antibacterial substances and hosts immunity responses. Therefore, the possession of the cell wall in prokaryotes like B. subtilis offers them an exceptional level of protection and stability, enabling their survival in diverse, often challenging, habitats. Hence, the presence or absence and the specific constitution of the cell wall, forms a key aspect in understanding the survival strategies of prokaryotes, their interaction with the environment, and their implication in antibacterial resistance. Figure 8 highlights important differences between Eukaryotes and Prokaryotes.

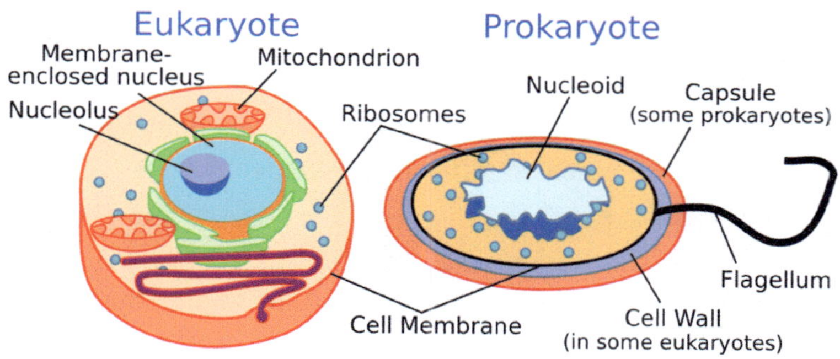

Figure 8. Difference between Eukaryotes and Prokaryotes.

Source: By Science Primer (National Center for Biotechnology Information). Vectorized by Mortadelo2005. - This file has been extracted from another file, Public Domain, https://commons.wikimedia.org/w/index.php?curid=2145991

External structures known as appendages are present in some prokaryotic cells (Sales-Gil et al., 2021; Salinas-Luypaert et al., 2021; Schornig & Taverna, 2021). Flagella are long, whip-like structures that enable cell movement, while pili are shorter, hair-like structures involved in adherence to surfaces and the exchange of genetic material through a process called conjugation. Indeed, some prokaryotic cells feature external structures known as appendages which serve various specialized purposes, contributing to the bacterium's survival and performance in diverse environments. To illustrate this point, consider the case of flagellated bacteria, such as Escherichia coli. Flagella are whip-like appendages protruding from the cell surface, facilitating motility— a vital function that allows the bacteria to move towards favorable environments or away from harmful substances, a phenomenon scientifically termed as 'chemotaxis'. These elongated appendages, composed of a protein called flagellin, rotate like propellers driven by a molecular motor at the base, propelling the bacterium forward. In addition to flagella, some prokaryotes possess shorter, bristle-like appendages called pili or fimbriae that aid in attachment to surfaces or other cells. These structures play a crucial role in bacterial conjugation, a form of genetic exchange that enhances genetic diversity and can contribute to antibiotic resistance. Taking E. coli as an example again, the F-pilus is a critical structure that enables the transfer of plasmids carrying antibiotic-resistance genes from one bacterial cell to another. Thus, through these diverse appendages, prokaryotic organisms are endowed with several capabilities, such as motility and genetic exchange, which not only are critical for their survival and adaptability but also have profound implications for human health and disease.

Prokaryotic cells perform essential functions that sustain life and contribute to their ecological significance (Skoniecka et al., 2021; Smith et al., 2022; Smith et al., 2023).

1. Genetic Information: The genetic material in prokaryotic cells consists of a single circular DNA molecule that contains the instructions for the cell's structure and functions (Skoniecka et al., 2021; Smith et al., 2022; Smith et al., 2023). The genetic material is responsible for the cell's growth, reproduction, and response to the environment. The genetic material in prokaryotic cells indeed comprises a single circular DNA molecule, which holds the critical instructions necessary for the cell's structure and function. Putting this idea into context, one can refer to the living microorganism Escherichia coli, a resident bacterium in the human gut. As typical of prokaryotes, the genome of E. coli is organized into a singular, circular DNA molecule situated in an area known as the nucleoid. This genetic circle, a complex network of coiled strands, is not enveloped within a nucleus, representing a significant departure from eukaryotic cells. The E. coli chromosome spans about 4.6 million base pairs and is responsible for encoding approximately 4,400 proteins. These proteins, the final product of genetic expression, are implicated in all aspects of the bacterium's life, encompassing processes such

as replication, transcription, translation, metabolic pathways, formation of cellular structures, and response to environmental stimuli. For instance, the lac operon present on the E. coli chromosome, a distinctive section of DNA that houses genes required for lactose metabolism, is a classic example of gene regulation in response to environmental change. It explains how E. coli, under certain conditions, can utilize lactose as a source of energy. Therefore, the circular DNA in prokaryotes, such as E. coli, underpins the bacterium's structure, function, survival, and adaptability, illuminating the potency of this singular molecule in orchestrating life's manifold processes.

2. Protein Synthesis: Prokaryotic cells have ribosomes, molecular machines responsible for protein synthesis (Wu & Liu, 2021; Yang et al., 2021; Xu & Li, 2021; Xie et al., 2022; Xue et al., 2022). These ribosomes are smaller than those in eukaryotic cells and play a critical role in translating the genetic code into functional proteins. Indeed, prokaryotic cells contain ribosomes, essential molecular machines tasked with the process of protein synthesis. A classic case study to illustrate this concept can be drawn from the bacterium, Escherichia coli. E. coli and other prokaryotic cells contain 70S ribosomes, a term referring to their size as measured through a unit called Svedberg units, reflecting their rate of sedimentation during ultracentrifugation. The prokaryotic ribosome, smaller than its eukaryotic counterpart, comprises two subunits, a larger 50S and a smaller 30S, both instrumental in the process of protein synthesis known as translation. The mRNA transcript, synthesized during transcription, binds to the small ribosomal subunit. As translation proceeds, transfer RNA (tRNA) molecules carrying specific amino acids recognize and bind to the codon sequence on the mRNA in the ribosome, thereby adding the appropriate amino acid to the growing polypeptide chain. The large subunit catalyzes the formation of the peptide bond between these amino acids, facilitating the creation of the protein. Upon completion, the polypeptide chain is released, marking the end of translation. Thus, through this fundamental process, the genetic code is translated into functional proteins, underlying the functioning and survival of prokaryotic cells. This ubiquitous molecular machine, the ribosome, underpins all of

life's functions, lending credence to its recognition as the bedrock of biology.

3. Energy Production: Prokaryotic cells generate energy through various mechanisms, such as aerobic respiration, anaerobic respiration, and fermentation (Shen et al., 2021; Siu et al., 2021; Shi et al., 2022; Shoup & Priola, 2023). They utilize different electron acceptors and donors to produce ATP, the cell's primary energy currency. Indeed, prokaryotic cells utilize a range of mechanisms to generate energy, underpinning their adaptability to diverse environmental conditions. These mechanisms include aerobic respiration, anaerobic respiration, and fermentation. A suitable case study that perfectly encapsulates this concept is the bacterium Escherichia coli, a notable member of the human gut flora. E. coli is a facultative anaerobe, an organism that thrives in both oxygen-rich and oxygen-poor environments by shifting its energy-generating mechanism. When residing in environments with ample oxygen, E. coli, like most living beings, prefers to generate energy through aerobic respiration. This process involves the breakage of glucose into carbon dioxide and water, releasing considerable energy, which is then captured in ATP molecules. However, when oxygen supply is scarce or absent, E. coli seamlessly transitions to either anaerobic respiration or fermentation to meet its energy needs. In anaerobic respiration, alternate electron acceptors like nitrate or fumarate substitute oxygen in the electron transport chain to achieve ATP synthesis albeit with less efficiency. Conversely, during fermentation, E. coli converts glucose into substances like lactic acid or ethanol, which comparatively yields significantly less ATP. Through these power-generation strategies, E. coli can optimize its survival and continue to perform essential functions, illustrating the remarkable adaptability harbored within prokaryotic cells.

4. Nutrient Uptake: Prokaryotic cells acquire nutrients from their environment through processes such as diffusion, facilitated diffusion, and active transport (McCutcheon, 2021; Masur et al., 2021; Marvin et al., 2022). Many prokaryotes can also perform photosynthesis to generate energy from sunlight. Indeed, prokaryotic cells, such as bacteria, harness nutrients from their surrounding environment via several processes, namely diffusion, facilitated diffusion, and active transport. As an emblematic

example, consider Escherichia coli, an extensively studied gram-negative bacterium. For simple nutrients, such as water or gases like oxygen and carbon dioxide, E. coli primarily relies on simple diffusion, where these substances move from regions of high concentration to places of lower concentration across the cell membrane. In contrast, for the case of more complex nutrients, such as glucose or amino acids, E. coli employs facilitated diffusion, which requires specific transporter proteins embedded in the cell membrane that guide these nutrients into the cell along the concentration gradient. Furthermore, E. coli, much like many other prokaryotic cells, uses active transport to uptake important nutrients, particularly when they are scarce in the environment. Herein, nutrients are transported against their concentration gradient, necessitating the energy primarily derived from ATP hydrolysis. An illustrative example is the utilization of lactose by E. coli, where a membrane protein, known as lactose permease, facilitates the entry of the lactose molecule into the cell in an energy-dependent manner. Through this wide repertoire of nutrient acquisition strategies, prokaryotic cells elegantly display their remarkable adaptability, thus enabling their survival and proliferation in a variety of ecological niches. In essence, the nutrient gathering capacity of prokaryotes such as E. coli underscores the multifaceted interactions between cells and their environments, emphasizing its criticality in understanding cellular and ecological dynamics.

5. Reproduction: Prokaryotic cells reproduce through a process called binary fission, where a single cell divides into two identical daughter cells (Stancliffe & Patti, 2022; Stratman et al., 2022; Stowers, 2023). This rapid reproduction contributes to the abundance and diversity of prokaryotic organisms. Prokaryotic cells indeed reproduce through a process known as binary fission, a remarkable and efficient mode of reproduction wherein a single cell splits into two identical daughter cells. This process is central to the growth, survival, and adaptability of prokaryotic organisms. To illustrate this concept, an exemplary case can be drawn from the study of Escherichia coli, a well-researched bacterium known for its swift multiplication rate in favorable conditions. In binary fission, E. coli begins with the replication of its single circular chromosome, allowing it to disperse the precise genetic blueprint

to each daughter cell. Following replication, the cell elongates while concurrently directing the duplicated chromosomes towards opposite poles of the cell. When the cell has approximately doubled its size from the initial stage, a division septum forms in the middle, gradually constricting the cell until it splits into two separate cells. Each of these daughter cells is endowed with an identical set of chromosomes, hence creating genetic clones of the parent cell. Presuming optimal conditions, E. coli, utilizing binary fission, can replicate roughly every 20 minutes, which marks its impressive proliferative capacity. The process of binary fission therefore carries profound implications in areas such as microbial ecology, infectious disease progression, and antibiotic resistance.

6. Ecological Significance: Prokaryotic cells play critical roles in ecosystems. They are involved in nutrient cycling, decomposing organic matter, and establishing symbiotic relationships with other organisms (Ma & Hoffmann, 2021; Lu et al., 2022; Lynes et al., 2022). Some prokaryotes, like nitrogen-fixing bacteria, are vital for nutrient availability in the environment. Indeed, prokaryotic cells, encompassing bacteria and archaea, play fundamental roles in ecosystems, including nutrient cycling, decomposing organic matter, and establishing symbiotic relationships with other organisms. This wide-ranging influence can be aptly illustrated by the critical role played by prokaryotes in soil ecosystems. Soil-dwelling bacteria and archaea drive essential biogeochemical cycles, like the nitrogen cycle, where certain bacteria convert atmospheric nitrogen into a form that plants can utilize, a process known as nitrogen fixation. On the other hand, decomposer prokaryotes in soil play a crucial role in breaking down organic matter, hence contributing to nutrient recycling and soil fertility. Furthermore, prokaryotes often establish symbiotic relationships with other organisms, which can significantly impact ecosystem dynamics. A classic example of such a relationship is found between leguminous plants and nitrogen-fixing bacteria from the Rhizobium genus. The bacteria inhabit nodules in the roots of the plants, where they convert atmospheric nitrogen into a usable form for the plant. In return, the plant provides the bacteria with organic compounds for their nourishment. Through these diverse contributions, prokaryotic cells considerably influence

the structure, functioning, stability, and diversity of ecosystems, thereby underscoring their indispensability to life on earth.

In summary, prokaryotic cells, with their relatively simple structure and essential functions, represent a diverse and ecologically significant group of microorganisms. Understanding the intricacies of prokaryotic cells is fundamental in the field of cell microbiology and provides insights into the evolution, ecology, and adaptation of these abundant and ancient life forms. The final section in this chapter will discuss membrane-bound organelles and their roles in eukaryotic cells.

1.5 EUKARYOTIC CELLS: MEMBRANE-BOUND ORGANELLES AND THEIR ROLES

Eukaryotic cells are characterized by their complexity and contain membrane-bound organelles that compartmentalize different cellular functions (Prazak et al., 2021; Pressly et al., 2022; Qazi et al., 2022). These organelles play crucial roles in various cellular processes, contributing to the efficiency and organization of eukaryotic cells. Understanding the structure and functions of these organelles is essential in the field of cell microbiology. Eukaryotic cells, noted for their intrinsic complexity, are defined by the presence of membrane-bound organelles that facilitate the compartmentalization of varied cellular functions. Central to understanding the significance of this compartmentalization is the phenomenon observed in the multicellular organism Paramecium caudatum, a protozoan ciliate known for being a eukaryote par excellence. The cellular structure of Paramecium caudatum, with multiple internal membrane-bound compartments, astutely reflects the intricate complexity of eukaryotic cells. For instance, the nucleus, one of the most prominent organelles within the membrane enclave of this organism, can distinctly segregate functions such as replication and transcription to modulate genetic processes, further contributing to the whole cellular network. Moreover, mitochondria, often termed the powerhouse of the cell, underscores another layer of functional compartmentalization as it enables aerobic respiration by housing the necessary enzymes for oxidative phosphorylation. Furthermore, the presence of Golgi bodies and endoplasmic reticulum aids in protein and lipid synthesis demonstrating yet another level of task assignment and isolation within the cell. This compartmentalization illustrates a profound degree of sophistication and efficiency, thereby enhancing the survival and proliferation of eukaryotic cells. Each of these organelles, working in a well-coordinated manner within their bounded milieu, perform specialized tasks. This modality of operation essentially enables eukaryotic cells to carry out a far wider range of functions, with higher efficiency and precision in comparison to their prokaryotic counterparts. Therefore, the presence of membrane-bound organelles compartmentalizing cellular functions, as exhibited in Paramecium caudatum, is fundamental to the inherent complexity that typifies eukaryotic cells.

The most prominent organelle in eukaryotic cells is the nucleus, which houses the cell's genetic material, DNA, organized into multiple linear chromosomes (Jones et al., 2021; Karlsson & Draber, 2021; Ju et al., 2022;). The nucleus controls cellular activities by regulating gene expression and synthesizing RNA. It is surrounded by a double membrane called the nuclear envelope, which contains nuclear pores that enable the movement of molecules between the nucleus and the cytoplasm. A salient case study exploring the most distinguishable organelle in eukaryotic cells, the nucleus, reveals a fascinating world of intricate cellular mechanisms. The nucleus, which is the controlling center of the cell, is encircled by the nuclear envelope that safeguards the cell's genetic material wrapped inside it. This genetic material, known as Deoxyribonucleic Acid (DNA), is the blueprint of life, dictating all cellular processes and inheriting traits from generation to generation. In the structure of eukaryotic cells, the DNA is organized into complex entities called chromosomes, which are distinctively linear compared to their circular counterparts in prokaryotic cells. Each chromosome contains a steady number of genes, varying between different organisms – for instance, a human cell consists of approximately 20,000 to 25,000 genes organized into 46 chromosomes. These packages of genes, with their distinct order, facilitate a systematic and uniform inheritance of traits and maintain the fundamental structure of the organism. The transference of these genes during cell division is a meticulously orchestrated process, with the chromosomes being visible during only the metaphase and anaphase stages of mitosis. Two identical replicas of each chromosome, or sister chromatids, are pulled apart to opposite sides of the cell, as a prelude to cellular fission. This process regulates the equal distribution of identical genetic data to each daughter cell, ensuring the uniformity and continuity of life forms. The organization of DNA into multiple linear chromosomes in the nucleus of eukaryotic cells is, hence, a pivotal system guaranteeing the reproduction, growth, healing, and evolution of all higher forms of life. Figure 9 is organelles and their functionality.

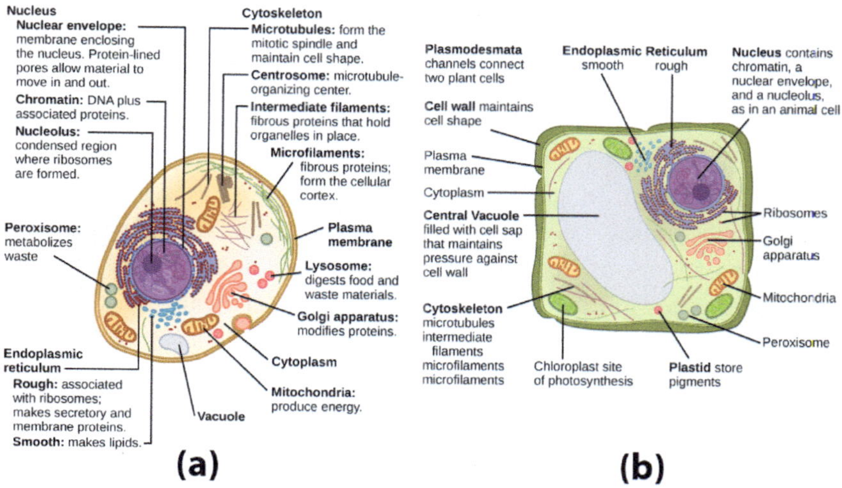

Figure 9. A selection of organelles and their functionality.

Source: Lumen Learning (https://courses.lumenlearning.com/suny-wmopen-biology1/chapter/outcome-organelles/)

Another important organelle is the endoplasmic reticulum (ER), a network of membranes extending from the nuclear envelope throughout the cell (Bettegazzi et al., 2021; Bennison et al., 2022; Beppler et al., 2022). The rough ER, studded with ribosomes on its surface, is involved in protein synthesis, with ribosomes translating mRNA into proteins that are either secreted or integrated into the cell's membranes. The smooth ER, lacking ribosomes, plays a role in lipid metabolism, detoxification of drugs, and storage of calcium ions. Delving deeper into the magnificent wonders of eukaryotic cell physiology reveals the unique role and complexity of an essential organelle – the endoplasmic reticulum. Seemingly an extension of the nuclear envelope that encloses the nucleus – the control center of the cell – this vast network of membranes undulates and meanders throughout the cell, providing a substantial area for chemical reactions. The endoplasmic reticulum exists in two distinct forms, variegated by their structural attributes: the rough endoplasmic reticulum and the smooth endoplasmic reticulum. Covered with numerous embedded ribosomes, the rough endoplasmic reticulum, named due to its rough appearance under the electron microscope, is principally associated with the manufacturing and packaging of proteins. The nascent proteins enter the rough endoplasmic reticulum, where they are folded and chemically modified to achieve their functional three-dimensional conformations. Additionally, this organelle plays a critical function in the glycosylation of proteins, where specific carbohydrate molecules are attached to proteins – a necessary modification for their stability and functional versatility. Furthermore, it serves as the site for the synthesis of phospholipids, which plays a critical role in cell membrane formation. In stark contrast, the smooth endoplasmic reticulum – devoid of ribosomes – is chiefly involved in the metabolism of lipids, the biosynthesis network of steroids in certain organs, and the detoxification of harmful substances in liver cells. It also holds the critical role of regulating the cellular concentration of calcium ions, which is imperative for multiple cellular activities. Evidently, from the synthesis of proteins and lipids to the regulation of calcium ion concentration, the endoplasmic reticulum is indispensable for upholding the cell's overall functionality.

The Golgi apparatus is a stack of flattened, membrane-bound sacs that receive and modify proteins synthesized in the ER (Soon et al., 2021; Smole et al., 2022; Spangler et al., 2022). It processes, sorts, and packages proteins into vesicles for transportation to various cellular destinations. The Golgi apparatus is essential for the secretion of proteins outside the cell and the formation of lysosomes, specialized organelles for intracellular digestion. A fascinating insight into the realm of cellular microbiology reveals the considerable intricacies and indispensable role of the Golgi Apparatus. This organelle, identified by Camillo Golgi in 1897, is often portrayed as a stack of delicately flattened, membrane-bound sacs, or cisternae, skillfully arranged in a polarized manner. There is the 'cis' face, located near the endoplasmic reticulum, aptly named the entry face, and on the opposing end, the 'trans' face dubbed the exit face, from which the processed proteins are transported. The protein saga commences at the endoplasmic reticulum, where lipid and protein synthesis occurs. These newly synthesized proteins, often equipped with a specific signal peptide, get transferred into the Golgi Apparatus via vesicular transport. Upon the proteins' arrival at the Golgi Apparatus, an elaborate sequence of modifications ensues.

It is within these distinctive cisternae that the proteins undergo a systematic process of 'post-translational modifications'. These modifications can range from the removal, addition, or alteration of certain functional groups, including phosphate (phosphorylation), acetate (acetylation), various sugars (glycosylation) to form glycoproteins, and lipids (lipidation) to form lipoproteins. These modified proteins are then sorted and packaged into transport vesicles at the trans-Golgi network. These vesicles bud off from the 'trans' face and are trafficked to their specific destinations, which could be areas within the cell or outside the cell through a process known as exocytosis. From the internalization of proteins from the endoplasmic reticulum to their requisite modifications, sorting, and eventual dispatch, entering the extraordinary world of the Golgi Apparatus reveals the beauty and complexity inherent in the diligent orchestra directed by this crucial organelle within the cell. Figure 10 is an illustration of cell compartment.

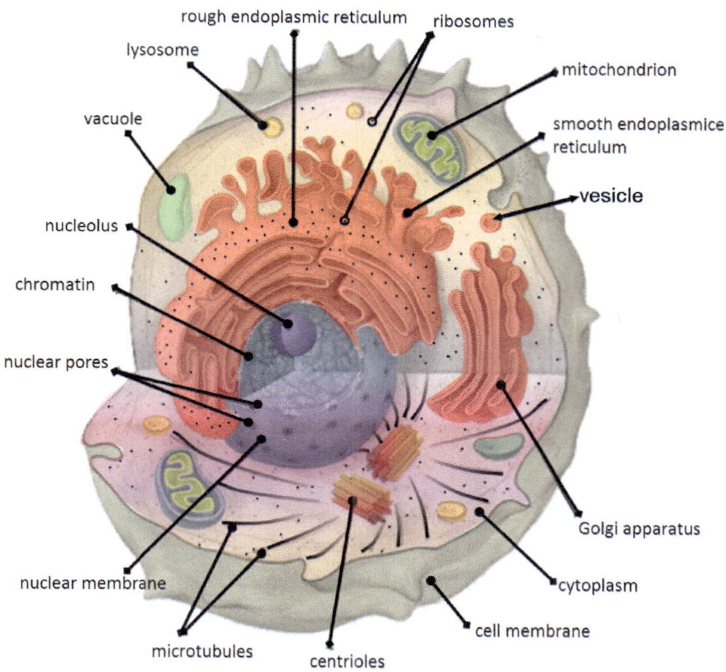

Figure 10. Illustration of cell compartments.

Source: Wikipedia (https://en.wikipedia.org/wiki/Cellular_compartment)

Lysosomes are membrane-bound organelles containing enzymes that break down cellular waste, damaged organelles, and ingested materials through a process called autophagy (Choo et al., 2022; Cibelli et al., 2022; Chohra et al., 2023). They play a crucial role in recycling cellular components and maintaining cellular homeostasis. In the vast arena of cellular microbiology, the importance of maintaining a balanced and organized intracellular environment is paramount. Among the armamentarium of eukaryotic cells exists a particular set of specialized, membrane-bound organelles known as lysosomes. These microscopic entities, primarily responsible for the digestion and recycling of various cellular constituents, were first identified by the Belgian cytologist Christian de Duve in the mid-twentieth century. Functioning as de facto waste disposal systems, lysosomes have an array of acidic hydrolases – enzymes specifically designed to dismantle macromolecules into their respective subunits. These enzymes, manufactured in the endoplasmic reticulum and matured in the Golgi apparatus, are adept at degrading a spectrum of substances, including proteins, nucleic acids, polysaccharides, and lipids. With an internal pH of about 5 maintained by proton pumps, optimal for the activity of these enzymes, lysosomes are well-equipped for their task. Autophagy, a term coined by de Duve himself and derived from the Greek words for 'self-eating', is a process by which these organelles engulf and decompose damaged, non-functional cellular elements. This not only rids the cell of potential hazards but also enables the recycling of the resulting molecular fragments for renewed cellular activities. Similarly, they manage extracellular materials that have been internalized through processes such as endocytosis and phagocytosis, thereby protecting the cell from potential exogenous threats. Hence, lysosomes, with their powerful array of hydrolytic enzymes, play a critical role in the controlled decomposition and recycling of cellular waste, the elimination of damaged organelles, and the management of ingested materials, thereby maintaining the integrity and health of the cell.

Mitochondria are often referred to as the "powerhouses" of the cell due to their role in cellular respiration, where they produce ATP, the cell's energy currency (Arber et al., 2021; Ardalan et al., 2022; Arhontoulis et al., 2022). They contain their own DNA and can replicate independently within the cell. Mitochondria are believed to have originated from symbiotic bacteria engulfed by ancient eukaryotic cells, a process known as endosymbiosis. In the intriguing realm of cellular microbiology, mitochondria are often hailed as "powerhouses" of the cell, and this appellation is ascribed to their cardinal role in cellular respiration – an energy-producing process vital for cell survival and function. Mitochondria are unique organelles typified by their double membrane structure with the outer membrane being smooth and the inner membrane being deeply invaginated, resulting in numerous infoldings known as cristae. This distinctive structure significantly increases the surface area of the inner mitochondrial membrane where the electron transport chain, the final and most productive phase of cellular respiration, occurs. Cellular respiration, a critical metabolic pathway, involves the systematic breakdown of glucose in the presence of oxygen to produce adenosine triphosphate (ATP) – the primary energy currency of cells. This biological process commences in the cytosol with glycolysis, followed by the Krebs cycle and culminates in the mitochondria with oxidative phosphorylation. It is this latter phase where the organic compound's potential chemical energy is fully harnessed and translated into an accessible form – ATP. Mitochondria also house their own DNA, which encodes essential components of the electron transport chain and is independently replicated and transcribed, suggesting an evolutionary origin from symbiotic bacteria. Therefore, mitochondria, through their facilitation of cellular respiration, play an unequivocally central role in driving cellular activity by providing the necessary energy currency, ATP, which perpetuates the myriad cellular processes necessary for life.

Chloroplasts are found in plant cells and some protists, where they carry out photosynthesis, converting light energy into chemical energy (glucose) and releasing oxygen as a byproduct (Yasuhiko et al., 2021; Ye et al., 2021; Yla-Anttila, 2021; Yoshimatsu et al., 2022). Like mitochondria, chloroplasts have their own DNA and are believed to have originated from endosymbiotic cyanobacteria. Diving into the realm of cellular microbiology, one finds a unique organelle, the chloroplast, which plays a central role in the survival and energy production of plant cells and some protists. These double-membrane bound organelles, exclusive to the photosynthetic members of the biological world, are distinctive for their high concentration of pigments known as chlorophyll. These pigments, enclosed within membrane-bound flattened sacs known as thylakoids, are the primary actors responsible for the absorption of light energy from the sun, initiating a marvel of biological engineering – photosynthesis. This crucial process can be broadly categorized into two stages: light-dependent reactions that occur in the thylakoid membrane and light-independent reactions, or the Calvin cycle, unfolding in the stroma, a fluid-filled space within the chloroplast. The absorbed light energy excites electrons within the chlorophyll, propelling them down an electron transport chain, resulting in the synthesis of ATP, the universal energy currency, and NADPH, an electron carrier. These products then fuel the Calvin cycle, where carbon dioxide from the atmosphere undergoes a series of enzymatic reactions, ultimately resulting in the conversion of inorganic carbon into organic glucose molecules. This miraculous transformation of solar energy into chemical energy drives life on Earth, providing not only the base of the food chain but also a significant portion of atmospheric oxygen through the splitting of water molecules during light-dependent reactions. Thus, chloroplasts play an instrumental role in harnessing light energy, converting it into usable chemical energy in the form of glucose, and releasing life-sustaining oxygen as a byproduct, thereby energizing and oxygenating our planet.

Peroxisomes are membrane-bound organelles that contain enzymes involved in various metabolic processes, including the breakdown of fatty acids and the detoxification of harmful substances (Stancliffe & Patti, 2022; Stratman et al., 2022; Stowers, 2023). They play a crucial role in protecting cells from oxidative stress. Navigating the intricacies of cellular microbiology, attention must be turned to peroxisomes, small yet mighty organelles encapsulated by a single membrane and present in nearly every type of eukaryotic cell. Discovered by Christian de Duve, the same cell biologist who discovered lysosomes, peroxisomes are vital for multiple metabolic processes. These include, notably, the breakdown of long-chain fatty acids through a process known as beta-oxidation. Beta-oxidation, encompassing a cyclic series of enzymatic reactions, cleaves two carbon atoms from the fatty acid chain per cycle, generating acetyl-coenzyme A, which can be further processed in the mitochondria to produce ATP. Additionally, peroxisomes aid in the synthesis of plasmalogens, crucial lipid components of certain cell membranes, notably in brain and nerve cells. Beyond lipid metabolism, peroxisomes are also cellular guardians, protecting the cell from toxic substances, particularly reactive oxygen species (ROS) like hydrogen peroxide. Enzymes such as catalase, housed in peroxisomes, neutralize these potentially damaging substances, converting them into harmless water and oxygen. Interestingly, peroxisomes can also produce some of these reactive oxygen species as byproducts of oxidative reactions, thereby embodying both the generation and decomposition of these molecules. Thus, despite their diminutive size, peroxisomes play a multifaceted role in cellular metabolism, from the breakdown of fatty acids to the synthesis of lipids, and in biochemical defense, neutralizing toxic substances to maintain the health of the cell.

The cytoskeleton is a dynamic network of protein filaments that provides structural support to the cell and facilitates intracellular transport and movement (MacArthur & Dawlaty, 2021; Ma et al., 2022; Ma & Wu, 2022). It is composed of microtubules, microfilaments, and intermediate filaments, each with specific roles in cell division, movement, and mechanical support. In the complex landscape of cellular microbiology, the cytoskeleton emerges as a dynamic and intricate network of protein filaments, orchestrating a multitude of critical functions that maintain cell integrity and facilitate cellular processes. Constituting what can be perceived as the cellular architecture, the cytoskeleton is a meticulous assembly of three distinct types of protein fibers: microtubules, intermediate filaments, and actin filaments. Each component has unique features, functions, and dynamics that contribute to the overall versatility of this network. Microtubules, composed of tubulin protein, are the largest and serve as essential tracks for intracellular transport, aided by motor proteins such as dynein and kinesin. They also form the mitotic spindle during cell division, ensuring the even distribution of chromosomes to daughter cells. Intermediate filaments, having a diverse protein composition, provide mechanical strength, preventing the cell from being deformed under physical stress. Lastly, actin filaments, made from actin protein, have both structural and motility roles, influencing cell shape and enabling cell movement. Furthermore, they contribute to muscle contraction in a cooperative interaction with myosin, another motor protein. Apart from serving as the cellular backbone, the cytoskeleton also aids in the movement of organelles, vesicles, and cellular contents, intracellular signaling and the maintenance of cell polarity, thereby ensuring optimal cell functionality. In essence, the cytoskeleton is a versatile and dynamic network that intricately weaves structural support, intracellular transport, movement, and several other functions into the elaborate tapestry of cellular life.

In summary, eukaryotic cells are characterized by the presence of membrane-bound organelles, each with specific roles in cellular processes. The nucleus houses the genetic material, while the endoplasmic reticulum, Golgi apparatus, and vesicles work together in protein synthesis and transport. Lysosomes, mitochondria, and chloroplasts are involved in digestion, energy production, and photosynthesis, respectively. Peroxisomes handle metabolic reactions, and the cytoskeleton provides structural support and facilitates intracellular movement. The complexity and coordination of these organelles contribute to the diverse functions of eukaryotic cells, making them the basis of life's most complex organisms. Let us now close this first chapter with a summary.

CHAPTER 1 SUMMARY

Chapter 1 of the book "Introduction to Living Cells: The Building Blocks of Life" delves into the historical discoveries and foundational principles of cell theory. It explores the contributions of prominent scientists such as Robert Hooke, Matthias Schleiden, Theodor Schwann, and Rudolf Virchow, who collectively laid the groundwork for our understanding of cells.

- Section 1.1 focuses on the establishment of cell theory. Robert Hooke, in the 17th century, made significant contributions with his microscope observations, coining the term "cell" when he examined cork tissue. Matthias Schleiden and Theodor Schwann, in the 19th century, proposed that all living organisms consist of cells and that cells are the basic building blocks of life. Rudolf Virchow's concept of "omniscellula e cellula" emphasized that cells arise from pre-existing cells, completing the principles of cell theory.
- Section 1.2 delves into the advancements in cell biology through the development of microscopes. Microscopes have revolutionized our ability to observe and study cells at a microscopic level. Technological advancements in microscopy have led to the discovery of various cellular structures and organelles, paving the way for deeper insights into cellular functions and processes.
- Section 1.3 explores the fundamental distinction between unicellular and multicellular organisms. Unicellular organisms are self-sufficient entities, capable of performing all life functions within a single cell. In contrast, multicellular organisms comprise specialized cells that collaborate to support the overall function of the organism. Understanding cell diversity between these two categories provides insights into the evolution and complexity of life forms on Earth.
- Section 1.4 focuses on prokaryotic cells, which are characterized by their simple structure and lack of membrane-bound organelles. These single-celled microorganisms, such as bacteria and archaea, have played a crucial role in shaping the planet's ecology. The chapter delves into the unique structure and functions of prokaryotic cells, including their genetic organization, cell wall, and roles in energy production and nutrient uptake.
- Section 1.5 delves into eukaryotic cells, which represent a higher level of cellular organization. Eukaryotic cells contain membrane-bound organelles, such as the nucleus, endoplasmic reticulum, Golgi apparatus, lysosomes, mitochondria, chloroplasts (in plant cells), and more. Each organelle has specific roles in cellular processes, including DNA regulation, protein synthesis, energy production, and cellular transport. Understanding the functions of these organelles enhances our understanding of the complexity and efficiency of eukaryotic cells.

In conclusion, Chapter 1 provides a comprehensive overview of the historical milestones that led to the establishment of cell theory and the subsequent advancements in cell biology. It introduces readers to the distinct characteristics and functions of prokaryotic and eukaryotic cells, setting the stage for a deeper exploration of living cells as the essential building blocks of life.

Chapter

2

CELL STRUCTURE AND FUNCTION

CONTENTS

In this chapter, we delve into the intricacies of cell structure and function, exploring the remarkable architecture that underlies the diverse and essential processes of life. Cells, as the fundamental units of life, exhibit remarkable complexity, housing an array of specialized organelles, each contributing to specific cellular functions.

The understanding of cell structure and function is crucial for comprehending how living organisms maintain their vitality and carry out vital processes. At the center of cell organization lies the plasma membrane, the gatekeeper of the cell. The plasma membrane regulates the passage of molecules into and out of the cell, enabling it to maintain a controlled internal environment. We examine the mechanisms of transport, such as diffusion and active transport, that govern the movement of essential substances across this vital boundary.

Moving deeper within the cell, we encounter the nucleus, the command center of genetic information. Within the nucleus lies the cell's DNA, holding the blueprint for the cell's activities and determining its characteristics. The processes of DNA replication, transcription, and translation allow the cell to synthesize proteins and carry out its specialized functions. As we venture further into the cellular landscape, we encounter the endoplasmic reticulum and Golgi apparatus, crucial players in protein synthesis and modification. The endoplasmic reticulum facilitates protein synthesis, while the Golgi apparatus modifies, sorts, and packages these proteins for distribution within and outside the cell.

The mitochondria come into focus as the powerhouses of cellular respiration. Here, cellular energy production takes place, providing the cell with adenosine triphosphate (ATP), the currency of cellular energy. We explore the processes of glycolysis, the Krebs cycle, and the electron transport chain that culminate in ATP synthesis. Finally, we turn our attention to chloroplasts, which perform photosynthesis and produce energy in plant cells. Through the capture of light energy and the conversion of carbon dioxide and water into glucose, chloroplasts enable plants to sustain themselves and serve as the foundation of terrestrial food chains.

Throughout this chapter, we unravel the inner workings of these essential organelles and the interconnectedness of their functions. Understanding the cellular machinery is pivotal not only for biology students and researchers but also for appreciating the intricate beauty of life itself. By examining the mechanisms and structures underlying cell function, we gain insight into the remarkable complexity and harmony of living systems. Let us open this second chapter by exploring the role of the plasma membrane as the gatekeeper of the cell.

2.1 PLASMA MEMBRANE: THE GATEKEEPER OF THE CELL

The plasma membrane is a dynamic and vital structure that serves as the gatekeeper of the cell (Cheung et al., 2022; Choezom & Gross, 2022; Chhabra et al., 2023). It is a selectively permeable barrier that separates the internal environment of the cell from the external surroundings, allowing for the regulation of molecular exchange. This delicate membrane is composed of a phospholipid bilayer with embedded proteins and plays a pivotal role in various cellular processes. The plasma membrane, a vibrant and essential structure in cell biology, invariably performs the role of a gatekeeper, controlling the ingress and egress of various molecules to maintain internal cellular equilibrium. Figure 11 illustrates how the cell membrane is a gatekeeper element.

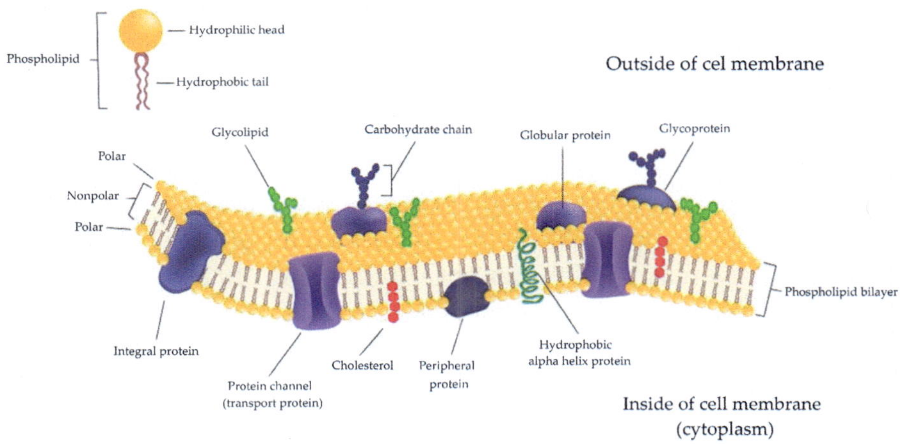

Figure 11. Illustration of the cell membrane as a gatekeeper.

Source: Da Jiménez-Jiménez C, Manzano M, Vallet-Regí M. Nanoparticles Coated with Cell Membranes for Biomedical Applications. Biology. 2020; 9(11):406. https://doi.org/10.3390/biology9110406

A well-documented case that exemplifies this daunting role is evident within human neurons (Arber et al., 2021; Ardalan et al., 2022; Arhontoulis et al., 2022). Here, the plasma membrane, with its complex lipid bilayer interspersed with various proteins, underpins the functionality of neurons by regulating the movement of charged ions, creating the essential electrical currents needed for nerve impulse transmission. Sodium ions, normally concentrated outside the cell, and potassium ions, present in higher quantities inside the cell, are capable of passing through their respective ion channels embedded within the membrane. However, their movement is only initiated upon receiving an appropriate signal. For example, during an action potential, the membrane initially becomes more permeable to sodium ions via the opening of sodium channels, prompting their diffusion into the cell, and as a result, creating a change in the electrical charge across the membrane. Once the membrane potential reaches its peak, these channels close while potassium channels open, allowing potassium ions to exit the cell, returning the membrane to its resting state. Thus, the well-timed orchestration of these ion movements is largely directed by the selective permeability of the plasma membrane - an attribute that is crucial to the neurons' ability to generate and conduct electrical signals efficiently. Highlighting the integral role of the plasma membrane in a neuron's functionality underscores the profound relevance of the plasma membrane in preserving the interactive dynamism of cellular processes. The plasma membrane's role as the gatekeeper is vividly illuminated, revealing its capacity to maintain the delicate internal architecture of the cell while negotiating its relationship with the exterior environment.

The phospholipid bilayer forms the backbone of the plasma membrane, with hydrophilic phosphate heads facing outward towards the aqueous environments and hydrophobic lipid tails oriented inward, away from water (Efstathiou & Draviam, 2021; Duncan, 2022, 2023; Dunkley et al., 2022). This unique structure allows the membrane to maintain its integrity while permitting the passage of certain molecules through its hydrophobic interior. The structure of the phospholipid bilayer, with its hydrophilic phosphate heads oriented outward towards aqueous environments and hydrophobic lipid tails facing inward away from water, serves as the backbone of the plasma membrane in cell microbiology. This delicate arrangement of components is the fundamental unit of all cellular life, and its importance can be comprehensively appreciated through the case of a generic eukaryotic cell. In this context, the plasma membrane encapsulates the cell, demarcating it from its surroundings and carefully segregating and protecting its internal environment. The phosphate heads, due to their hydrophilic nature, naturally align themselves towards both the intracellular cytosol and the extracellular fluid, forming a stable interface with these aqueous environments. Meanwhile, the lipid tails, being hydrophobic, congregate in the middle of the bilayer, sheltered from the surrounding water. This amphipathic architecture of the phospholipid bilayer allows the cell to maintain crucial differences in the concentration of molecules and ions inside and outside of the cell. It further acts as a permeability barrier for many substances, enabling selective diffusion of specific molecules like gasses and small, uncharged polar molecules, whereas blocking others. Simultaneously, proteins embedded in the bilayer facilitate the transport of ions and larger, charged molecules. Through these mechanisms, the phospholipid bilayer performs as the gatekeeper, navigating the controlled traffic of substances to ensure cellular functionality. The central role played by the phospholipid bilayer in the architecture and performance of the eukaryotic cell's plasma membrane vividly illustrates how the unique characteristics of its components - hydrophilic phosphate heads and hydrophobic lipid tails - are meticulously utilized to uphold the tenets of life at a microscopic level.

Integral and peripheral membrane proteins are embedded within the lipid bilayer and play diverse roles in cellular function (Vakonakis, 2021; Veerasamy et al., 2021; van den Dries et al., 2022). Integral proteins traverse the entire membrane, serving as channels, carriers, or receptors. These proteins enable the movement of specific substances across the membrane, facilitate signal transduction, and mediate cell-cell communication. Peripheral proteins are loosely attached to the membrane's surface and participate in various cellular processes, such as cell adhesion and cytoskeletal support. Integral and peripheral membrane proteins that are embedded within the lipid bilayer significantly contribute to cellular function, with roles as multifaceted and diverse as the cells themselves. For example, in the neuronal cells of the nervous system, the symbiotic relationship between these proteins and the lipid bilayer becomes explicitly clear. Integral proteins, designed to span the entire membrane with portions exposed both inside and outside the cell, serve a myriad of roles. Among these roles, some integral membrane proteins form ion channels. These channels, like the voltage-gated sodium and potassium channels, are essential for establishing and changing the membrane potential, which is critical for the passage of electrical signals along the neuron. Another subset of integral proteins acts as transporters, such as the sodium-potassium pump, which actively maintains the concentration gradient of these ions across the membrane, critical for the neuron's functionality. In contrast, peripheral membrane proteins, attached to the inner or outer surface of the membrane, serve different roles. They often act as enzymes catalyzing specific reactions, or as a part of the cytoskeleton in maintaining cell shape and facilitating movement. Some peripheral proteins even function as signal transduction molecules, helping the cell respond to changes in its environment. Therefore, by scrutinizing the neuron, it is easy to recognize the contributions of integral and peripheral membrane proteins to cell operation and how this harmonious alignment of diverse functionalities embedded within the lipid bilayer orchestrates the proper functioning of cells, thereby ensuring the healthy performance of more complex biological systems.

The plasma membrane's selective permeability is essential for maintaining cellular homeostasis and facilitating specific cellular functions (Das et al., 2021; de Albuquerque et al., 2021; De Camilli, 2021; de Torres-Jurado et al., 2022). Small, non-polar molecules like oxygen and carbon dioxide can diffuse freely through the lipid bilayer, while larger, polar molecules and ions require the assistance of transport proteins for passage. The selective permeability of the plasma membrane, a vital attribute in cell microbiology, is indispensable for establishing cellular homeostasis and facilitating particular cellular functions. This is aptly exemplified in the erythrocytes, or red blood cells (RBCs), that circulate in the human bloodstream. The plasma membrane of an RBC, akin to other cells, exhibits a selective permeability that allows specific substances to pass through while restricting others. Saliently, it permits the free diffusion of gasses like oxygen and carbon dioxide. Oxygen, carried from the lungs to cells via the bloodstream, diffuses from the high concentration in the blood into the cells where its concentration is lower, facilitating cellular respiration. Conversely, carbon dioxide, a metabolic waste product, diffuses out of the cells into the bloodstream, from where it is transported to the lungs for exhalation. Notwithstanding this, the membrane prevents the loss of vital molecules like proteins and ions that are crucial for cellular processes. Besides such passive movements, active transport mechanisms led by membrane-embedded proteins play a role in maintaining ionic gradients that regulate cell volume and contribute to the membrane potential. Furthermore, receptor proteins on the RBC membrane enable the recognition and binding of specific substances, such as hormones, which mechanically triggers intracellular responses. Thus, through this exploration of the erythrocyte, the pivotal, multifunctional role of selective permeability in maintaining cellular homeostasis and mediating specific cellular functions becomes apparent. This underlines the exceptional orchestration at the microscopic level that meticulously balances cellular integrity and function while elaborately interacting with the external environment. Figure 12 compares the membrane in animal and plant cells.

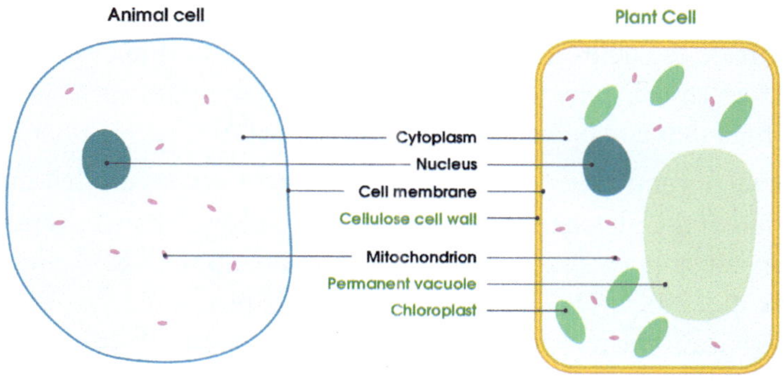

Figure 12. Comparing plasma membranes in animal and plant cells.

Source:https://commons.wikimedia.org/wiki/File:Differences_between_simple_animal_and_plant_cells_%28en%29.svg

The membrane's role as a gatekeeper is particularly significant in active transport and endocytosis/exocytosis processes (Magalhaes et al., 2021; Maib & Murray, 2022; Mahaldashtian et al., 2023). Active transport involves the movement of molecules against their concentration gradient, requiring energy from ATP. This energy-driven process allows cells to maintain ion concentrations, essential for nerve impulses, muscle contractions, and nutrient uptake. The role of the cell membrane is fundamental in regulating cellular processes, serving as a gatekeeper that facilitates or hinders the transport of substances in and out of the cell. Active transport and endocytosis/exocytosis mechanisms exemplify this vital role, which is particularly prominent in the functioning of neurons. Active transport involves transporting substances across the cell membrane against their concentration gradient, a process that requires energy, hence referred to as "active". Neurons utilize this process to manage the concentrations of sodium and potassium ions on either side of their cell membrane. This is achieved through the sodium-potassium ATPase pump, which expels three sodium ions from the neuron in exchange for two potassium ions, thereby establishing a resting potential in preparation for the conduction of electrical signals. Simultaneously, neurons also engage in endocytosis and exocytosis to govern the release and uptake of neurotransmitters, chemical messengers vital to nerve impulse transmission. Neurotransmitters are initially packaged into vesicles, which during exocytosis, fuse with the cell membrane, providing an exit route for the neurotransmitters. Conversely, endocytosis can be employed by neurons to retrieve neurotransmitters or recycle membrane components, by letting the membrane engulf the desired molecules. This results in the formation of a vesicle inside the neuron containing the engulfed substances. Collectively, these mechanisms display the pivotal role of cell membranes as gatekeepers, maintaining a delicate balance that underpins cellular survival and function.

Endocytosis and exocytosis are bulk transport processes that involve the invagination of the plasma membrane to internalize substances (endocytosis) or release substances (exocytosis) (Doostmohammadi & Ladoux, 2022; Drozdz et al., 2022; Du et al., 2023). These processes are essential for nutrient uptake, waste removal, and the secretion of cellular products like hormones and enzymes. Endocytosis and exocytosis, bulk transport processes that represent manifestations of the plasma membrane's dynamic adaptability within cell microbiology, have transformative implications for the internalization or release of substances. These mechanisms are illustrated cogently in the context of a macrophage, a type of white blood cell integral to the immune system. Macrophages are essentially the sanitation workers of the body, responsible for identifying, engulfing, and digesting cellular debris and foreign pathogens. For this purpose, macrophages exploit endocytosis in a process known as phagocytosis. On encountering a pathogenic intruder, the macrophage extends pseudopodia or "false feet," that revolve around the pathogen. Subsequently, the pseudopodia merge, causing an invagination of the plasma membrane that encapsulates the pathogen inside a phagosome, essentially a large vesicle. This internalization of the pathogen through the endocytic process enables the macrophage to contain and neutralize the potential threat effectively. Parallelly, exocytosis plays a pivotal role in the macrophage's functionality. Once the pathogen is digested within the phagosome, the resultant waste material needs to be expelled. Here, the phagosome fuses with the plasma membrane, enabling the release of waste material into the extracellular environment, thereby completing the process of exocytosis. Thus, an in-depth analysis of a macrophage attests to the critical functional duality of endocytosis and exocytosis. The processes together illustrate the inherent flexibility and responsiveness of the plasma membrane, driving the cell's dynamism and fortifying its robust interaction with the environment.

Furthermore, the plasma membrane is involved in cell recognition, immune response, and cell signaling (Baluska & Yokawa, 2021; Aridor, 2022; Arjmand et al., 2023). Cell surface receptors interact with specific ligands or signaling molecules, initiating intracellular cascades that regulate gene expression and cellular responses. The plasma membrane, beyond its role as a physical barrier, is instrumental in cell recognition, immune response, and cell signaling. A compelling example can be drawn from the intricate workings of the human immune system, particularly the interaction between a macrophage and a T cell. The membrane of a macrophage, or any antigen-presenting cell (APC), contains specialized molecules called Major Histocompatibility Complex II (MHC II) proteins. On ingestion of a pathogen, the macrophage digests it into fragmented antigens that bind to the MHC II molecules. The antigen-MHC II complexes are subsequently expressed on the cell surface, making the antigens visible to T cells - a process crucial for cell recognition and immune response initiation. A helper T cell recognizes this complex through its antigen-specific T cell receptor (TCR), leading to the activation of the T cell. This interaction is further strengthened by other membrane proteins that serve as co-receptors, enhancing the stability of the T cell-APC interaction. Once activated, the T cell proliferates, releasing cytokines that initiate and coordinate further immune responses, thereby amplifying the fight against the pathogen. This interaction also exemplifies the plasma membrane's role in cell signaling, facilitating communication not only within the cell but between different cells. Hence, the macrophage-T cell interaction case underlines the plasma membrane's critical function in ensuring cells effectively communicate, identify each other, and coordinate their responses - a role that ultimately safeguards the well-being of the entire organism.

The fluid mosaic model aptly describes the dynamic nature of the plasma membrane (Patriarca et al., 2021; Pauletto et al., 2021; Passanha et al., 2022). It exhibits lateral movement of lipids and proteins, allowing for adaptability and flexibility. Membrane fluidity is influenced by factors such as temperature and lipid composition, impacting cellular function. The fluid mosaic model, a conception of the plasma membrane's organization, aptly encapsulates its dynamic nature – an aspect that can be vividly illustrated through the functioning of a specific type of white blood cells known as B-cells. This model encapsulates the idea that the plasma membrane's semi-viscous constitution allows its components, including lipids, proteins, and carbohydrates, to intersperse and move within the membrane. Combining the fluidity of movement with the varied assortment of these intercalated components, the fluid mosaic model provides a sound framework that explains the sculpting and functioning of B-cells and, by extension, their role in the immune response. B-cells employ the fluidity provided by the phospholipid bilayer to perform receptor-mediated endocytosis of antigens. When an antigen connects with the B-cell receptor, the fluid structure of the membrane allows the antigen-receptor complex to aggregate and form an invagination. Subsequently, the antigen is internalized by the process of endocytosis. Additionally, these B-cells use the mosaic nature of the membrane, characterized by high protein content, to present these antigens to helper T-cells, in a process spurring the production of antibodies. Therefore, as demonstrated by the workings of B-cells, the fluid mosaic model befittingly describes the dynamic nature of the plasma membrane, bringing forth a deeper understanding of the intricate processes integral to cellular life.

In conclusion, the plasma membrane serves as the gatekeeper of the cell, governing the passage of molecules in and out of the cell. Its selective permeability, aided by diverse membrane proteins, ensures the maintenance of cellular homeostasis and facilitates essential cellular processes. The intricate organization and functions of the plasma membrane are crucial aspects of cell microbiology, offering insights into the fascinating mechanisms that govern cellular life. The next section will focus on the nucleus as the command center of genetic information.

2.2 THE NUCLEUS: COMMAND CENTER OF GENETIC INFORMATION

The nucleus is a central and vital organelle within eukaryotic cells, acting as the command center of genetic information (Vakonakis, 2021; Veerasamy et al., 2021; van den Dries et al., 2022). It is an intricate and dynamic structure that plays a crucial role in regulating cellular activities and transmitting genetic instructions to guide cell functions. The nucleus occupies a central, vital status within eukaryotic cells, acting as the command center of genetic information - a fact quintessentially demonstrated within human somatic cells. Being referred to as the cell's "control center," this complex double-membraned organelle houses the cell's most valuable asset – its DNA, which is organized into chromosomes. Each human somatic cell nucleus contains 46 chromosomes that consist of genes, the basic units of heredity that guide the synthesis of proteins necessary for the proper functioning of the body. Through a meticulous process known as transcription, specific segments of the DNA's sequence are copied into mRNA (messenger RNA) within the nucleus. This mRNA, carrying the genetic instructions, is then transported out of the nucleus into the cytoplasm via nuclear pores, structures within the nuclear envelope, which selectively allow the transit of molecules between the nucleus and the rest of the cell. Within the cytoplasm, ribosomes read these instructions and translate them into corresponding proteins – a process termed translation. Furthermore, the nucleus takes part in regulating the cell cycle, playing a decisive role in cell division, as it dictates when DNA duplication and cell division should occur. Therefore, the commanding function of the human somatic cell nucleus perfectly embodies how eukaryotic cells safeguard and make use of their genetic profiles, underscoring the nucleus's pivotal role. This role enables the complexity of form and function observed within multicellular organisms, bringing to the fore the marvels of cell microbiology, whereby intricate processes are carried out efficiently and accurately. Figure 13 illustrates the nucleus of an animal cell.

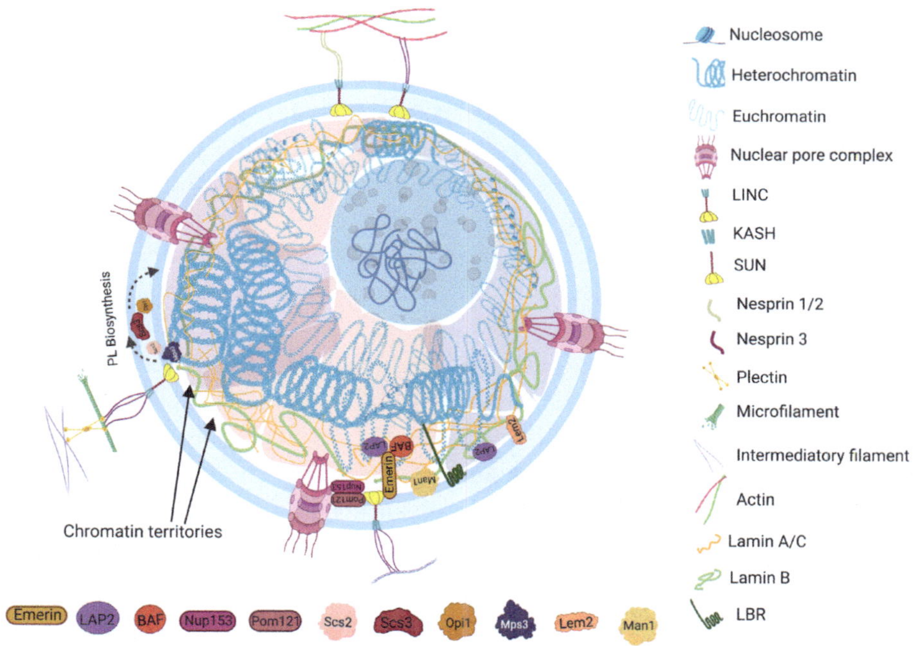

Figure 13. The nucleus of an animal cell.

Source: Manda NK, Golla U, Sesham K, Desai P, Joshi S, Patel S, Nalla S, Kondam S, Singh L, Dewansh D, et al. Tuning between Nuclear Organization and Functionality in Health and Disease. Cells. 2023; 12(5):706. https://doi.org/10.3390/cells12050706

Enclosed within a double-membrane nuclear envelope, the nucleus houses the cell's genetic material in the form of chromatin (Tolkin & Hubbard, 2021; Tomoda & Kime, 2021; Tuieng et al., 2021). Chromatin consists of DNA wrapped around histone proteins, forming a complex and organized structure. During cell division, chromatin condenses into visible chromosomes, allowing for the accurate segregation of genetic material into daughter cells. Encased within the formidable double-membrane nuclear envelope, the nucleus guards the cell's genetic material, which resides in the form of chromatin, a function that particularly comes into prominence in the process of cellular division as observed in a human skin cell. Human skin, with its constant exposure to external stressors and the resulting high turnover rate of cells, is an extraordinary site for observing how the nucleus, and the chromatin within it, play a role in maintaining the body's largest organ.

Within the life cycle of a skin cell, when it is ready to divide and proliferate, the chromatin housed in the nucleus, which generally exists in a diffuse, uncondensed state for regular transaction of genetic information, undergoes a transformation. It condenses into tightly packed, discrete chromosomes, thereby protecting the genetic information during the tumultuous process of cell division. This process, known as mitosis, starts with the nuclear envelope breaking down, allowing spindle fibers to interact with the chromosomes. Each chromosome is then pulled apart into two identical sets, each destined for one of the two new cells. Following this, the nuclear envelope reassembles around each set of chromosomes, in effect, creating two new nuclei in two new cells. Hence, the case study of a human skin cell underscores the importance of the nucleus in protecting and organizing genetic material. This ability is not only pivotal in responding to day-to-day functioning, but also in ensuring the perpetuity of the cell's genetic legacy, thereby allowing the organism to survive, grow, and adapt.

The nucleus functions as the repository of the cell's genetic information, which controls the synthesis of proteins essential for cellular processes (Shen et al., 2021; Siu et al., 2021; Shi et al., 2022; Shoup & Priola, 2023). This genetic information is stored in the form of genes, which are specific segments of DNA that encode instructions for producing proteins. The process of transcription occurs in the nucleus, where the DNA sequence of a gene is transcribed into a complementary RNA molecule called messenger RNA (mRNA). The nucleus acts as a repository of genetic information in a cell and this inherent attribute is integral to the synthesis of proteins essential for various cellular processes. A case study that brings this critical function to the fore concerns insulin production in pancreatic cells known as β-cells. Each β-cell contains a nucleus with hundreds of copies of the gene encoding insulin, a hormone instrumental in the regulation of glucose levels in the body. Upon detection of increased blood glucose levels, the nucleus initiates the process of gene expression, converting the information contained in the insulin gene into a format that can guide the synthesis of the insulin protein. The first step involves the transcription of the insulin gene into a messenger RNA (mRNA) molecule, a process mediated by RNA polymerase, an enzyme housed within the nucleus. This mRNA molecule, a characteristic copy of the gene sequence, thereafter migrates from the nucleus to the cytoplasm through pores in the nuclear membrane, a critical interaction that testifies to the dynamic nature of the nuclear membrane. Following this, a cellular structure known as the ribosome reads the sequence of the mRNA in a process termed translation and assembles the insulin protein based on this information. Subsequently, the insulin protein is packaged and dispatched to fulfill its pivotal role in glucose regulation. Hence, the nucleus of the β-cell, with its control over the cell's genetic material, lies at the heart of this complex yet intriguingly precise and efficient biological mechanism. The orchestration of protein synthesis within the cell reaffirms the nucleus's fundamental role in governing cellular functionality and well-being.

After transcription, the mRNA exits the nucleus through nuclear pores, which are protein channels that traverse the nuclear envelope (Pinson & Huttner, 2021; Postler, 2021; Pratt et al., 2021). The mRNA then travels to the cytoplasm, where it binds to ribosomes for translation into proteins. This translation process is fundamental for synthesizing enzymes, structural proteins, and regulatory molecules that govern cellular functions. In order to elucidate the statement that posits that after transcription, the mRNA exits the nucleus through nuclear pores, which are essentially protein channels that traverse the nuclear envelope, it would be profoundly beneficial to draw one's attention toward a case study in the realm of cell biology. Consider, for this discussion, the human cell, characterized by its intricate multi-layered structure, hosting an elaborate set of processes that ensure the continuance of life itself. Scholarly research has shown that, melded within this complex structure, lies the noteworthy process of RNA transcription. This very operation commences within the nucleus with the unwinding of the molecule of DNA. The unwound DNA strand, acting as a template, is read by the enzyme RNA polymerase to build a transcript of messenger RNA (mRNA), a complementary strand that mirrors the sequence of the DNA but incorporates uracil (U) in lieu of thymine (T). Following its synthesis, the crucial process of the mRNA's egress from the nucleus takes place. This, as alluded to in the statement, occurs via nuclear pores, infinitesimal gateways composed of multiple proteins known as nucleoporins bedded into the double-layered nuclear envelope, which incidentally delineates the nucleus from the cytoplasm. The nuclear pores, integral players in the nucleocytoplasmic traffic, permit the passage of macromolecules like mRNA from the nucleus to the cytoplasm. The mRNA export is not random or passive; instead, it is, as research validates, a highly regulated and specific process. The exported mRNA, bearing the genetic instructions sourced from the DNA, then sways towards the ribosomes present in the cytoplasm. Here, the narrative of the mRNA continues, transforming from a mere sequence of nucleotides to the precisely folded, physiologically active protein, fulfilling the prophecy inaugurated by the DNA. Consequently, the statement enshrines an indispensable facet of cellular function, signifying the trail of life that follows from the heart of a cell's nucleus to its periphery.

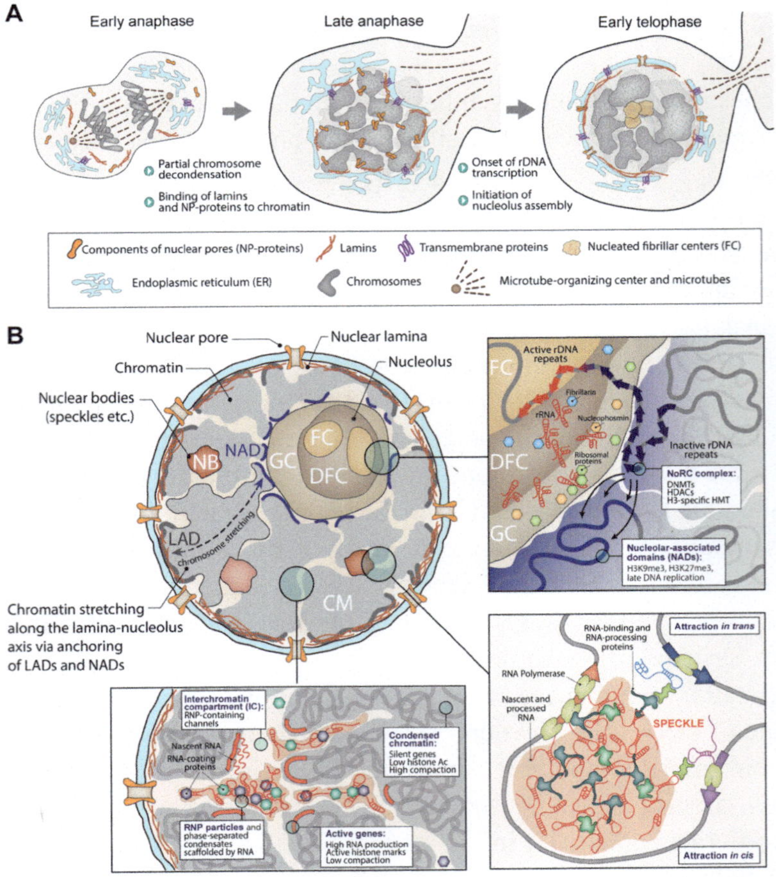

Figure 14. The structure and functionality of a cell nucleus.

Source: Razin SV, Ulianov SV. Genome-Directed Cell Nucleus Assembly. Biology. 2022; 11(5):708. https://doi.org/10.3390/biology11050708

The nucleus also houses the nucleolus, a distinct region within the nucleus responsible for ribosome biogenesis (Yu et al., 2022; Yuan et al., 2021; X. J. Zhang et al., 2023; X. Y. Zhang et al., 2023). The nucleolus contains DNA sequences that encode ribosomal RNA (rRNA), which combine with proteins to form ribosomes. Ribosomes are essential cellular components involved in protein synthesis, highlighting the nucleus's central role in controlling the cell's protein production machinery. In order to deepen our understanding of the assertion that the nucleus houses the nucleolus, a specific region specialized in ribosome biogenesis, it is essential to explore a representative case study within the field of cell microbiology. Here, a cell from the robust organism Saccharomyces cerevisiae, commonly known as yeast, is selected due to its relative simplicity and its extensive contribution to our understanding of eukaryotic cell biology. Contained within the confines of its nuclear envelope, the yeast cell possesses a well-organized nucleus wherein the nucleolus, the notable organelle of interest, is prominently situated. This specialized sub-nuclear compartment, the nucleolus, distinguished from the surrounding nucleoplasm, serves as the epicenter for the complex process of ribosome biogenesis (see figure 14). This procedure begins with the transcription of ribosomal RNA genes hosted by the nucleolus.

Employing the enzyme RNA polymerase I, rRNA precursors of 35S and 5S are synthesized (He et al., 2021; Henderson et al., 2022; Herdman et al., 2022). These nascent molecules are then subjected to a meticulous and intricate course of processing, which includes cleavage events and chemical modifications. This sequence of these operations results in the constituent rRNAs of the mature ribosome, namely 18S, 5.8S, and 25S. Meanwhile, ribosomal proteins, synthesized in the cytoplasm and imported into the nucleus, travel into the nucleolus to congregate with the aforementioned rRNAs. This crucial assembly ties the previously fragmented rRNAs and proteins into the intricate yet harmonious structures of the small and large ribosomal subunits. Remarkably, the accomplished subunits are exported via the nuclear pores to the cytoplasm, where they are united to form functioning ribosomes, as the situation demands. Hence, the nucleolus, residing in the nucleus, fulfills the profound role of a master orchestrator, directing the symphony of ribosome biogenesis responsible for the essential protein synthesis of a cell. This monumental manifestation of cellular strategy tucked within the nucleus underscores the truth value of the initial statement, reinforcing the fundamental principles of cell biology. Beyond protein synthesis, the nucleus plays a critical role in cell cycle regulation

(Chen et al., 2021; Chen et al., 2023; Chen et al., 2022). During interphase, the cell prepares for cell division by replicating its DNA in the nucleus. This process ensures that each daughter cell receives an identical copy of the genetic material. The nucleus also monitors cell cycle checkpoints, ensuring the fidelity of DNA replication and repair processes before proceeding to the next phase of the cell cycle. To fully grasp the assertion that the nucleus plays a pivotal role in cell cycle regulation, beyond its well-known function in protein synthesis, it might be beneficial to examine a case study in the realm of cell microbiology.

Consider, for instance, a somatic human cell, which provides an excellent model for understanding these dynamics. Within the intricately arranged interior structure of the cell, inside the protective double layer of the nuclear envelope, occur sequences of events that are tightly regulated and distinctly choreographed. These procedures, which form the essence of the eukaryotic cell cycle, constituting iterative cycles of growth and division, are governed by the nuclear instructions. The nuclear control of the cell cycle commences with the DNA, which performs as the primary custodian of cell cycle regulation by encoding key functional proteins integral to cell cycle progression. These include cyclins, cyclin-dependent kinases (CDKs), and their inhibitors. The harmonized activity of these proteins is integral to the progression of the cell through various well-defined phases of the cell cycle: G1, S, G2, and M. During G1, the cell prepares for DNA replication by producing the necessary enzymes for the synthesis of DNA and closely assessing the state of the cellular environment. Once these preparatory actions are deemed satisfactory and the cell progresses past the G1 checkpoint, it enters the S phase.

Here, the process of DNA replication ensures that each daughter cell receives a complete copy of the parent cell's genetic material. This replication, conceived and governed by the nucleus, forms the core of the cell cycle processes. The synthesis phase is followed by the G2 phase, a period during which the cell prepares for division by synthesizing proteins required for building the mitotic spindle. As the final phase of the cell cycle, the M phase, ensues, mitosis leads to the segregation of identical sets of chromosomes to each daughter cell in the process of cell division. This intricate sequence of events in the life cycle of a cell is the outcome of the elegant regulatory mechanisms dwelling within the cell nucleus. Therefore, the assertion that the nucleus plays a fundamental role in cell cycle regulation rings true, with its influence traversing far beyond protein synthesis and casting a significant impact on the cell's life cycle.

Furthermore, the nucleus is involved in gene regulation, determining which genes are expressed and which are repressed in response to internal and external signals (Lewis & Wells, 2021; D. Li et al., 2021; Lerit, 2022). Transcription factors and other regulatory proteins control gene expression, allowing cells to adapt to changing environmental conditions and respond to developmental cues. Deepening our appraisal of the complexities within cellular biology prompts examination of the assertion that the nucleus is fundamentally involved in the regulation of gene expression, navigating the delicate balance between genes being expressed or repressed in reaction to a compendium of internal and external signals. For this exploration, an instance from the mammalian species, particularly the human cell, provides a rich tableau of detailed case study for research.

Within the layered enceinte of the nuclear envelope, lies the nucleus, a command center of sorts, harboring the prime genetic material, aka DNA, the helical structure of which consists of multiple sequences of nucleotides representing individual genes. These genes are potential recipes for the synthesis of proteins, considered as the workhorses of cellular function. Logically then, the synthesis of a protein from a gene, a process known as gene expression, is a momentous affair within a cell's existence. The nucleus, in this context, acts as the primary decision-maker, directing which genes shall be transcribed into mRNA, and subsequently translated into proteins, as well as which genes shall lie dormant, refraining from such expression. This determination is based upon an array of sophisticated signaling pathways responding to a variety of internal and external stimuli. Internal signals may include feedback loops from metabolic pathways or signals from cellular checkpoints during the cell cycle, while external cues can span diverse environmental stimuli like temperature, light, or the presence of certain molecules. The mechanism of gene regulation, supervised by the nucleus, involves intricate systems of activator and repressor proteins that bind to specific sites within the DNA, modulating the rate of gene transcription, and hence, its expression or repression. Further complexity is introduced through processes such as DNA methylation or histone modification, both integral to the field of epigenetics. Consequently, viewed from the perspective of a human cell, the nucleus undeniably performs an integral role in the regulation of gene expression, governing the nuanced, dynamic interplay between activating and repressing genes in response to a diverse palette of inducements. This resonates firmly with the core of the initial assertion, underscoring the nucleus's status as the master regulator within the architecture of life's most elemental unit.

In summary, the nucleus is the command center of genetic information in eukaryotic cells, playing a pivotal role in regulating cellular activities. It houses the cell's DNA, orchestrates the synthesis of proteins through transcription and translation, and governs cell cycle progression and gene regulation. Understanding the nucleus's intricate functions provides insights into the complex mechanisms that underlie cell microbiology and the fundamental processes that sustain life. The next section will explore protein synthesis and modification.

2.3 ENDOPLASMIC RETICULUM AND GOLGI APPARATUS: PROTEIN SYNTHESIS AND MODIFICATION

The Endoplasmic Reticulum (ER) and Golgi Apparatus are two crucial organelles involved in the synthesis and modification of proteins in eukaryotic cells (Ford et al., 2021; Forveille et al., 2021; Gan, 2021; Gallusser et al., 2022). These organelles work in tandem to ensure the proper folding, processing, and distribution of proteins throughout the cell. The Endoplasmic Reticulum (ER) and the Golgi Apparatus, central components of the eukaryotic cell's endomembrane system, are essential for the complex processes of protein synthesis and modification.

The ER, a continuous network of flattened sacs and tubules, is categorized into two types: the rough ER (characterized by ribosomes studded on its surface) and smooth ER, each contributing to different aspects of protein synthesis and maturation. For instance, in the case study examining Saccharomyces cerevisiae, a simple eukaryotic model organism, the rough ER was observed to have an active role in the early stages of protein synthesis through the attachment of ribosomes that translate the mRNA into nascent polypeptides. Concurrently, the ER lumen serves as an ideal environment for protein folding, ensuring correct tertiary structures are formed, and where necessary, disulfide bonds are frequently introduced within the cysteine residues. Post-translational modifications such as glycosylation also take place in the ER, adding further refinement to the nascent proteins. Following their formation and initial modifications within the ER, these proteins are then encapsulated within vesicles - small enclosed sacks - for transport to the Golgi Apparatus. Regarded as the 'post office' of the cell, the Golgi Apparatus, through its cisternae, further modifies these proteins through processes such as phosphorylation or removal of certain oligosaccharides. Detailed investigation of cell protein markers in the S.

cerevisiae again illuminates the Golgi's role in packaging the modified proteins into secretory vesicles targeting various cellular locations or even for exocytosis. Therefore, through a combined and sequential effort, the ER and the Golgi Apparatus ensure that proteins within eukaryotic cells like S. cerevisiae are accurately synthesized, correctly folded, adequately modified, and suitably transported, reinforcing the intricate and vital role these organelles play in cell biology.

The Endoplasmic Reticulum is an extensive network of membranous tubules and sacs located near the nucleus (Donato et al., 2021; Don & Kalebic, 2022; Domcke & Shendure, 2023). It can be classified into two regions: rough ER and smooth ER. The rough ER is studded with ribosomes on its cytoplasmic surface, giving it a rough appearance. Ribosomes on the rough ER synthesize proteins that are destined for secretion, incorporation into the cell membrane, or transport to other organelles. As the nascent polypeptide chain emerges from the ribosome, it enters the ER lumen, where it undergoes co-translational protein folding. Situated in close proximity to the nucleus, the Endoplasmic Reticulum (ER) is an expansive network of membranous tubules and sacs that perform key functions pertaining to protein synthesis, modification and transport in eukaryotic cells. A closer examination of this claim can be substantiated by studying the yeast, Saccharomyces cerevisiae - a model eukaryotic organism in cell biology. S. cerevisiae demonstrates the intricate workings of the ER in protein synthesis and modification through its two distinctive forms - the Rough Endoplasmic Reticulum (RER) and the Smooth Endoplasmic Reticulum (SER).

The rER, distinguished by the presence of ribosomes on its exterior cytosolic surface, facilitates the primary phase of protein synthesis. The transcribed mRNA from nuclear genes is translated into nascent polypeptides by ribosomes docked on rER. These new proteins then enter the ER lumen where they fold correctly into their functional forms under chaperone supervision. Concurrently, primary modifications like glycosylation occur, where oligosaccharide groups are added to these nascent proteins. Conversely, the SER, lacking ribosomes, orchestrates lipid synthesis and calcium storage. Once the primary protein synthesis and modification processes conclude, these proteins are packed into vesicles, which bud off from the transitional ER, a region often considered as a connection between the ER and the Golgi apparatus. These transitional ER vesicles then deliver the newly synthesized and modified proteins to the Golgi apparatus, wherein subsequent protein processing steps occur, and eventual protein targeting takes place. This coordinated operation of the Endoplasmic Reticulum demonstrates its fundamental role in cellular protein synthesis (see figure 15), modification, and transport, thereby underscoring its significance in cellular function.

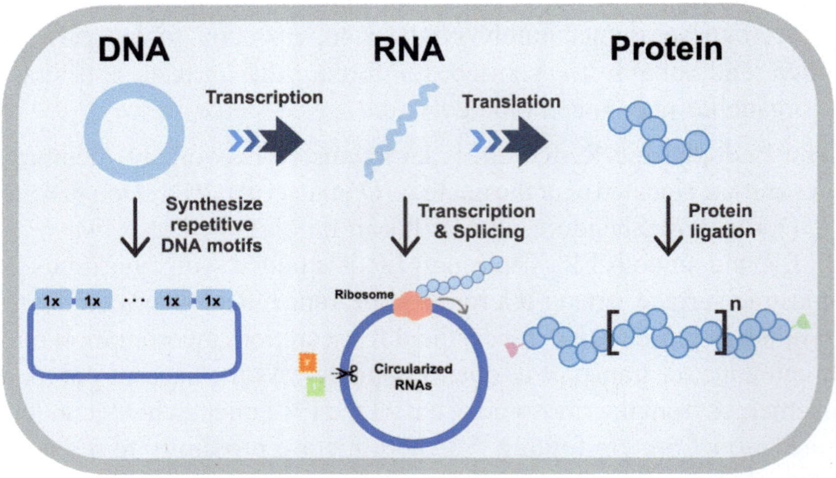

Figure 15. Illustration of protein synthesis.

Source: P Jeon J, Subramani SV, Lee KZ, Jiang B, Zhang F. Microbial Synthesis of High-Molecular-Weight, Highly Repetitive Protein Polymers. International Journal of Molecular Sciences. 2023; 24(7):6416. https://doi.org/10.3390/ijms24076416

The process of co-translational translocation ensures that the growing polypeptide chain is appropriately folded and stabilized within the ER (Donato et al., 2021; Don & Kalebic, 2022; Domcke & Shendure, 2023). Chaperone proteins within the ER lumen assist in protein folding, preventing misfolding or aggregation. Proteins that require post-translational modifications, such as glycosylation or disulfide bond formation, undergo these modifications within the ER lumen. These modifications are crucial for protein stability, function, and proper targeting. The process of co-translational translocation represents a pivotal mechanism in protein synthesis, where the nascent polypeptide chain is concurrently translated and translocated into the Endoplasmic Reticulum (ER). This mechanism ensures appropriate folding and stabilization of the polypeptide, deemed essential for protein function. Explicated through a case study of Saccharomyces cerevisiae, a eukaryotic model organism, it becomes evident how co-translational translocation ensures protein fidelity. Herein, ribosomes synthesize polypeptides following the decoding of messenger RNA (mRNA). During this synthesis, the translocation process commences upon identification of the signal sequence, a distinctive peptide sequence found at the N-terminal of the nascent polypeptide, by the signal recognition particle (SRP). The SRP then halts translation temporarily to direct the ribosome-mRNA-polypeptide complex to the SRP receptor located on the ER membrane. Subsequent to SRP detachment, translation resumes concurrently with translocation; the nascent chain passes through the translocon pore into the lumen of the ER. Entrapped in the ER, the nascent polypeptide undergoes simultaneous folding into its native tertiary structure, often with the assistance of various chaperone proteins such as clostriplast. Moreover, the ER environment facilitates the formation of disulfide bridges, a process steered by protein disulfide isomerase, to stabilize the protein structure. Therefore, through the mechanism of co-translational translocation, the ER ensures the correct folding and stabilization of the nascent polypeptide, reinforcing its critical role in maintaining protein homeostasis in eukaryotic cells.

The smooth ER lacks ribosomes and is involved in various metabolic processes (He et al., 2021; Henderson et al., 2022; Herdman et al., 2022). It plays a significant role in lipid synthesis, including the synthesis of phospholipids and cholesterol, essential for the formation of cell membranes and steroid hormones. The smooth ER is also involved in detoxification, where it detoxifies harmful substances by adding functional groups to make them more water-soluble and easily excretable. The Smooth Endoplasmic Reticulum (SER), visibly differentiated from its counterpart, the Rough Endoplasmic Reticulum, by the distinctive absence of ribosomes on its surface, plays an indispensable role in an array of metabolic processes vital for cellular functioning. A case analysis of a liver hepatocyte, a cell abundantly populated with smooth ER, illustrates its broad-spectrum functions. One of the key metabolic engagements of the SER in hepatocytes involves the biotransformation of lipophilic xenobiotics, which are potentially deleterious to the cell. The SER carries an array of enzymes including cytochrome P450 monooxygenases that conduct phase I detoxification reactions to render these xenobiotics more hydrophilic and, therefore, more accessible for further conversion and eventual excretion. Additionally, the SER assumes a crucial role in lipid metabolism, catalyzing the synthesis of phospholipids and cholesterol, molecules that serve as critical components of cell membranes. The cholesterol synthesized in the SER of hepatocytes gets converted into bile acids, aiding in fat digestion. Furthermore, the SER, being the primary site for the synthesis, modification, and packaging of lipoproteins, governs the transport of lipids to different parts of the body. Another function observed particularly in muscle cells includes the regulation of intracellular calcium levels, where the SER functions as a calcium ion reservoir, releasing or sequestering the ions as dictated by the cell's physiological demands. Thus, through its varied metabolic involvements, as elucidated in the liver hepatocyte example, the Smooth Endoplasmic Reticulum substantiates its intrinsic role in several crucial cellular operations. Figure 16 is an illustration of the Golgi apparatus.

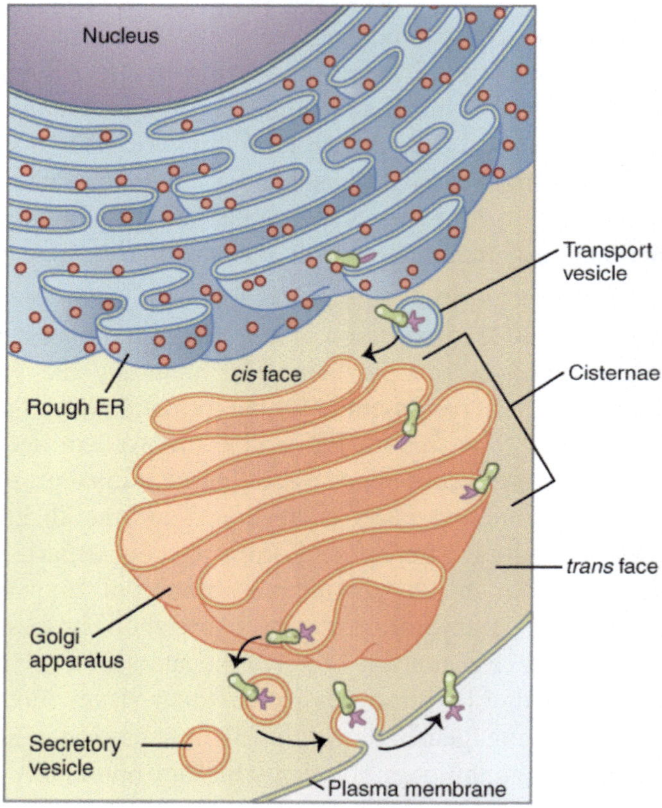

Figure 16. Illustration of the Golgi apparatus.

Source: By OpenStax - https://cnx.org/contents/FPtK1zmh@8.25:fEI3C8Ot@10Version 8.25 from the TextbookOpenStax Anatomy and PhysiologyPublished May 18, 2016, CC BY 3.0, https://commons.wikimedia.org/w/index.php?curid=64286585

After the initial synthesis and modification in the ER, proteins destined for secretion or cell membrane incorporation move to the Golgi Apparatus

(Gowrishankar et al., 2021; Giolito & Plateroti, 2022; Gordeeva et al., 2022). The Golgi Apparatus is composed of flattened membranous sacs called cisternae and is responsible for further processing and sorting of proteins. Once the stage of initial synthesis and modification within the Endoplasmic Reticulum (ER) is complete, a protein destined for secretion or destined for integration within the cell membrane is transported to the Golgi Apparatus. A detailed examination of this process can best be understood by studying the hormone-producing pituitary cells. In these cells, precursors of peptide hormones are initially synthesized in the rough ER, where they undergo glycosylation, an essential modification process. The newly synthesized and modified peptide hormone precursors are then packaged in COPII-coated vesicles, which bud off from the ER and are subsequently targeted to the cis-Golgi network, the entrance compartment of the Golgi Apparatus. The COPII coating is removed once the vesicles are fused with the membranes of the Golgi's cis face, releasing the enclosed proteins. Within the confines of the Golgi Apparatus, these proteins undergo further modifications such as phosphorylation and sialylation, which effectively change the charge and structure of the glycoprotein, ensuring they are not reabsorbed by the ER. Through successive stages within the Golgi stack, the modified proteins are sorted based on their respective destinations: secretion, incorporation in the cell membrane, or internal cellular organelles. These stages are characterized by changes in pH and enzymatic activity designed to aid in the molecules' 'maturation' and to enable the molecules to reach their functional state. Proteins earmarked for secretion or cell membrane incorporation are packaged within clathrin-coated vesicles, bud off from the trans-Golgi network, the exit face of the Golgi Apparatus, and are teleported to their final destination through exocytosis or incorporated within the plasma membrane. The intrinsic role of the Golgi Apparatus in protein trafficking within and outside of the cell therefore underlines its essential function in cellular protein homeostasis.

The Golgi Apparatus receives proteins from the ER in its cis face and processes them as they move through its cisternae (Mangeat et al., 2021; Manils et al., 2022; Maricchiolo et al., 2022). This processing includes additional glycosylation, phosphorylation, and proteolytic cleavage, among other modifications. These modifications help to define the final structure and function of the proteins. The Golgi Apparatus, an essential organelle of the cell, receives proteins from the Endoplasmic Reticulum (ER) at its cis face for further processing and transportation through its stacks of cisternae. This dynamic functionality is exemplified through a case study of the pancreatic acinar cells, key constituents in the production and secretion of digestive enzymes. Primarily synthesized in the ER, these proteins are encapsulated into COPII-coated vesicles that transit to the Golgi Apparatus. The vesicles fuse with the membrane at the cis face of the Golgi Apparatus, depositing the digestive enzymes into the cisternal space for additional modification. As these proteins traverse the Golgi stack, sequential post-translational modifications occur. These modifications include processes such as the addition and removal of carbohydrate groups, phosphorylation, and sulphation, which are crucial for proper protein maturity, stability and function. It is within the different cisternae of the Golgi stack that these modifications occur, due to varying enzymes specific to each cisterna. Upon reaching the trans face, the processed proteins are sorted for their respective destinations; in this case, for the pancreatic acinar cells, these proteins are targeted to specific secretory vesicles called zymogen granules. The granules then fuse with the cell membrane, releasing their enzymatic content into the small intestine for digestive purposes upon specific physiological stimuli. Therefore, through this orchestrated sequence of events within pancreatic acinar cells, the integral role of the Golgi Apparatus in protein modification, processing, and targeted transport is elucidated.

The Golgi Apparatus also sorts proteins into different vesicles, depending on their final destination (Masur et al., 2021; McCutcheon, 2021; Marvin et al., 2022). Some proteins are packaged into secretory vesicles and transported to the cell surface for release through exocytosis. Others are sorted into lysosomes, where they will be used for intracellular digestion or recycling of cellular components. The Golgi Apparatus functions not only as an assembly line for protein processing and modification but also as a sorting station that categorizes proteins into apt vesicles, as per their final destination. This nuanced, multi-level sorting task can be comprehensively understood by closely examining the workings within a neuron, the core unit of the nervous system. Neurons synthesize specialized proteins, such as neurotransmitter receptors and ion channels, essential for the cell's communication function. Following initial synthesis and rough processing in the Endoplasmic Reticulum, these proteins reach the Golgi Apparatus where they undergo further refinement. Upon reaching the trans-Golgi network, the final compartment of the Golgi cisternae, the processed proteins are sorted and allocated for packaging into distinctive vesicles. This sorting process is driven by the recognition of signal patches on the protein molecule by Golgi-resident sorting receptors. Proteins destined for secretion, cell membrane incorporation, or conveyance to lysosomes are dispatched in separate vesicles generated at the trans-Golgi network. For instance, neurotransmitter receptors and ion channels are packaged into vesicles that fuse with the cell membrane, allowing them to embed within the membrane and carry out their functions. Other proteins are targeted to the lysosomes or other organelles within the cell. Neurons also especially demonstrate the sending of vesicles to long distances, such as down the axon to the synaptic terminals, necessitating special regulation over their delivery. Thus, the Golgi Apparatus, through its multifaceted approach to protein shipping in neurons, underlines its critical role as a pivotal sorting and dispatch station within the eukaryotic cell.

In summary, the Endoplasmic Reticulum and Golgi Apparatus work collaboratively in protein synthesis and modification. The ER ensures proper co-translational folding and post-translational modifications, while the Golgi Apparatus further processes and sorts proteins before their final destination within the cell. These organelles play critical roles in maintaining cellular homeostasis, supporting cellular functions, and ensuring the proper functioning of the cell's protein machinery. Understanding the functions of the ER and Golgi Apparatus provides valuable insights into the intricate and dynamic processes of cell biology. In the next section, we consider the power and functions of mitochondria.

2.4 MITOCHONDRIA: POWERHOUSES OF CELLULAR RESPIRATION

Mitochondria are remarkable organelles often referred to as the "powerhouses of cellular respiration (Doostmohammadi & Ladoux, 2022; Drozdz et al., 2022; Du et al., 2023)." These double-membraned structures are crucial components of eukaryotic cells, where they play a central role in energy production through the process of cellular respiration. Mitochondria excel as exceptional organelles within the cell, often commended as the "powerhouses of cellular respiration," due to their essential role in energy production. This principle becomes evident when considering the case of the cardiac myocyte, the main cell type in the heart. The heart is a highly metabolic organ that requires consistent, copious amounts of energy supplies to maintain its relentless contractile activity, hence the substantial prevalence of mitochondria in cardiac myocytes. These organelles independently carry out the process of oxidative phosphorylation as part of cellular respiration, converting carbon-based fuels into adenosine triphosphate (ATP), the chief energy currency of the cell. Cardiac myocytes are highly reliant on fatty acid metabolism, whereby long-chain fatty acids are transported to the mitochondrial matrix and become substrates for beta-oxidation, a metabolism process that yields high-energy molecules, NADH and FADH2. These high-energy molecules then deliver their electrons to the electron transport chain located in the inner mitochondrial membrane, a process that culminates in a substantial generation of ATP. The ATP generated is then used to power the contractile machinery of the cardiac myocytes, enabling the constant beating of the heart. Remarkably, mitochondria also maintain their own DNA (mtDNA), RNA, and ribosomes, and demonstrate a double-membrane architecture, suggesting a level of autonomy within the cell and reaffirming the hypothesis of their bacterial origin. This manifestation underlines the extraordinary role that mitochondria play in both cellular energy supply and overall cell health (see figure 17), underscoring the genuine significance of these organelles within the cell's complex machinery.

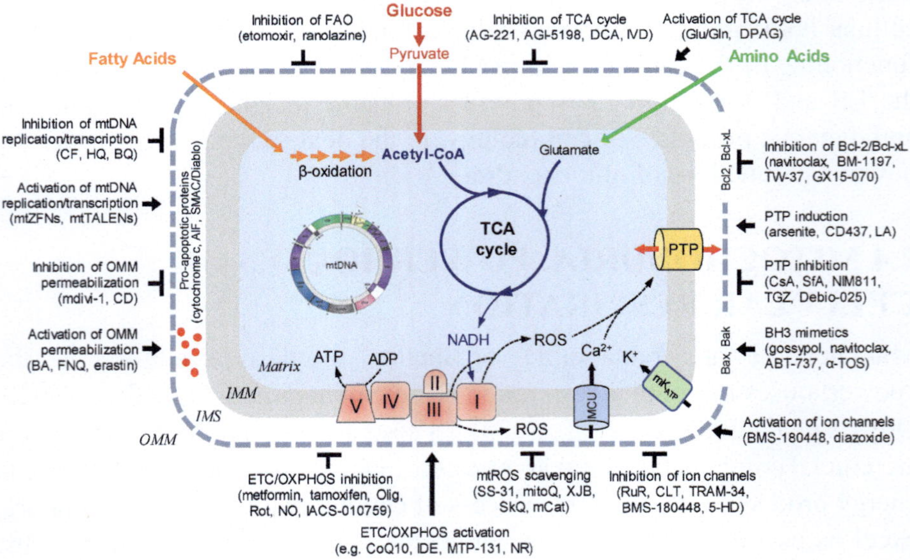

Figure 17. The role of mitochondria in cellular respiration.

Abbreviations: AIF, apoptosis-inducing factor; BA, betulinic acid; BQ, p-benzoquinone, mCAT, mitochondrial-targeted catalase; CD, chlorodiazepam; CF, ciprofloxacin; CLT, clotrimazole; CsA, cyclosporin A; DCA, dichloroacetate; DPAG, dipyruvyl-acetyl-glycerol; FAO, fatty acid oxidation; FNQ furanonaphthoquinone; 5-HD, 5-hydroxydecanoate; HQ, hydroquinone; Glu/Gln glutmate/glutamine; IDE, idebenone; IMM, inner mitochondrial membrane; IMS, intermembrane space; IVD, ivosidenib;

mK_{ATP}, mitochondrial ATP-sensitive potassium channel; LA, lonidamine; MCU, mitochondrial calcium uniporter; NR, nicotinamide riboside; Olig, oligomycin, OMM, outer mitochondrial membrane; PTP, permeability transition pore; Rot, rotenone; RuR, ruthenium red; SfA, sanglifehrin A; mtTALENs, mitochondrially targeted transcription activator like effectors (TALE) fused with a Fok1 nuclease; α-TOS, α-tocopheryl succinate; TGZ, troglitazone; mtZFNs mitochondrially targeted zinc finger nucleases. The names of representative compounds are shown in brackets. Source: Javadov S, Kozlov AV, Camara AKS. Mitochondria in Health and Diseases. Cells. 2020; 9(5):1177. https://doi.org/10.3390/cells9051177

Mitochondria are unique among cellular organelles as they contain their own DNA and ribosomes, reminiscent of bacterial cells (Y. Zhang et al., 2022; Y. F. Zhang et al., 2022; Zhao et al., 2023). This suggests that mitochondria evolved from ancient symbiotic relationships between early eukaryotic cells and aerobic bacteria, leading to the development of an energy-efficient symbiosis. Mitochondria retain an exclusive status among cellular organelles, not merely through their substantial contribution to cellular energy production but also due to their hallmark attributes: possession of their own DNA and ribosomes.

Comparatively, this trait is parallel to bacterial cells, giving credence to the endosymbiotic theory that asserts the origins of eukaryotic mitochondria from engulfed prokaryotic organisms. A detailed analysis of a typical eukaryotic cell in the context of disease progression, the HeLa cell, can further underscore this distinct feature. HeLa cells, derived from a cervical cancer patient, have been extensively used in medical research due to their unique ability to proliferate indefinitely. Within these cells, mitochondria manifest themselves as double-membraned organelles equipped with their own circular DNA (mtDNA) and 70S ribosomes, the latter being a feature primarily associated with prokaryotes. The presence of mitochondrial DNA, distinctly apart from the nuclear genome, enables mitochondria to self-replicate and carry out transcription and translation processes independently within their boundary membranes, demonstrating a level of autonomy within the cell. The mitochondrial genome in HeLa cells encodes 13 proteins, all of which are integral parts of the oxidative phosphorylation system, along with 22 tRNAs and 2 rRNAs necessary for intramitochondrial protein synthesis. Thus, in the backdrop of HeLa cells, it is clear that the biomechanics of mitochondria are quite faithfully reminiscent of bacterial cells, thus setting them apart from other eukaryotic organelles and reinforcing their exceptional status within the cell.

The main function of mitochondria is to generate adenosine triphosphate (ATP), the primary energy currency of the cell (Donato et al., 2021; Don & Kalebic, 2022; Domcke & Shendure, 2023). This process occurs through cellular respiration, a series of interconnected reactions that involve the breakdown of glucose and other organic molecules to release energy. Mitochondria, often hailed as the "powerhouses" of the cell, primarily function to generate adenosine triphosphate (ATP), the cardinal energy currency of the cell. This function is pivotal for the maintenance and survival of virtually all eukaryotic cells, as ATP fuels numerous cellular processes. The fundamental role of mitochondria in ATP production can be illuminated through a case study of skeletal muscle cells during exercise. Fast-twitch skeletal muscle fibers, which are geared towards rapid, powerful contractions, rely heavily on the ATP generated by mitochondria. These cells contain plentiful mitochondria that efficiently execute aerobic respiration to meet their high metabolic demand. Mitochondria break down glucose and fatty acids, obtained from the food we consume, to produce ATP via a series of reactions known as the Krebs cycle (or citric acid cycle) and oxidative phosphorylation. In the Krebs cycle, which occurs within the mitochondrial matrix, the products of glucose and fatty acid breakdown are further oxidized, and the energy released is captured in NADH and FADH2. These high-energy carriers then donate their electrons to the electron transport chain, embedded in the inner mitochondrial membrane, a process coupled with the pumping of protons across the membrane. The resulting proton gradient drives ATP synthesis as protons flow back across the membrane through the ATP synthase enzyme. Thus, under conditions of strenuous exercise, skeletal muscle fibers' massive demand for immediate ATP is satisfied by the efficient operation of mitochondria, underscoring their crucial role in securing the cell's energy finances.

Cellular respiration consists of three main stages: glycolysis, the citric acid cycle (Krebs cycle), and oxidative phosphorylation (Bonilha, 2022; Boukhatmi, 2021; Boumard & Bardin, 2021). Glycolysis, which occurs in the cytoplasm, breaks down glucose into two molecules of pyruvate and produces a small amount of ATP and reduced coenzymes, NADH, and FADH2. Cellular respiration, the process through which cells extract energy from nutrients, typically unfolds through three principal stages: glycolysis, the citric acid cycle (or Krebs cycle), and oxidative phosphorylation. A detailed probe into the yeast Saccharomyces cerevisiae, used commonly in brewing and baking, offers a concisely observable representation of cellular respiration. Glycolysis, the initial step of this energy releasing process, occurs in the cytosol. During glycolysis, a single molecule of glucose, a six-carbon sugar, splits into two three-carbon molecules through a series of enzymatic reactions, yielding two molecules of ATP and two molecules of NADH. These three-carbon sugar molecules, known as pyruvate, then enter the mitochondria, which marks the beginning of the citric acid cycle. Once inside the mitochondria's matrix, the pyruvate undergoes a complex progression of enzymatic reactions known as the Krebs cycle. The Krebs cycle oxidizes the pyruvate, generating carbon dioxide as a by-product and transferring the electrons to NAD+, creating several molecules of NADH and FADH2. Further, a modest yield of ATP is also produced directly in the cycle. Following the Krebs cycle, oxidative phosphorylation, the final stage of cellular respiration, transpires across the inner mitochondrial membrane. This phase consists of the electron transport chain and chemiosmosis. In these reactions, the NADH and FADH2 produced earlier donate their electrons to a chain of protein complexes in the mitochondrial inner membrane. As electrons move along the chain, energy is used to pump protons across the membrane, generating a gradient. The potential energy thus created propels the synthesis of ATP as protons flow back across the membrane through ATP synthase, an enzyme that catalyzes the formation of ATP from ADP and inorganic phosphate. Thus, through the intricate coordination of glycolysis, the citric acid cycle, and oxidative phosphorylation, cellular respiration achieves the crucial transformation of nutrient energy into a form useful to the cell: ATP.

The pyruvate then enters the mitochondria and undergoes further breakdown in the citric acid cycle (Mangeat et al., 2021; Manils et al., 2022; Maricchiolo et al., 2022). In this cycle, carbon atoms are released as carbon dioxide, and high-energy electrons are captured by NAD+ and FAD to form NADH and FADH2. The function of pyruvate within the context of cellular metabolism extends beyond glycolysis, the process in which a glucose molecule is broken down into two pyruvate molecules. The true potential of pyruvate as a conduit for energy release is actualized when it enters the mitochondria and undergoes further metabolism in the citric acid cycle. To elucidate this process further, the human liver cell offers a relevant vantage point. The human liver, an organ involved in numerous metabolic processes, possesses an abundance of mitochondria and plays a crucial role in glucose homeostasis. Following glycolysis in the cytosol of these hepatocytes, each pyruvate molecule is transported into the mitochondrial matrix via the mitochondrial pyruvate carrier. Once within the mitochondria, pyruvate is promptly decarboxylated in a reaction catalyzed by the pyruvate dehydrogenase complex, culminating in the formation of an acetyl-CoA molecule. This new molecule is then introduced into the citric acid cycle, a sequence of enzyme-catalyzed reactions that contributes significantly to cellular respiration. Throughout the course of the cycle, each two-carbon acetyl group attached to coenzyme A is further oxidized to two molecules of carbon dioxide. Meanwhile, electrons are transferred to carrier molecules NAD+ and FAD, forming NADH and FADH2, respectively. This relay of electrons provides the energy required for the production of ATP in the third and final stage of cellular respiration, oxidative phosphorylation. Thus, through its participation in the citric acid cycle and subsequent metabolism within the mitochondria of liver cells, pyruvate sequesters and channels the potential energy within glucose, ultimately furnishing ATP to meet the cell's energy demands. Figure 18 illustrates the basic structure of mitochondria.

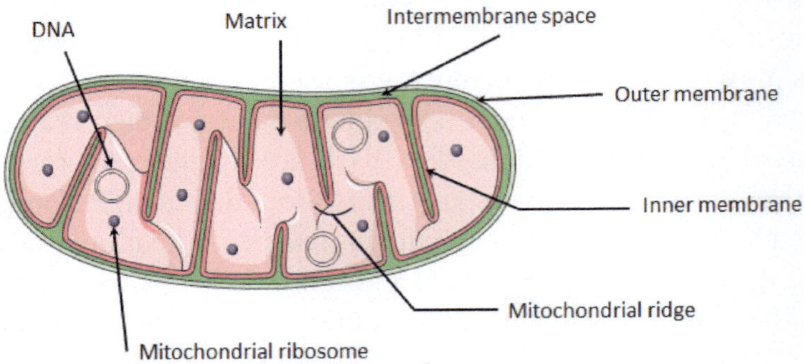

Figure 18. The structure of mitochondria.

Source: Andrieux et al. (2021)

Andrieux, P., Chevillard, C., Cunha-Neto, E., & Nunes, J. P. S. (2021). Mitochondria as a cellular hub in infection and inflammation. International journal of molecular sciences, 22(21), 11338.

The most energy-efficient phase of cellular respiration is oxidative phosphorylation, which takes place in the inner mitochondrial membrane (Carroll & Marqusee, 2022; Chang et al., 2023; Charpentier & King, 2021). Here, the NADH and FADH2 generated in glycolysis and the citric acid cycle donate their electrons to an electron transport chain (ETC). The ETC consists of a series of protein complexes that transfer electrons, releasing energy in the process. The phase of cellular respiration that gives maximal yield in terms of energy generation is oxidative phosphorylation, which largely occurs across the inner mitochondrial membrane. This process, integral to producing the cell's energy currency, ATP, is exemplified by the biological processes of the cardiac cells in the human heart. The incessant contraction and relaxation of cardiac cells, a process that enables the heart to pump blood throughout the body, places substantial, continual demands on cellular energy. Mitochondria in these cells meet this challenge through oxidative phosphorylation, the most ATP-productive stage of cellular respiration. During this process, the NADH and FADH2 molecules, generated during glycolysis and the citric acid cycle, donate their electrons to the electron transport chain, a series of protein complexes and electron carrier molecules located in the inner mitochondrial membrane. As electrons pass down this chain, energy is released and utilized to pump protons from the mitochondrial matrix into the intermembrane space, which creates a proton gradient or electrochemical potential. This potential represents a form of stored energy harnessed in the final step of oxidative phosphorylation, chemiosmosis. In chemiosmosis, protons from the intermembrane space flow down their gradient and into the matrix, coursing through ATP synthase, an enzyme embedded in the inner membrane. The translocation of protons through ATP synthase catalyzes the phosphorylation of ADP, generating ATP. Consequently, a human cardiac muscle cell, equipped with numerous mitochondria, can generate around 30 molecules of ATP per glucose molecule during oxidative phosphorylation. Thus, the adeptness of oxidative phosphorylation in the mitochondria allows the heart cells to meet their immense energy needs thereby allowing the heart to function efficiently.

As electrons move through the ETC, protons are pumped from the mitochondrial matrix into the intermembrane space, creating an electrochemical gradient (Yuan et al., 2021; Yu et al., 2022; X. J. Zhang et al., 2023; X. Y. Zhang et al., 2023). The flow of protons back into the mitochondrial matrix through ATP synthase results in the synthesis of ATP, effectively coupling electron transport with ATP production. The molecular processes that underpin cellular respiration serve as fundamental mechanisms in the generation of the energy currency of the cell, ATP. A critical component of this process involves the creation of an electrochemical gradient in the mitochondria—a process that can be elucidated through a case study of neuronal cells. In the densely packed neuronal mitochondria, which power the high energy demand of neurotransmission, energy derived from glucose metabolism is transduced into a proton gradient across the inner mitochondrial membrane via the electron transport chain (ETC). The ETC, composed of five protein complexes (Complexes I-V), accepts electrons from electron carriers NADH and FADH2. These electrons are then successively transferred through these complexes, set in the inner mitochondrial membrane. As electrons move through the ETC, from a region of high-energy to one of lower energy, the energy released is harnessed to pump protons (H+) from the mitochondrial matrix (inside the inner membrane) into the intermembrane space (between the inner and outer membranes). This active movement of protons against their concentration gradient metabolically couples the energy yielded from electron transport to the creation of an electrochemical gradient. This gradient, characterized by a high concentration of protons in the intermembrane space relative to the matrix, stores potential energy. This energy is ultimately coupled to ATP synthesis as protons flow down the gradient and into the matrix via the ATP synthase enzyme, thereby driving the phosphorylation of ADP to ATP. Through such meticulously orchestrated processes, neurons can fulfill their high metabolic demands and maintain their essential role in cellular communication and coordination.

This efficient process of oxidative phosphorylation generates a substantial amount of ATP, providing the cell with the energy it needs to carry out essential functions such as active transport, biosynthesis, and muscle contraction (Urmi et al., 2021; Usmani et al., 2021; Turner et al., 2022). Oxidative phosphorylation signposts an efficient process in the metabolic pathway, bearing the potential to generate a substantial yield of adenosine triphosphate (ATP), the primary energy currency for most cell functions. This functionality is observable in the human muscle cells during strenuous physical activity. The muscle cells, or myocytes, undergo rapid and repetitive contraction and relaxation, a physiological process demanding a continuous and copious supply of ATP. To cater to the high-energy requirements, mitochondria within these muscle cells effectively employ oxidative phosphorylation. During this process, electrons derived from the metabolic breakdown products of food, carried by electron carriers NADH and FADH2, traverse the electron transport chain located in the inner mitochondrial membrane. The inherent energy released in this electron flow propels the pumping of protons from the mitochondrial matrix into the intermembrane space, creating a substantial proton (or electrochemical) gradient. This gradient supports a pool of potential energy, which is harnessed by ATP synthase, the final enzyme complex in the electron transport chain. Protons flow back into the matrix through ATP synthase, an event that triggers the transformation of ADP to ATP. Thereby, through the functional mechanisms of oxidative phosphorylation, muscle cells manage to generate large amounts of ATP. This ATP molecule generation is an essential element, particularly during strenuous physical activity when ATP is rapidly consumed to fuel the active transport of ions across muscle cell membranes, biosynthetic processes for muscle repair and growth, and the direct mechanics of muscle contraction. Therefore, the oxidative phosphorylation process is emblematic of a highly efficient energy generation system that adequately fuels cellular functioning, especially in high energy-requiring cells such as muscle cells.

Beyond energy production, mitochondria also play critical roles in other cellular processes. They are involved in calcium storage and signaling, regulation of apoptosis (programmed cell death), and the synthesis of certain amino acids and lipids (Watson et al., 2021; Warecki & Sullivan, 2022; Weigert & Herhaus, 2023). A significant portion of the developing research in cell microbiology, supported by numerous laboratory investigations, underscores the elaborate roles of mitochondria that transcend their primary regulatory function in energy production. One such fascinating and decisive aspect is calcium homeostasis, the dynamic balance of calcium ion concentration within the cytosol and intracellular compartments - a role played by the mitochondria where they act as calcium buffers. They achieve this through the mitochondrial calcium uniporter, a protein complex highly selective for calcium ions, which facilitates the transfer of these ions from the cytosol into the mitochondria. This transfer is instrumental in preparing the cell for signal transduction, where calcium ions act as second messengers activating consequential biochemical cascades. The second rung of mitochondrial function is the regulation of apoptosis or programmed cell death, a crucial process that maintains cellular integrity and viability. Studies, such as the one by Kantari and Walczak from the National Center for Scientific Research, reveal mitochondria as key mediators of apoptosis via the release of pro-apoptotic proteins like cytochrome c from the intermembrane space into the cytosol, consequently leading to the activation of downstream caspases - the chief executioners of apoptosis. In addition, one must understand the role of the mitochondrion in the synthesis of particular amino acids and lipids. Tucked within the matrix of the mitochondria is the tricarboxylic acid cycle that aids in the production of non-essential amino acids like aspartate and glutamate. Moreover, the outer mitochondrial membrane houses enzymes involved in the construction of the lipid phosphatidylserine, which plays a crucial role in a minute assembly of things, from serving as a signal for phagocytosis of apoptotic cells to being a precursor for other cell membrane phospholipids. This panorama of cellular processes, steered by the mitochondria and backed by robust findings from the field of cell microbiology, positions the organelle as a ceaseless powerhouse that not only fuels the cell with energy but also steers critical cellular functions encompassing calcium signaling, programmed cell death, and molecular synthesis.

Mitochondrial dysfunction has been linked to various diseases and aging (Orr & Hynds, 2021; Ohi & Kenworthy, 2022; Okano et al., 2023). Mutations in mitochondrial DNA can lead to mitochondrial disorders, affecting energy production and leading to a wide range of symptoms, including muscle weakness, neurological problems, and organ dysfunction. Mitochondrial dysfunction, characterized by impaired energy production and aberrant signaling, has been associated with pathologies encompassing neurological disorders, metabolic syndromes, cancer, and the aging process. A prime example is the identification of mitochondrial dysfunction in neurodegenerative diseases such as Parkinson's and Alzheimer's, established through multiple studies. For instance, research by Loeffler, LeWitt, and colleagues from the University of Texas revealed reduced activities of mitochondrial respiratory chain components, especially complex I, in the substantia nigra region of the brains of patients with Parkinson's disease. This bioenergetic dysfunction leads to impaired neuronal functioning, contributing to the progressive deterioration witnessed in Parkinson's disease. Furthermore, investigations like those conducted by Moreira and colleagues found mitochondrial-related oxidative stress, impairing mitochondrial bioenergetics, in Alzheimer's disease. The oxidative damage leads to neuronal cell death, thereby contributing to the cognitive decline associated with this disorder. Similarly, aberrant mitochondrial metabolism is illustrated in type 2 diabetes. Petersen and colleagues at Yale University discovered through magnetic resonance spectroscopic studies that insulin resistance, a defining characteristic of type 2 diabetes, was associated with decreased mitochondrial ATP production rates and insulin-stimulated glucose disposal. Furthermore, recent research, such as studies led by Wallace of the Children's Hospital of Philadelphia, has elucidated the role mitochondrial dysfunction plays in the process of aging. The decline in mitochondrial function, partly due to the accumulation of mutations in mitochondrial DNA, results in increased production of reactive oxygen species leading to cellular damage. Additionally, the role of mitochondria in programmed cell death implies that mitochondrial dysfunction might contribute to an imbalanced apoptosis rate, a hallmark of aging. Making clear through these case studies, mitochondrial dysfunction is an undercurrent of a spate of diseases and the process of aging, underscoring the critical importance of mitochondrial health in cellular and organismal longevity. In summary, mitochondria are essential organelles responsible for energy production through cellular respiration. Through the highly efficient process of oxidative phosphorylation, they generate ATP, providing the cell with the energy needed to sustain life. The symbiotic origin and significance of mitochondria in cell microbiology underscore their critical role in the functioning and survival of eukaryotic cells. The final section in this second chapter will focus on chloroplasts and their role in photosynthesis as well as energy production among plants.

2.5 CHLOROPLASTS: PHOTOSYNTHESIS AND PLANT CELL ENERGY PRODUCTION

Chloroplasts are fascinating organelles found in plant cells and some photosynthetic protists (Clarke et al., 2021; Claude-Taupin et al., 2022; Colin et al., 2022). They are often referred to as the "kitchens" of the cell due to their role in photosynthesis and energy production. Through the process of photosynthesis, chloroplasts harness light energy and convert it into chemical energy in the form of glucose, which serves as the primary source of energy for plant cells. Chloroplasts, renowned as the sites of photosynthesis, emerge as enthralling organelles resident in plant cells and certain photosynthetic protists. These intricate entities house chlorophyll, a green pigment that traps light energy, thereby earning them their emerald hue. Their significance truly manifests in their unique capability to transduce solar energy into chemical energy in the form of adenosine triphosphate (ATP) and reduced nicotinamide adenine dinucleotide phosphate (NADPH). This transduction sets in motion the generation of glucose from water and carbon dioxide, a process triggering the production of oxygen - a distinctive feature that has sculpted the planet's atmosphere and supports most life forms. This comprehensive process occurs in two stages - the light-dependent reactions and the light-independent or Calvin cycle. Light-dependent reactions, occurring in the thylakoid membrane, harness light energy to generate ATP and NADPH, while the Calvin cycle, happening in the stroma, utilizes the ATP and NADPH to convert carbon dioxide into glucose. An exemplary case study that hones this understanding further is that of the unicellular green alga, Chlamydomonas reinhardtii, that researchers like Harris from Dartmouth College have investigated. Chlamydomonas reinhardtii possesses a single, cup-shaped chloroplast that occupies a substantial part of the cell's volume. This chloroplast exhibits regions specialized for capturing light energy (photosystems) and elaborate networks for carrying out the Calvin cycle (stroma lamellae). Through this research, the versatility of chloroplasts in energy conversion becomes distinguished – a principle that replicates in all photosynthetic organisms, from the unicellular Chlamydomonas to the towering redwood trees. Hence, the role of chloroplasts – from being the engines driving photosynthesis to being dynamic game-changers of Earth's atmosphere – is a testimony to their mind-boggling significance in biological thought and study.

Photosynthesis is a complex process that occurs in the thylakoid membranes of chloroplasts (Quintero-Carmona, 2021; Radulovic et al., 2022; Rao et al., 2022). These membranes are stacked in structures called grana, which maximize the surface area for capturing light energy. Within the thylakoid membranes, several pigments, including chlorophyll a and chlorophyll b, absorb light of various wavelengths. These pigments are crucial for the absorption of light energy during photosynthesis. Photosynthesis, a cornerstone process vital to the sustenance of life on earth, takes place in the thylakoid membranes of chloroplasts and involves a harmonious interplay between lights capturing reactions and carbon fixation reactions to convert solar energy into nutritive organic molecules. Initiated by the absorption of photons by chlorophyll molecules in the light-harvesting complex within the thylakoid membranes, this absorbing of light energy enables the chlorophyll to displace an excited electron onto the primary electron acceptor of Photosystem II (PS II). The displaced electron is then transferred through an electron transport chain, including plastoquinone, b6-f complex and plastocyanin, creating a proton gradient across the thylakoid membrane and yielding ATP via chemiosmosis - a process known as photophosphorylation. Subsequently, light energy absorbed by Photosystem I (PS I) enables the reinstatement of the "missing" electron in PS II and reduces NADP+ to NADPH. The eminent case of cyanobacterium Synechocystis, substantiated by research entities like Rögner from the University of Bochum, illuminates a holistic understanding of this sequence of reactions. Research studies using ultrafast spectroscopy and imaging techniques have ascertained the structure and spatial organization of thylakoid membranes and photosystems in Synechocystis to shed light on the efficient energy transfer processes. Furthermore, they have authenticated the crucial role of PS II and PS I, jointly operating to maximize energy conversion under varying light conditions, thus reflecting an ingenious, adaptive trait of photosynthetic organisms. Hence, photosynthesis is indeed an intricate event that evolved in the thylakoid membranes of the chloroplasts, embodying one of the universe's most efficient energy conversion systems, serving as the optimal role model of green energy. Figure 19 illustrates the basic structure of a chloroplast.

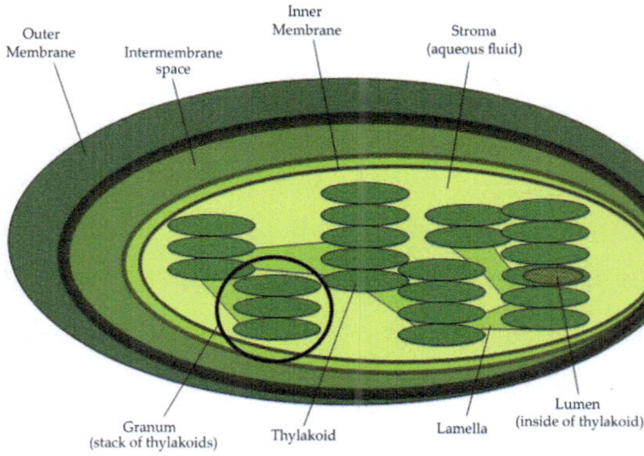

Figure 19. The basic structure of a chloroplast.

Source: Public Domain, https://commons.wikimedia.org/w/index.php?curid=434456

The light-dependent reactions of photosynthesis take place in the thylakoid membranes and involve two photosystems, aptly named Photosystem I and Photosystem II (Lanz et al., 2021; Larson & McFarlane, 2021; Le Goff et al., 2022). These photosystems work in tandem to harness light energy and convert it into chemical energy in the form of ATP and NADPH. The thylakoid membranes of the chloroplasts serve as the site for the light-dependent reactions of photosynthesis, with two distinguished complexes, Photosystem I and Photosystem II, at the helm of these reactions. Both systems harness light energy and pigments, but work differently. Photosystem II embarks on the route of photosynthesis, absorbing light primarily at a wavelength of 680 nm, thus also termed P680. Here, the energized photons cleave water molecules into oxygen and hydrogen ions, liberating electrons to fill the void created in the split chlorophyll a molecules. These electrons then journey down an electron transport chain - from plastoquinone to cytochrome b6f complex to plastocyanin, playing an instrumental role in generating ATP through a process known as photophosphorylation. On the other hand, energized by light primarily at 700 nm, Photosystem I, or P700, accepts electrons from plastocyanin, restores the electrons lost in Photosystem II and reduces NADP+ to NADPH in the stroma via ferredoxin-NADP+ reductase. This clear demarcation between the roles of Photosystem I and II elucidates how the light-dependent reactions operate, leading to a seamless transition to the light-independent reactions. A definitive example to reflect this is evidenced in the work by Jansson and colleagues detailing cyanobacterium Synechocystis photosynthetic processes. They validate the proficiency of Photosystem I and II in collaborating for efficient light harvesting and energy conversion, underlining the capability of these tiny reaction centers to fuel life on Earth using radiant energy from the sun. Therefore, the light-dependent reactions, powered by Photosystem I and II in the thylakoid membranes, form the critical first half of the photosynthetic process, indubitably marking them as central players in energy transformation.

During Photosystem II, light energy is absorbed by pigments, exciting electrons within the reaction center (Charpentier & King, 2021; Carroll & Marqusee, 2022; Chang et al., 2023). These high-energy electrons are then passed through a series of electron carriers in the thylakoid membrane, leading to the pumping of protons across the membrane and the creation of a proton gradient. This gradient drives ATP synthesis through the enzyme ATP synthase. Photosystem II, a central component of the light-dependent reactions, comprises light-absorbing pigments, primarily chlorophyll a, orchestrated to catalyze the photochemical phase of photosynthesis in the thylakoid membrane of the chloroplasts. Positioned in the core of Photosystem II, these pigments anticipate the absorption of light photons to instigate a chain of events. The light energy, mainly absorbed at a wavelength of 680 nm, incites electrons within the reaction center, specifically in the P680 chlorophyll a molecules, to reach an excited state. The elevated energy level causes these excited electrons to defect from the molecules, creating a positive charge or “hole” in the P680, which is immediately filled up by electrons derived from the photolysis of water. This series of reactions, known as the primary photochemical reaction, leads to the generation of molecular oxygen, a by-product of utmost importance to aerobic life. An exemplary case study that sheds light on the functionality of Photosystem II is established through numerous investigations on a single-celled, photosynthetic cyanobacterium Synechocystis. Research work by Nürnberg and colleagues from Imperial College London illuminated the fine details of the electron excited state, water-splitting process, and oxygen evolution in Photosystem II using ultrafast X-ray and optical spectroscopic methods. Their findings delineated the step-wise electron transfer processes, contributing to a more nuanced understanding of how the influx of light energy can trigger a series of finely tuned photochemical reactions in Photosystem II, a process essential to nearly all life on earth. Hence, the role of Photosystem II – from capturing light energy to sparking complex biochemical cascades – underscores the harmony of cellular processes and photosynthetic life's sustainment.

In Photosystem I, light energy is again absorbed by pigments, leading to the re-energization of electrons (Leao et al., 2023; Lee & Rompolas, 2022; Lefranc, 2021). These high-energy electrons are transferred to another electron carrier, which ultimately donates them to NADP+ to form NADPH. NADPH is a crucial reducing agent used in the subsequent reactions of photosynthesis. Once the electrons translocate through the electron transport chain originating from the Photosystem II, they reach Photosystem I, an intricate protein-pigment complex pivotal to the photosynthetic machinery operating in the thylakoid membranes. Functioning primarily on light absorbed at a wavelength of 700 nm, the chlorophyll a molecules within the Photosystem I reaction center, hence aptly named P700, reenergize these electrons. When pigments capture photons, the P700 center is excited, pushing electrons to a higher energy state and allowing them to be captured by the primary electron acceptor. These highly energized electrons then progress down a second electron transport chain, which primarily consists of ferredoxin molecules. This procession culminates into the conversion of NADP+ to NADPH through the enzyme ferredoxin-NADP+ reductase. The light-induced electron flow in Photosystem I exhibits the intriguing nature of the photosynthetic process, where light is transformed into chemical energy. A demonstrative case observing this process can be found in the algae Chlamydomonas reinhardtii. In a study by Drop and colleagues from the VU University in Amsterdam, the researchers used sophisticated structural biology techniques to unveil the architecture of the Photosystem I complex and its associated light-harvesting complex. This detailed understanding of the Photosystem I structure allows for deeper insight into the mechanisms governing the absorption of light energy and the process it set into motion, including electron flow and the re-energizing process. Therefore, Photosystem I, with its crucial role in photon absorption and electron re-energization, undeniably forms the heart of light-dependent photosynthetic reactions, consolidating the intricacy and efficiency of the energy conversion processes in photosynthesis.

The light-independent reactions, also known as the Calvin cycle, occur in the stroma of chloroplasts (Liao et al., 2021; Liedtke et al., 2022; K. Lim et al., 2023). In this cycle, carbon dioxide from the atmosphere is fixed into organic molecules, eventually leading to the synthesis of glucose. The Calvin cycle relies on the ATP and NADPH generated during the light-dependent reactions to power its chemical reactions. Under rigorous scientific examination, the voluminous world of cell biology unveils astounding microscopic processes, one of which is the light-independent reactions—or referred to in academic circles as the Calvin cycle—a process that takes place within the spacious confines of the stroma located in chloroplasts.

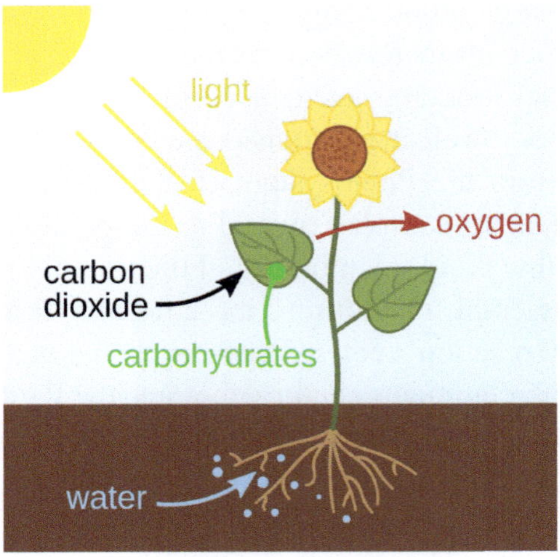

Figure 20. The process of photosynthesis.

Source: By At09kg, Wattcle, NefronusAt09kg: originalWattcle: vector graphicsNefronus: redoing the vector graphics - This file was derived from: Photosynthesis.gif:, CC BY-SA 4.0, https://commons.wikimedia.org/w/index.php?curid=49183032

Operating as the minuscule yet intensely intricate engine room for photosynthesis, the Calvin cycle proves critical for life as we know it by facilitating carbon fixation—a sequential process that takes carbon dioxide from the environment and integrates it into existing organic molecules within the cell. This meticulously executed fixation process—inextricably anchored to the cellulose fibers underlying plant structural strength—profiles the breadth of this reaction's significance. Pursuing a deeper probe into this phenomenon, numerous studies have elucidated an astute three-stage cycle that unfolds. The initial phase unveils carbon fixation, where the enzyme RuBisCO catalyzes the merger of carbon dioxide and Ribulose-1,5-bisphosphate, a five-carbon sugar, to form an unstable six-carbon compound that abruptly splits into a pair of three-carbon molecules of 3-phosphoglyceric acid (3-PGA). The middle phase magnifies as one where an array of enzymes harmoniously facilitate chemical reactions that convert 3- PGA into glyceraldehyde-3-phosphate, using ATP and NADPH subsequently produced during the prior light-dependent reactions. In the burgeoning terminal phase, the majority of the G3P molecules are recycled to regenerate RuBP, which aids in maintaining the process's perpetuation. The remaining G3P, however, becomes the framework for the consequent synthesis of glucose and other organic compounds. Therewith, by entrenched symbiosis with the concomitant light-dependent reactions, the calvin cycle commandeers molecular processes that embolden life on Earth, invigorating an exploration that bears witness to the paradox of simplicity in complexity within the cell's microscopic realm. Figure 20 shows the process of photosynthesis.

Through photosynthesis, chloroplasts play a critical role in capturing and transforming light energy into chemical energy stored in glucose (Spiers et al., 2021; Stacey & Healy, 2021; Stallaert et al., 2022). This energy is then utilized by plant cells for various purposes, including growth, reproduction, and cellular respiration. Chloroplasts, in the context of cell biology, occupy a distinctive position due to their function as pivotal agents in the process of photosynthesis, a method that proficiently captures and transmits light energy into a stored form of chemical energy, notably glucose. This specialized function of chloroplasts is far from trivial concerning the sustenance of life as we comprehend it. Spearheading the series of photosynthetic operations is a group of pigment molecules, famously associated with chlorophyll, residing within the thylakoid membranes of chloroplasts. In the presence of light, these pigments absorb photons primarily in the blue and red regions of the spectrum, inducing an excitation of their electrons to higher energy states. As the excited electrons make their hasty descent back to their ground state, they embark on a labyrinthine journey through a complex of proteins, known as the electron transport chain, where the energy derived from these electrons is harnessed to effect endergonic reactions that generate adenosine triphosphate (ATP) and nicotinamide adenine dinucleotide phosphate (NADPH). The generated ATP and NADPH molecules enter the complex arena of the Calvin cycle within the accommodating stroma of the chloroplast, where carbon dioxide is reduced and methodically stitched into the fabric of glucose. Notably, in the case of the production of glucose, two molecules of glyceraldehyde 3-phosphate (G3P), a product of the Calvin cycle, are committed and restructured via a series of enzymatic reactions into a singular molecule of glucose. Thus, through this dynamic dance of absorption, excitation, reduction, and synthesis, chloroplasts illustrate their irrefutable importance by converting mere fleeting sunlight into the tangible and versatile entity of glucose, consequently sustaining life across the biosphere.

In addition to photosynthesis, chloroplasts also have other important functions (Das et al., 2021; de Albuquerque et al., 2021; De Camilli, 2021; de Torres-Jurado et al., 2022). They are involved in the synthesis of various molecules, such as amino acids and lipids, that are essential for plant growth and development. Chloroplasts are also involved in the regulation of redox reactions and the detoxification of harmful oxygen radicals produced during photosynthesis. Situated on the front lines of cellular operations, chloroplasts execute a repertoire of paramount functions beyond photosynthesis, participating in the key synthesis of requisite molecules such as amino acids and lipids, which serve as fundamental substrates for plant growth and development. A considerable portion of these components originates from the photorespiratory pathway, an indispensable biochemical detour tethered to the Calvin cycle. Substantial scientific investigations anchor their focus on the glycolate pathway—a facet of photorespiration occurring within the chloroplast and peroxisomes—where RuBisCO's catalytic promiscuity with oxygen initiates the conversion of glycolate into glycine. This process later engages in a series of reactions within the mitochondria to eventually generate serine—an essential building block for protein synthesis. Moreover, the chloroplast, through the regulation of its resident enzyme acetyl-CoA carboxylase, also mediates the synthesis of fatty acids, the precursors for complex lipids. In spectral contrast to these anabolic activities, chloroplasts bear the significant responsibility of maintaining cellular redox homeostasis and the detoxification of pernicious oxygen radicals, residues left over from photosynthesis. More precisely, the chloroplast presents an array of enzymatic machinery, such as superoxide dismutase and ascorbate peroxidase, both implicated in the neutralization of reactive oxygen species. Alongside this, chloroplasts institute an indispensable role in the proliferation of the cellular antioxidant glutathione, as directed by the oxidative pentose phosphate pathway, indispensable for purging cells of damaging molecular interlopers. Thus, these revelations expose the chloroplast as a multifunctional organelle, interweaving a web of essential synthetic pathways and robust detoxification mechanisms to help orchestrate plant growth and development and herald a healthy cellular environment.

The presence of chloroplasts in plant cells is a defining characteristic of plants, setting them apart from other organisms that rely on different mechanisms for energy production (Mangeat et al., 2021; Manils et al., 2022;

Maricchiolo et al., 2022). The intricate process of photosynthesis carried out by chloroplasts is crucial for the survival and productivity of plant life, serving as the foundation of terrestrial food chains and ecosystems. The presence of chloroplasts within plant cells is a distinguishing component that sets them apart from other organisms, underpinning an entirely unique mode for energy production. Emphasizing the comprehensive capabilities of chloroplasts, both within and beyond the bounds of photosynthesis, this particular organelle is lauded by experts as the centerpiece of stunning biochemical choreography, rendering the plant cell and, by extension, the plant itself, a masterpiece of biological innovation. A standout feature of chloroplasts is their unique repertoire to convert sunlight into utilizable energy through photosynthesis - a process distinctly absent in other organisms that hinge upon varying energy production mechanisms. Fundamentally, the photochemical reactions within chloroplasts transform captured sunlight into chemical energy in the form of ATP and NADPH, which are then flaunted to power the synthesis of a vast range of vital organic molecules, most notably glucose. In stark contrast, other organisms such as animals and fungi draw upon cellular respiration, a process that leverages organic compounds as energy sources, usually in the form of glucose. These organisms rely on the oxidation of glucose in their mitochondria to produce ATP, fundamentally different from the energy-harvesting solar power system of plants. While some bacteria can employ photosynthesis, their versions are structurally and functionally distinct from the oxygenic photosynthesis occurring in the eukaryotic chloroplast, further demonstrating the unique nature of plant energy metabolism. Therefore, the existence of chloroplasts facilitates a unique bioenergetics toolkit that fundamentally sets plant cells apart from others, contributing to the rich tapestry of life's multifarious energy strategies.

In conclusion, chloroplasts are remarkable organelles responsible for photosynthesis and energy production in plant cells. Through their intricate machinery and pigment systems, they capture light energy and convert it into chemical energy, which is utilized by the cell for various purposes. The fascinating interplay between light-dependent and light-independent reactions in chloroplasts enables plants to harness the energy of the sun and sustain life on Earth. Understanding the functions of chloroplasts provides valuable insights into the intricate processes of cell microbiology and the vital role of plants in the ecosystem.

CHAPTER 2 SUMMARY

Chapter 2 of the book "Introduction to Living Cells: The Building Blocks of Life" delves into the intricate world of cell structure and function, exploring various organelles that play essential roles in cellular processes. The chapter covers five significant topics: the plasma membrane, the nucleus, the endoplasmic reticulum, the Golgi apparatus, mitochondria, and chloroplasts. The plasma membrane, known as the gatekeeper of the cell, serves as a selectively permeable barrier that separates the internal environment of the cell from the external surroundings. Composed of a phospholipid bilayer with embedded proteins, the membrane regulates molecular exchange, enabling essential substances to enter the cell while preventing unwanted materials from entering.

The nucleus, often considered the command center of genetic information, houses the cell's DNA within a double-membraned nuclear envelope. Genes, specific segments of DNA, encode instructions for producing proteins that govern various cellular functions. The nucleus is involved in transcription, where DNA is transcribed into messenger RNA (mRNA), which is then transported to the cytoplasm for translation into proteins. The endoplasmic reticulum (ER) and Golgi apparatus work collaboratively in protein synthesis and modification. The ER, consisting of rough and smooth regions, plays a vital role in co-translational protein folding and post-translational modifications. The Golgi apparatus further processes and sorts proteins, ensuring their proper distribution to different cellular destinations.

Mitochondria, often referred to as the powerhouses of cellular respiration, generate ATP through the process of oxidative phosphorylation. This intricate process involves the breakdown of glucose and other organic molecules to release energy, which is stored in ATP molecules, providing the cell with the energy it needs for various cellular processes.

Chloroplasts are fascinating organelles found in plant cells, responsible for photosynthesis and plant cell energy production. Through photosynthesis, chloroplasts capture light energy and convert it into chemical energy in the form of glucose. This process provides the primary source of energy for plant cells and plays a crucial role in sustaining life on Earth. Chapter 2 explores the remarkable complexity and functionality of these organelles, providing valuable insights into the dynamic and interdependent processes within cells. Understanding cell structure and function is essential for comprehending the intricacies of cell microbiology and the fundamental mechanisms that underlie life. By unraveling the roles of these organelles, readers gain a deeper appreciation for the sophistication of living organisms and the essential role of cells as the building blocks of life. The next chapter will focus on cellular transportation mechanisms.

Chapter

3 CELLULAR TRANSPORT MECHANISMS

CONTENTS

Chapter 3 of the book "Introduction to Living Cells: The Building Blocks of Life" delves into the intricate mechanisms of cellular transport, crucial for maintaining cellular homeostasis and enabling communication between cells. This chapter encompasses five significant topics, each exploring different aspects of cellular transport.

Section 3.1, Passive Transport, focuses on essential processes such as diffusion, osmosis, and facilitated diffusion. These mechanisms allow molecules to move across the cell membrane without the expenditure of cellular energy.

Diffusion is the random movement of molecules from regions of high concentration to low concentration, ensuring an even distribution within the cell. Osmosis, on the other hand, is the movement of water molecules across a semipermeable membrane, balancing solute concentrations on either side. Facilitated diffusion involves the assistance of specific transport proteins to move molecules across the membrane.

Section 3.2, Active Transport, explores the ATP-driven movement of molecules, which occurs against their concentration gradient, requiring cellular energy. Active transport mechanisms, such as the sodium-potassium pump, play a crucial role in maintaining ion gradients and cellular electrical potentials, essential for nerve impulses and muscle contractions.

Section 3.3 delves into Endocytosis and Exocytosis, which are bulk transport processes used to move large substances in and out of the cell. Endocytosis involves the invagination of the plasma membrane to engulf extracellular substances into vesicles. On the other hand, exocytosis releases intracellular substances by fusing vesicles with the plasma membrane. These processes are vital for the uptake of nutrients and secretion of cellular products.

Section 3.4 focuses on Cell Communication, highlighting signaling pathways and receptor proteins. Cells communicate through chemical signals, allowing coordination and integration of cellular activities. Signaling pathways involve a series of biochemical reactions triggered by ligand-receptor interactions, resulting in cellular responses. Receptor proteins on the cell surface or within the cell mediate the recognition of specific ligands, initiating the signaling cascade.

Section 3.5 addresses Cell Adhesion and the Extracellular Matrix, which are fundamental for maintaining tissue integrity and cellular organization. Cell adhesion involves interactions between cell surface molecules and the extracellular matrix or neighboring cells. The extracellular matrix provides structural support to tissues and influences cell behavior, including proliferation, migration, and differentiation.

In this chapter, readers will delve into the fascinating world of cellular transport mechanisms, unveiling the intricate processes that underlie cellular communication, nutrient uptake, and the maintenance of cellular and tissue integrity. Understanding these mechanisms is fundamental to comprehending how living cells maintain their functional and structural harmony, contributing to the overall complexity and diversity of life. Let us begin by exploring passive transport in terms of osmosis and diffusion.

3.1 PASSIVE TRANSPORT: DIFFUSION, OSMOSIS, AND FACILITATED DIFFUSION

Passive transport is a fundamental and intricate process in cell microbiology that ensures the efficient exchange of substances across cell membranes without the consumption of energy (McFall & Stites, 2021; Melo et al., 2022; Mendes, 2023; Mierke, 2023). This chapter explores three significant types of passive transport: diffusion, osmosis, and facilitated diffusion, and delves into their mechanisms and implications in cellular physiology. Passive transport, an essential and complex mechanism in the field of cell microbiology, performs a critical role in ensuring the seamless exchange of substances across cell membranes and remarkably, it manages to do so without expending any energy. To delve into the intricacies of this process, it is instructive to take the case of a human red blood cell (RBC) as the example. A human RBC, serving as a vessel for oxygen transport in the body, acts as an excellent case study for passive transport through its continuous movement of oxygen (O2) and carbon dioxide (CO2). The RBC, while traversing the capillaries, encounters areas of high oxygen concentration, especially in the lungs, where O2 molecules exceed those in the cytoplasm of RBCs. This difference invokes the passive diffusion process that drives the O2 molecules, impelled by their concentration gradient, to traverse the cell membrane and penetrate into the RBC interior. On the contrary, CO2 molecules accumulated in the RBC as the byproduct of cellular respiration are motivated to move out of the cell due to their higher concentration within the cell, compared to the surrounding blood plasma. Thus, these molecules diffuse out through the semipermeable cell membrane of the RBC. These examples of oxygen intake and carbon dioxide expulsion encapsulate the delicate balance maintained by the passive transport mechanism, which enables cells to function efficiently. The process of passive transport thereby illuminates the precision and harmony with which these exchanges take place, and the balancing act that happens without the expenditure of ATP (adenosine triphosphate) energy, which is further indicative of nature's sophistication and prudence in managing cellular resources. Therefore, the human RBC example serving as the case study here highlights the significance of passive transport in maintaining homeostasis in the cellular environment, a critical aspect of cell microbiology. Figure 21 compares active and passive transport.

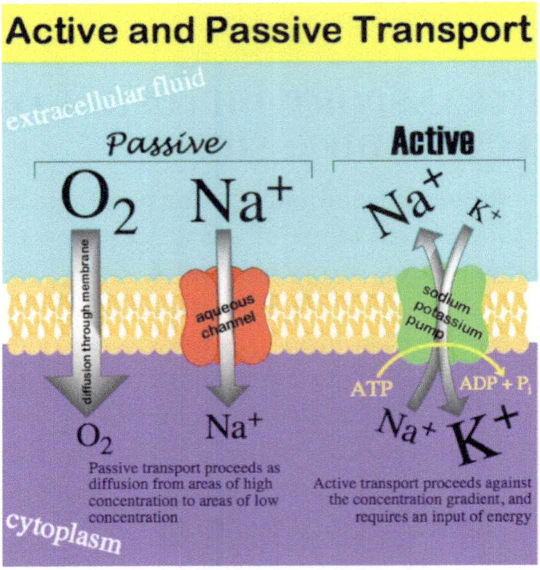

Figure 21. Comparing active and passive transport.

Source: https://upload.wikimedia.org/wikipedia/commons/4/44/Active-passive_transport.jpg

Diffusion is an elegant process driven by the inherent random motion of molecules (Montagnoli et al., 2021; Migliano et al., 2022; Montgomery et al., 2023). When there is a concentration gradient across a cell membrane, with higher concentrations on one side and lower concentrations on the other, molecules will spontaneously move from regions of high concentration to regions of low concentration. This molecular dance is a result of countless collisions and interactions between molecules, leading to their even distribution over time. Diffusion is indeed a beautiful and elegant mechanism largely driven by the random yet precisely directed motion of molecules – a phenomenon that becomes starkly evident when we delve into the microscopic world of cells. A comprehensive understanding of this process becomes possible with an exploration of the role of diffusion in a neuron or nerve cell, an integral component of the nervous system. In the neuron, diffusion enables the seamless transmission of signals through the electrochemical process known as an action potential. As sodium and potassium ions permeate the neuron's cell membrane, they create an electrical charge that travels along the neuron. However, the manner of their movement is not quite as random as one might expect. Sodium ions are concentrated outside the neuron, while potassium ions are concentrated inside. When an electrical signal needs to be transmitted, the negative charge inside the neuron becomes positive due to the influx of sodium ions via diffusion, following their concentration gradient from a region of higher concentration to that of a lower one. This rapid change in charge travels down the neuron as the action potential. Following this, potassium ions flow out of the neuron to restore the negative charge inside, again moving from a region of higher to lower concentration, driven by diffusion. The coordinated diffusion of these ions, from regions of higher concentration to lower, powered by their random kinetic energy and potential disequilibrium, leads to what is acknowledged as the fundamental process of impulse transmission in the nervous system – an elegant display of the inherent random motion of molecules resulting in an essential biological function. Consequently, the case of neuron signal transmission accentuates the fascinating role of diffusion as a key driver of cellular operation, further underscoring the remarkable elegance inherent in the seemingly disorganized randomness of molecular motion.

In cell microbiology, diffusion is ubiquitous and plays a pivotal role in various cellular processes (Doostmohammadi & Ladoux, 2022; Drozdz et al., 2022; Du et al., 2023). For instance, in gas exchange, oxygen and carbon dioxide diffuse across cell membranes, allowing cells to respire and maintain internal oxygen concentrations essential for metabolic reactions. Additionally, small, non-polar molecules, such as lipids, can freely diffuse through the hydrophobic interior of the lipid bilayer, contributing to cell membrane fluidity. The ubiquity of diffusion in cell microbiology testifies to its pivotal role in driving an array of cellular processes, a concept that is perfectly encapsulated in the widely studied and understood case of osmosis in plant cells. Osmosis, a form of diffusion involving the movement of water molecules from an area of low solute concentration to one of high concentration, is instrumental in maintaining the structural integrity and functional capabilities of plant cells. Taking a closer look at an individual plant cell, one finds an enclosed structure surrounded by a rigid cell wall, within which lies the semi-permeable cell membrane that encapsulates the cytosol, a water-based solution where the cell's organelles are suspended. Upon placing the plant cell in a hypotonic environment, where the concentration of solutes external to the cell is lower than that inside the cell, water molecules start diffusing into the cell through the semipermeable cell membrane. The water molecules traverse the cell membrane, propelled by their inherent kinetic energy and the concentration gradient, moving to an area where the concentration of water is lower due to the presence of solutes. As water fills the cell, it creates turgor pressure against the rigid cell wall, giving the cell its shape and structure, and enabling it to stand upright against the force of gravity. This ultimately sets the stage for processes such as photosynthesis by allowing for optimal exposure to sunlight. This instance of osmosis in a plant cell firmly emphasizes the immense significance of diffusion, demonstrating how its ubiquitous presence in cell -microbiology influences the complex processes involved in life sustenance. It augments the central theme that, in the microscopic world, the random motion of molecules made predictable by the guiding principles of diffusion is what lays the foundation for the determined functioning and very survival of the cells.

Osmosis is a specialized form of diffusion that pertains specifically to the movement of water across semi-permeable membranes (Donato et al., 2021; Don & Kalebic, 2022; Domcke & Shendure, 2023). When there is a concentration gradient of solute particles across the membrane, water molecules will move from regions of lower solute concentration to regions of higher solute concentration, seeking to balance the solute concentrations on both sides. Osmosis, a specialized type of diffusion exclusive to the transportation of water molecules across semi-permeable membranes, governs a variety of cell-microbiological processes with enviable efficiency. This selective random motion of water is indeed a defining principle of life at the cellular level, as exemplified in the case study of paramecia, a group of unicellular ciliates. These freshwater organisms, with permeable cellular membranes, are often found in environments where the cytosol, the liquid found within their cells, has a higher concentration of solutes compared to the surrounding water. This discrepancy in concentration, or gradient, sets the stage for osmosis as water molecules naturally move from areas of lower concentration, i.e., the freshwater environment, to areas of higher concentration, i.e., the internal environment of the paramecium. This continuous stream of water into the cell, although crucial for nutrient uptake, also requires careful controlling to prevent the potential risk of cellular rupture due to over-inflating. The paramecium ingeniously solves this dilemma through the presence of contractile vacuoles. These dedicated organelles function as effective bilge pumps, continually expelling water to maintain osmotic balance - a beautifully orchestrated biological response to ensure the survival of the paramecium in its aquatic habitat. This detailed case study of diffusion in the form of osmosis in paramecium thus reinforces the significance of this specialized process in cell biology. It underscores the untold narratives of survival at the microscopic level, demonstrating how life has evolved intricate cellular mechanisms, guided by the selective motion of water molecules, to thrive even in the most challenging environments.

In cell biology, osmosis is of utmost importance in maintaining cell integrity and functionality (LeGuennec et al., 2021; Leidal & Debnath, 2021; Lendahl, 2022). In plant cells, for instance, the uptake of water through osmosis results in turgor pressure, providing the cell with structural support and maintaining its shape. In animal cells, osmosis influences cell volume and contributes to the regulation of cell osmolarity and stability. The principle of osmosis, integral in cell biology, is of utmost importance in preserving cell integrity and functionality, and this can be perfectly illustrated through a case study focusing on human red blood cells (RBCs). These biconcave structures, distinguished by their cytosolic content rich in hemoglobin, are contained within a semipermeable membrane. They exist in a dynamic environment, the isotonic plasma, that features an equilibrium in the concentration of solutes both external and internal to the cells. However, any alteration in this balance has direct implications on cellular function. For instance, placing RBCs in a hypertonic environment, where the concentration of solutes outside the cell exceeds that inside, triggers a process known as crenation. During crenation, osmosis drives water molecules to move from their higher concentration inside the cell, across the membrane, to the lower concentration outside. As a consequence of this efflux of water, the cells shrink and wrinkle, disrupting their normal functionality and shape. Conversely, placing RBCs in a hypotonic solution, in which the concentration of solutes is greater inside the cell, leads to a rapid influx of water via osmosis, causing the cells to swell and potentially burst in a process called hemolysis. Both of these scenarios illustrate the devastating impact of osmotic imbalance on cell functionality and integrity. Fundamental cellular processes like gas exchange are compromised, threatening the overall survival of the organism. Hence, by highlighting the extreme consequences of osmotic imbalance in RBCs, the critical role of osmosis in maintaining cell functionality and structural integrity in cell biology is clearly underscored. Thus, this reinforces the understanding that the constant, diligent regulation of the osmotic environment is crucial for the proper functioning of these microscopic building blocks of life. Figure 22 illustrates facilitative, passive, and active transport in cell microbiology.

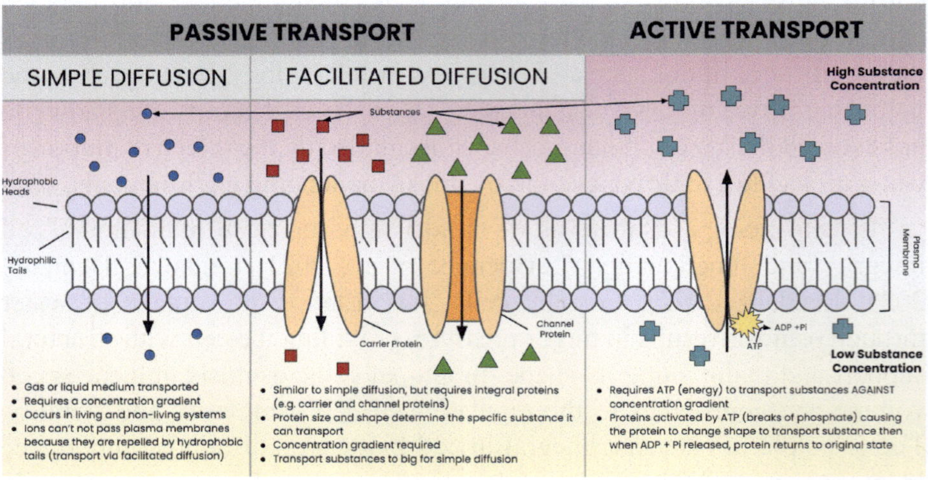

Figure 22. Facilitative, passive, and active transport.

Source: https://commons.wikimedia.org/wiki/File:Passive_vs_Active_Membrane_Transport.svg

Facilitated diffusion involves the assistance of specific transport proteins to enable the passage of larger or charged molecules across the cell membrane (Ma & Hoffmann, 2021; Lu et al., 2022; Lynes et al., 2022). These transport proteins act as selective gates, allowing molecules to move down their concentration gradients. Facilitated diffusion, a critical strategy employed within cell biology, involves the aid of specialized transport proteins to permit the traversal of larger or charged molecules across the cell membrane. A commonly referenced case study that highlights the mechanistic intricacy and biological importance of facilitated diffusion is the process of glucose transportation in human cells. Glucose, a vital fuel molecule for cellular energy production, is a large hydrophilic substance that cannot easily diffuse through the lipophilic cell membrane. To overcome this obstacle, cells harness a class of transport proteins named glucose transporters (GLUT). Each GLUT protein embedded in the cell membrane operates like a meticulously engineered gateway, specifically designed to recognize and bind to glucose molecules outside the cell. On binding, these proteins undergo a sophisticated conformational transformation, effectively flipping their overall architecture to allow the release of the glucose molecule into the intracellular environment, whilst simultaneously shielding it from the surrounding lipophilic core of the membrane during its journey. This fascinating molecular relay race enables cells to effectively harvest glucose from their environment, irrespective of the concentration of glucose inside the cell. The orchestrated performance of GLUT proteins, which seamlessly combine recognition and transport, underscores the elegance and efficiency of facilitated diffusion. Therefore, this intricate mechanism of glucose transportation, underlining the widespread biological necessity of facilitated diffusion, exemplifies how cell biology has fine-tuned the interface between the cell and its surroundings, enabling larger molecules or charged particles to cross the cell membrane barrier effortlessly.

In cell biology, facilitated diffusion is crucial for the uptake of essential molecules, such as glucose and amino acids (Montagnoli et al., 2021; Migliano et al., 2022; Montgomery et al., 2023). Glucose transporters facilitate the movement of glucose molecules into the cell, ensuring a steady supply of energy for cellular processes. Similarly, ion channels facilitate the movement of ions across the membrane, contributing to nerve impulse transmission and muscle contractions. Facilitated diffusion plays a critical role in cell biology, particularly in the uptake of vital molecules like glucose and amino acids, the significance of which becomes apparent in the case of human muscle cells, which require both glucose and amino acids to perform optimally. The muscle cells, like a well-structured entity, utilize the help of a dedicated set of transport proteins that meticulously assist in the reception and transportation of these nutrients. The glucose transport proteins (GLUT4), for instance, facilitate the passage of glucose across the cell membrane. Upon stimulation (such as during exercise), GLUT4 transporters translocate to the cell surface. Acting like an exquisitely programmed gateway, these proteins bind to glucose molecules in the extracellular fluid, undergo a conformational shift, and release the glucose molecule into the cell. This series of events effectively bypasses the energy-requiring active transport methods and hinges purely on the existing glucose concentration gradient. Simultaneously, the cell furnishes itself with the specialized transporters SLC38A9 and LAT1 to permit the entry of necessary amino acids. These transporters recognize and escort various amino acids across the cell membrane, once again without consuming ATP energy but rather exploiting the existing concentration gradient between extracellular and intracellular environments. In addition to their role in protein synthesis, these amino acids can contribute to glucose homeostasis, thereby impacting cellular energy management. Ultimately, these examples of facilitated diffusion in human muscle cells underline its importance in maintaining the uptake of essential molecules. The concerted efforts in transporting glucose and amino acids via facilitated diffusion underscore the complex yet efficient handling of nutrient transport in cell biology, playing a pivotal role in cellular function and, indeed, survival.

In conclusion, passive transport mechanisms, including diffusion, osmosis, and facilitated diffusion, are fundamental to cell biology, enabling the precise regulation of cellular processes and maintaining cellular homeostasis. The intricacy of these processes is a testament to the remarkable adaptation and precision of cellular mechanisms. Understanding the mechanisms of passive transport is vital for grasping the intricacies of cellular physiology and the functioning of microorganisms, forming the foundation for further exploration into the fascinating world of cell biology. The next section explores active transport in terms of ATP-driven movement of molecules.

3.2 ACTIVE TRANSPORT: ATP-DRIVEN MOVEMENT OF MOLECULES

Active transport is a fundamental cellular process that enables the movement of molecules against their concentration gradient, from areas of low concentration to areas of higher concentration (Collins et al., 2021; Costa & Sousa, 2022; Colin et al., 2023). Unlike passive transport, which relies on diffusion and does not require cellular energy, active transport consumes energy, usually in the form of ATP (adenosine triphosphate). This process is essential for maintaining specific ion concentrations inside and outside the cell and for performing vital functions like nerve impulse transmission and muscle contraction. Active transport mechanisms, bestowed upon cells as an essential biological process with paramount efficiency, engage in the task of transporting molecules against their concentration gradients from regions characterized by a low level of these particles, to zones with higher levels of concentration. The operation executed by cellular components known as transport proteins embedded within cell membranes delineates this directive of active transport mechanisms. Figure 23 demonstrates how active transport moves against the concentration gradient.

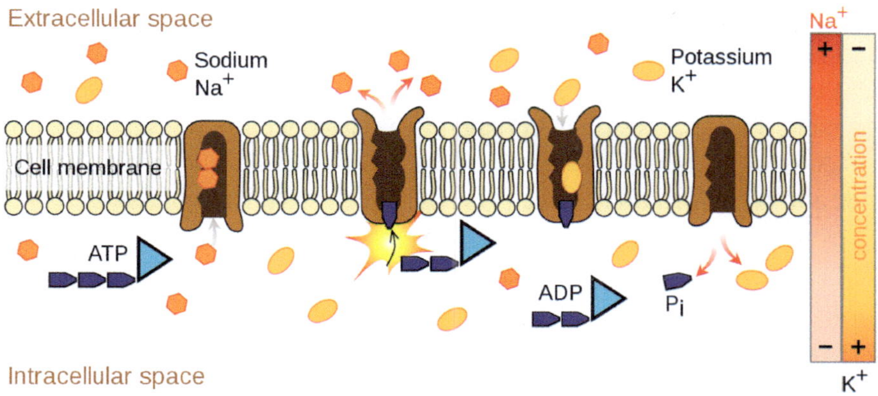

Figure 23. Illustration of how active transport moves against the concentration gradient.

Source: By LadyofHats Mariana Ruiz Villarreal - Own work, Public Domain, https://commons.wikimedia.org/w/index.php?curid=3981038

A primary case study of active transport finds its existence in the realm of microbiology within the bacterium Escherichia coli (E.coli), known for its adaptiveness in a variety of environments due to its highly functional active transport systems (Das et al., 2021; de Albuquerque et al., 2021; De Camilli, 2021; de Torres-Jurado et al., 2022). A key transport system within E.coli, the phosphoenolpyruvate-dependent phosphotransferase system, or PTS, effectively channels glucose from the extracellular milieu to the cell's interior, a region of higher concentration, requiring energy in the form of phosphoenolpyruvate molecules. The inherent complexity and regulation of the PTS make it one of the most understood and elaborately regulated systems within E.coli, functioning not only as a transport system but also as a signaling cascade modulating responses to glucose availability. Structurally, this system comprised four different protein components: enzyme I (EI), HPr, Enzyme II (EII), and the EII complexes which are specialized for individual sugars. Through a stepwise trans-phosphorylation cascade, where phosphate groups are sequentially transferred from PEP to EI, HPr, and then to EII, it facilitates the simultaneous phosphorylation and translocation of glucose against its gradient into the cell, thereby underscoring the far-reaching potential of active transport. In conclusion, the case of E.coli stands out as a testament to the intricacies and vitality of active transport within the broader orchestration of cellular processes, fostering adaptability and survivability amid fluctuating environments.

One of the most well-known examples of active transport is the sodium-potassium pump (Hynds et al., 2021; Jardin et al., 2022; Iachettini et al., 2023). This pump actively transports sodium ions (Na+) out of the cell and potassium ions (K+) into the cell against their concentration gradients. It is crucial for maintaining the proper balance of sodium and potassium ions inside and outside the cell, which is vital for various physiological processes, including nerve cell communication and muscle contractions.

One of the most recognized exemplars of active transport within the interdisciplinary domain of cellular biology is unequivocally the sodium-potassium pump, a distinctive carrier protein that is ubiquitously present in animal cell membranes. It operates as a form of primary active transport and plays an imperative role in maintaining the electrochemical gradient indispensable for myriad biological functions, including cell volume regulation, nutrient absorption and generation of action potential in neurons. The pump executes this function by transporting three sodium ions out of the cell and two potassium ions into the cell during each cycle of its operation using energy derived from the hydrolysis of ATP (Adenosine Triphosphate). This consistent and directional movement against the concentration gradients of both ions establishes the interior of the animal cells to be negatively charged with respect to the exterior, a critical characteristic that is foundational to the physiological functions they support. The initial step in the cycle starts when three intracellular sodium ions bind to the protein pump, stimulating the phosphorylation of an ATP molecule and leading to the release of adenosine diphosphate (ADP), which in turn induces a change in the pump's conformation, causing it to expose these sodium ions towards the extracellular side. Once these three sodium ions are ejected, two extracellular potassium ions bind to the protein, which concurrently catalyzes the dephosphorylation of the pump and initiates another conformational change, thereby causing the protein to revert to its original state and securing the potassium ions within the cell. This exemplification of the sodium-potassium pump, upon close examination, indeed, affirms the paramount importance and astounding efficiency of active transport in the perpetuation and governance of life processes within cellular structures.

The sodium-potassium pump is an integral membrane protein located in the cell membrane (Gowrishankar et al., 2021; Giolito & Plateroti, 2022; Gordeeva et al., 2022). It is an enzyme called Na+/K+-ATPase, which hydrolyzes ATP to ADP (adenosine diphosphate) and inorganic phosphate, releasing energy in the process. This energy is used to move three sodium ions out of the cell and two potassium ions into the cell for every ATP molecule hydrolyzed. The Sodium-Potassium pump serves as an integral membrane protein that occupies a central position in the cell membrane, playing a critical role in the cellular mechanism to maintain a distinct homeostatic balance. This pump, known scientifically as Na+/K+-ATPase, is a transmembrane complex and found in the plasma membrane of almost all animal cells where it constitutes nearly 20% of the total membrane protein. Its multisubunit structure includes alpha, beta, and gamma subunits.

The alpha unit provides the key functional sites for ATP binding, sodium and potassium ion translocation, and the enzymatic capability to hydrolyze ATP, whereas the beta and gamma subunits participate in structurally stabilizing the complex and modulating its activity. A distinctive feature of this pump is its ability to ferry ions in and out of the cell against their respective concentration gradients, consequently creating an ionic imbalance imperative for cellular functionality, including maintaining intracellular pH, regulating cellular volume, influencing membrane potential and driving secondary active transport. During operation, the tightly orchestrated cycle of Na+/K+-ATPase transports three molecules of sodium ions out of the cell while concurrently pulling in two molecules of potassium ions into the cell. This differential pumping leads to a lower concentration of sodium ions and a higher concentration of potassium ions within the cells than outside. Also noteworthy is its ATPase activity, which harnesses the energy required for ion movement from the hydrolysis of a molecule of ATP during each cycle of action, thereby transforming chemical energy into ionic gradients. The functional complexities of the Sodium-Potassium pump, hence, underline its role in biological systems as an integral participant in cellular metabolism and machinery. Thus, the Na+/K+ pump not only emphasizes the potency and precision of active transport systems but also underscores their vital contribution to the proper functioning and survival of cells.

Another example of active transport is the calcium pump (Lefranc, 2021; Lee & Rompolas, 2022; Leao et al., 2023). This pump actively transports calcium ions (Ca2+) out of the cytoplasm to maintain low calcium levels inside the cell. Intracellular calcium concentration is tightly regulated because calcium ions play essential roles in cell signaling, muscle contractions, and other cellular processes. The calcium pump, similar to the sodium-potassium pump, utilizes ATP to move calcium ions against their concentration gradient. Another quintessential representation of active transport in cellular biology pertains to the calcium pump, formally known as the calcium ATPase and commonly found in both the plasma membrane (PMCA) and the sarcoplasmic reticulum (SERCA) of cells. For this text, reference shall primarily be confined to the SERCA in skeletal muscle cells. Governed by ATP hydrolysis, the calcium ATPase is invaluable for the transportation of calcium ions from the cytoplasm where calcium concentrations are low, to the lumen of the sarcoplasmic reticulum, a site of high calcium concentration. Functionally with two forms, SERCA1 prevalent within mature skeletal muscle, and SERCA2 widespread in all cell types, the calcium ATPase performs an indispensable role in muscle contraction and relaxation. In a resting muscle cell, the concentration of calcium in the cytoplasm is kept low, maintained by active transport by SERCA into the sarcoplasmic reticulum. Upon neuronal stimulation, calcium is released into the cytoplasm, triggering muscle contraction. To return the cell to its resting state, calcium ATPase counters the flood of calcium ions by active transport back into the sarcoplasmic reticulum, thus enabling muscle relaxation. The structure of calcium ATPase includes a transmembrane domain and a large cytoplasmic domain where ATP hydrolysis occurs. For each cycle of ATP hydrolysis, two calcium ions are transported from the cytoplasm to the sarcoplasmic reticulum against their concentration gradient. Consequently, the calcium pump, pivotal for the regulation of intracellular calcium, sets the stage for coordinated muscle movement, demonstrating the profound power of active transport. Figure 24 highlights the role of protein pumps in active transport.

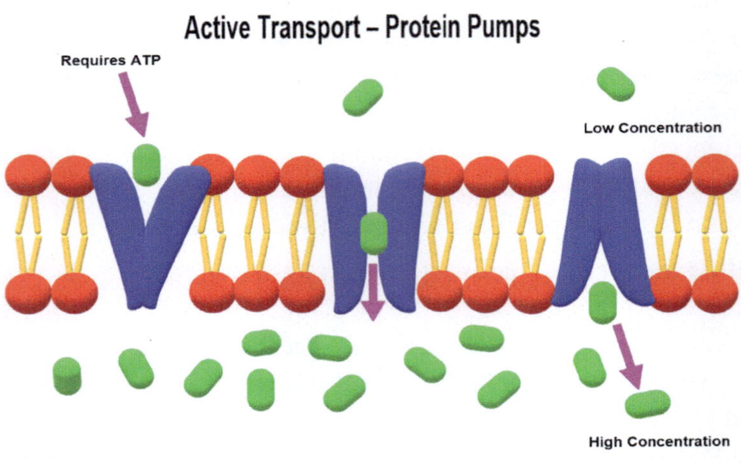

Figure 24. The role of protein pumps in active transport.

Source: Wiki Lectures (https://www.wikilectures.eu/w/Active_transport)

Active transport is not limited to ions; it also facilitates the uptake of essential nutrients by cells (Sales-Gil et al., 2021; Salinas-Luypaert et al., 2021; Schornig & Taverna, 2021). For example, in the intestinal epithelial cells, active transport mechanisms enable the absorption of nutrients such as glucose and amino acids from the gut lumen into the bloodstream. These nutrients are transported against their concentration gradients to ensure sufficient uptake and distribution throughout the body. In active transport, this cellular process certainly does not confine itself to the trafficking of ions; it rather expands its action to ensure the assimilation of essential nutrients possible for cells. A quintessential illustration of this phenomenon manifests in the bacterial species Escherichia coli, which employs an array of diverse specialized active transport systems for the intake of glucose, an indispensable nutrient for its survival and proliferation. One such prominent transporter system is the phosphoenolpyruvate: sugar phosphotransferase system (PTS). This system facilitates the uptake of glucose and a host of other sugars, allowing them to traverse the bacterial cell membrane against their concentration gradient from the extracellular milieu into the intracellular compartment, even when extracellular concentrations are considerably low. The PTS, a complex multi-protein system, is unmatched in its dual efficiency, which simultaneously phosphorylates and translocates the glucose molecule. Upon the uptake of glucose, the PTS system catalyzes the transfer of a phosphate group from phosphoenolpyruvate, an intermediate of glycolysis, to a specific histidine residue on enzyme I (EI). This phosphate group is subsequently transferred to HPr, enzyme II (EII), and finally to glucose as it is translocated into the cell. The imported glucose-phosphate can subsequently enter catabolic pathways, primarily glycolysis, thereby fueling the bacterium's metabolic processes. The efficacy regardless of nutrient availability of the PTS vividly illustrates the broader implication of active transport in nutrient assimilation. Active transport systems, therefore, find themselves interwoven into the fabric of cellular life, functioning as indispensable arbiters of survival and facilitators of growth.

In addition to primary active transport, cells also employ secondary active transport (Tolkin & Hubbard, 2021; Tomoda & Kime, 2021; Tuieng et al., 2021). In secondary active transport, the movement of one molecule against its gradient is coupled to the movement of another molecule down its concentration gradient. This coupling allows the transport of specific molecules across the cell membrane, even though they are moving against their own concentration gradient. Indeed, the metabolic versatility of cells is not limited to primary active transport systems alone. Cellular mechanisms also incorporate tools of secondary active transport, where the movement of a substance against its concentration gradient is coupled with the movement of another substance down its concentration gradient; thereby creating an energetically favorable scenario. An illustrative example of secondary active transport can be found in the intestinal absorption of glucose, amino acids, and various ions in epithelial cells. In the epithelial cells of the small intestine, a prime example of secondary active transport is demonstrated by the sodium-glucose transport proteins (SGLTs). These integral membrane proteins harness the downhill sodium ion gradient established by the primary active transport systems, most notably the sodium-potassium pump, to drive the uphill transport of glucose molecules. In essence, cellular energy in the form of ATP is indirectly used to facilitate the transport of glucose. The specific manner in which this works is through the SGLTs initially binding a sodium ion, and subsequently a glucose molecule. The combined presence of these two entities instigates a conformational change in the SGLT, causing it to deposit the sodium ion and glucose molecule into the cell. In this manner, the glucose concentration in the epithelial cells is increased, allowing glucose to diffuse into the bloodstream down its concentration gradient via GLUT2 transporters. Thus, highlighting the interconnected roles of both primary and secondary active transport mechanisms in cellular physiology, with secondary active transport mechanisms proving to be indispensable in nutrient absorption.

One well-studied example of secondary active transport is the sodium-glucose cotransporter (SGLT) in the kidney and intestinal cells (Jaspersen, 2021; Jaufmann et al., 2021; Ji et al., 2021; Jeziorski et al., 2023). The SGLT uses the energy stored in the sodium ion gradient established by the sodium-potassium pump to actively transport glucose against its concentration gradient into the cell. This mechanism enables the efficient absorption of glucose from the renal tubules and the intestines, preventing its loss through urine or feces. Scientific exploration within the realm of cell biology has furnished an extensive understanding of secondary active transport systems, with the Sodium-glucose Co-transporter (SGLT) present within renal and intestinal epithelial cells surfacing as a well-studied example. This co-transporter operates on the fundamental principle of coupling the movement of one molecule against its concentration gradient with the simultaneous facilitation of another molecule along its favored gradient. In the specific case of the SGLT, the 'downhill' movement of sodium ions from the lumen into epithelial cells, which is driven by the concentration gradient established by the sodium-potassium pump on the basolateral membrane, is concurrent with the 'uphill' transport of glucose from the lumen into these cells. The SGLT, an integral membrane protein, employs the potential energy of the sodium ions to drive the transport of glucose molecules, against their concentration gradient, into the kidney or intestinal cells where glucose concentration is significantly higher. The binding of a sodium ion to the SGLT causes a conformational modification in the protein, enabling it to bind to a glucose molecule subsequently. The consequent interaction causes another conformational alternation that releases both sodium and glucose into the cell's cytoplasm. The glucose molecules are then expelled into the bloodstream through facilitated diffusion by GLUT2 transporters present on the basolateral membrane. Thus, through the efficient coupling of favorable and unfavorable transport processes, the SGLT showcases how secondary active transport mechanisms play a pivotal role in the absorption and regulation of essential nutrients within cellular entities.

In summary, active transport is a vital cellular process that enables the movement of molecules against their concentration gradients. It requires energy, usually in the form of ATP, and is essential for maintaining ion concentrations, nutrient uptake, and cellular functions like nerve impulses and muscle contractions. Active transport mechanisms play a critical role in ensuring cellular homeostasis and the proper functioning of multicellular organisms. The next section considers bulk transport processes in terms of endocytosis and exocytosis.

3.3 ENDOCYTOSIS AND EXOCYTOSIS: BULK TRANSPORT PROCESSES

Endocytosis and exocytosis are essential bulk transport processes that enable cells to move large substances in and out of the cell, respectively (Katunin et al., 2021; Keenan et al., 2021; Khampang et al., 2023). Both processes involve the formation and fusion of membrane-bound vesicles, allowing the cell to regulate its internal environment, uptake nutrients, and eliminate waste materials. Endocytosis and exocytosis are salient processes in the milieu of bulk transport, enabling cells to ingest and expel large substances, respectively, surpassing the capability of simple diffusion or active transport mechanisms. One remarkable example of endocytosis is the internalization of low-density lipoprotein (LDL) particles, which carry cholesterol, into animal cells via receptor-mediated endocytosis. LDL particles in the extracellular fluid bind to specific LDL receptors on the plasma membrane, causing the membrane to invaginate inward and form a coated pit with the assistance of clathrin proteins. As the pit deepens, it pinches off from the plasma membrane, encapsulating the LDL particles into a vesicle inside the cell. This vesicle then combines with a lysosome, which contains hydrolytic enzymes that break the LDL particle down to release the cholesterol for cellular usage. On the other hand, neurotransmission at synapses in the nervous system represents an important instance of exocytosis. In response to an electrical signal, synaptic vesicles carrying neurotransmitters fuse with the presynaptic cell membrane, ejecting the neurotransmitters into the synaptic cleft. These neurotransmitters bind to specific receptors on the postsynaptic cell, thereby relaying the signal. This process is rapid and regulated, facilitated by SNARE proteins, which are responsible for the docking and fusion of the synaptic vesicles. Both endocytosis and exocytosis exemplify how cells manage the trafficking of large substances across the membrane, underpinning the dynamic nature of cells and their responsiveness to intricate and variable cellular demands. Figure 25 illustrates bulk transport.

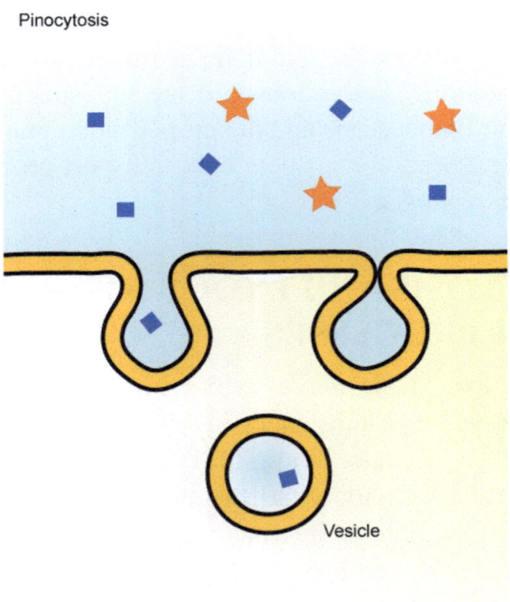

Figure 25. Illustration of bulk transport.

Source: Mariana Ruiz Villareal: https://bio.libretexts.org/Bookshelves/Introductory_and_General_Biology/General_Biology_1e_%28OpenStax%29/2%3A_The_Cell/5%3A_Structure_and_Function_of_Plasma_Membranes/5.4%3A_Bulk_Transport

Endocytosis is a process by which cells engulf extracellular substances, forming vesicles to bring those substances into the cell's interior (Cheung et al., 2022; Chhabra et al., 2023; Choezom & Gross, 2022). There are three main types of endocytosis: phagocytosis, pinocytosis, and receptor-mediated endocytosis.

1. Phagocytosis: In phagocytosis, specialized cells called phagocytes, such as macrophages and neutrophils, engulf large particles, such as bacteria, cell debris, or other foreign materials (Cheung et al., 2022; Choezom & Gross, 2022; Chhabra et al., 2023). The cell extends its membrane around the particle, forming a phagosome. The phagosome then fuses with lysosomes, which contain digestive enzymes, leading to the breakdown of the engulfed material. Phagocytosis is characterized by the active engulfing of large particles by specialized cells called phagocytes, chief among which include macrophages and neutrophils. This dynamic cellular process is instrumental in defense mechanisms and tissue homeostasis, particularly in the removal of foreign substances such as bacteria, dead cells, or cellular debris. A key example of this can be found within the mammalian immune system: when a bacterium enters the body, it can trigger an inflammatory response, which prompts these phagocytes to migrate to the site of infection. Notably, phagocytes possess surface receptors that can detect and bind to specific molecules on the surface of the foreign particle, such as bacterial cell wall components or antibodies, which are part of the body's immune response. Following the detection and attachment to foreign material, the phagocyte wraps its membrane around the particle, forming a phagosome. This intracellular vesicle subsequently fuses with a lysosome to create a phagolysosome. The enzymes and acidic environment within the lysosome facilitate the digestion of the ingested particle, producing residual bodies that contain indigestible material, which are ultimately expelled from the cell by exocytosis. The comprehensive interplay of this sequence, initiated by phagocytes and culminating in the destruction of foreign entities, emphasizes the crucial role of phagocytosis in maintaining physiological integrity, fortifying the body's defense mechanisms, and regulating cellular interaction.
2. Pinocytosis: Pinocytosis, also known as "cell drinking," involves the non-specific uptake of extracellular fluid and solutes (Jones

et al., 2021; Karlsson & Draber, 2021; Ju et al., 2022). The cell continuously forms small vesicles at its membrane to internalize fluid and dissolved substances, which are then delivered to endosomes for processing. Pinocytosis, colloquially termed as "cell drinking," represents another key aspect of adaptive cellular function, entailing the non-specific ingestion of extracellular fluid and solutes. This act of vesicle-mediated endocytosis enables cells to sample their surrounding environment and thus maintain their intracellular fluid composition. As an illustration, consider the function of capillary endothelial cells, particularly pertinent given their role in tissue nourishment and waste removal. These cells adopt pinocytosis as a routine characteristic of their operation, building small vesicles called caveolae that bud inward from the plasma membrane, encapsulating a portion of the extracellular fluid and the solutes therein. As caveolae are to special lipid rafts rich in cholesterol and glycolipids providing stability, they serve to perform transcytosis - a process where vesicles transport molecules across the cell, from one side to the other. This process, triggered by particular cues such as changes in cellular hydration status or certain molecular signals, results in the transcellular conveyance of nutrients, hormones, and plasma proteins from the bloodstream into the interstitial fluid, further facilitating the delivery of these substances to adjacent cells. Thereby, pinocytosis, through its role in maintaining cellular and tissue homeostasis and enabling cells to dynamically respond to environmental changes, highlights the remarkable adaptability and versatility found in cellular function and processes.

3. Receptor-Mediated Endocytosis: This process is highly specific and involves the uptake of specific molecules that bind to specific receptors on the cell surface (Hernandez-Gonzalez et al., 2021; Herranz-Iturbide et al., 2021; Hiatt et al., 2021). Ligand-receptor complexes form, leading to the formation of clathrin-coated vesicles, which are then internalized into the cell. Once inside the cell, the vesicles fuse with endosomes, and the ligands dissociate from the receptors. The receptors are then recycled back to the cell surface, while the ligands are delivered to lysosomes for degradation or utilized by the cell. Receptor-mediated endocytosis is an elevated, specific form of cellular transport that facilitates cellular uptake of specific extracellular molecules, known as

ligands, which bind to distinct receptors on the cell surface. One compelling illustration of receptor-mediated endocytosis is found in the cellular intake of low-density lipoprotein (LDL) particles. LDL, a courier of cholesterol throughout the body, contributes to cell membrane biosynthesis, and excessive concentrations of circulating LDL can pose cardiovascular risks. In receptor-mediated endocytosis of LDL, LDL particles bind to LDL receptors on the cell surface. Recognition and binding initiate inward budding of the cell membrane to create a coated vesicle within the cytoplasm. Concomitantly, clathrin, a protein on the cytoplasmic side, forms a lattice-like polygonal network, facilitating the transformation of the membrane into a vesicle while capturing the receptors and their bound ligands within. As the vesicle pinches off into the cytoplasm, dynamin, a GTPase, facilitates fission. Subsequently, the clathrin coat disassembles, freeing the vesicle to fuse with an early endosome, where the LDL particles release from their receptors due to a lower pH. Released LDL particles are then trafficked to lysosomes, where hydrolytic enzymes disassemble LDL to liberate cholesterol for cellular uses. This highly selective process ensures cells absorb the necessary amounts of specific molecules that sustain cell survival and contribute to cell proliferation, demonstrating the precision and adaptability of cellular mechanisms responsive to ever-changing internal and external environments.

In contrast, exocytosis (see figure 26) is the process by which intracellular substances, such as proteins, hormones, and neurotransmitters, are released from the cell (Bomont, 2021; Biermann & Reya, 2022; Bohnert & Johnson, 2022). Exocytosis is vital for secretion, cell communication, and the release of cellular waste materials. Exocytosis occurs when vesicles containing substances move toward the cell membrane and fuse with it, releasing their contents into the extracellular space. There are different types of exocytosis, depending on the specific substances being released.

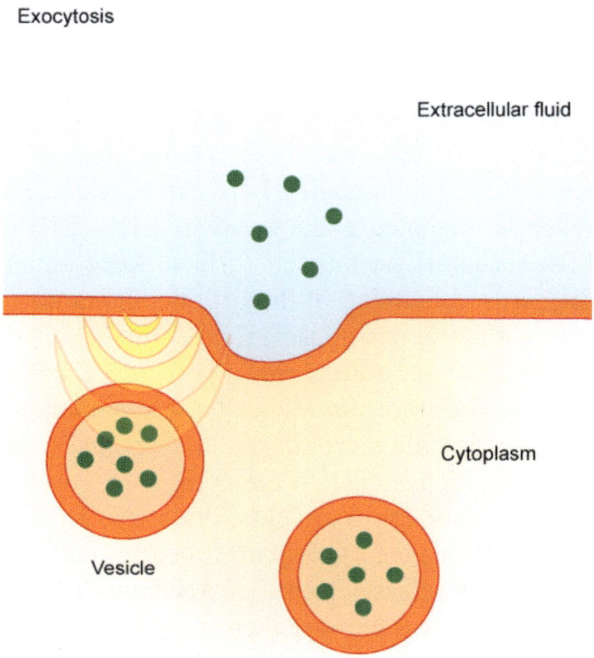

Figure 26. Illustration of exocytosis.

Source: credit: modification of work by Mariana Ruiz Villareal)- https://bio.libretexts.org/Bookshelves/Introductory_and_General_Biology/General_Biology_1e_%28OpenStax%29/2%3A_The_Cell/5%3A_Structure_and_Function_of_Plasma_Membranes/5.4%3A_Bulk_Transport

1. Constitutive Exocytosis: In constitutive exocytosis, vesicles continuously fuse with the cell membrane and release their contents into the extracellular space (Liang et al., 2021; Liang et al., 2022; Li et al., 2023). This process is essential for the continuous secretion of certain proteins and membrane components. Constitutive exocytosis delineates a timely yet incessant, non-regulated route of cellular export, in which vesicles coalesce persistently with the cell membrane and extrude their contents into the extracellular space. A clear example of this unfolds in fibroblast cells, crucial for tissue repair and extracellular matrix fabrication, as they continuously expel proteins such as collagen into the extracellular milieu to maintain tissue architecture and function. Within the fibroblast, collagen molecules destined for secretion are synthesized in

the endoplasmic reticulum, post-translationally modified, and then transported to the Golgi apparatus where they are further processed and packaged into vesicles. These vesicles, driven by microtubule networks, travel to and fuse with the cell membrane through a process facilitated by SNARE proteins. The merging of the vesicle and plasma membranes unveils a channel for the exported proteins to exit the cell. Thus, through this process of constitutive exocytosis, fibroblasts continuously secrete collagen molecules, which assemble extracellularly into fibers, forming an integral part of the connective tissue. This showcases the pivotal role of constitutive exocytosis in the persistent and harmonious production and release of molecules essential for maintaining the biological infrastructure and cellular communication.

2. Regulated Exocytosis: Regulated exocytosis is more specific and tightly regulated (Bettegazzi et al., 2021; Bennison et al., 2022; Beppler et al., 2022). It occurs in response to external signals or specific stimuli, leading to the release of specific substances. For example, neurotransmitters are released from nerve cells via regulated exocytosis in response to action potentials. Regulated exocytosis underlies a more regulated and precise mode of vesicular release, chiefly operational in cells tasked with secreting specific molecules in response to particular stimuli. A vivid illustration of such a highly controlled process is found in neurosecretory cells, for instance, neurons, orchestrating their unique neurotransmission role. Neurons store their neurotransmitters, the chemical messengers, within synaptic vesicles stacked in the nerve terminal awaiting anionic triggers. These neuronal signals, or action potentials, stem from the neuron's cell body and progress along the axon to reach the nerve terminals. Upon their arrival, voltage-gated calcium channels entrusted in the terminal plasma membrane open, permitting an influx of calcium ions from the extracellular space. This sudden surge in calcium concentration within the nerve terminal fosters the translocation of the neurotransmitter-laden vesicles toward plasma membrane docking sites. Guided by a complex of SNARE proteins, these vesicles dock and prime at the presynaptic membrane. The final fusion of vesicular and plasma membranes, mediated by the calcium-sensitive protein synaptotagmin, then results in the expulsion of the neurotransmitter into the synaptic

cleft. The neurotransmitter molecules bind to specific receptors on the adjacent postsynaptic neuron, thereby transmitting the signal from the first neuron. Hence, this elaboration on regulated exocytosis accentuates the sheer sophistication with which cells can conduct their secretory activities, thereby underscoring the precision and adaptability of cellular processes to cater to specialized functional needs.

Overall, endocytosis and exocytosis are crucial processes that enable cells to communicate with their environment, regulate internal conditions, and perform vital functions. These bulk transport mechanisms play critical roles in immune response, nutrient uptake, neurotransmission, and the maintenance of cellular homeostasis. Through endocytosis, cells can internalize substances for digestion or utilization, while exocytosis allows them to secrete substances for intercellular communication and waste elimination. Understanding these processes is fundamental to comprehending the complexity of cellular functions and their relevance in cell biology. The next section focuses on cell communications in terms of signaling pathways and receptor proteins.

3.4 CELL COMMUNICATION: SIGNALING PATHWAYS AND RECEPTOR PROTEINS

Cell communication is a fundamental process that allows cells to sense and respond to changes in their environment, as well as coordinate their activities with other cells (Sudhof, 2021; Summers et al., 2022; Sundaram & Chen, 2022). Signaling pathways and receptor proteins play a pivotal role in mediating cell communication and ensuring cellular responses to specific signals. Intercellular communication stands as a fundamental regulatory process, facilitating cells to sense, interpret, and adapt to modifications in their milieu, alongside orchestrating collaborative tasks with neighboring cells. One prominent example elucidating this process is the bacterial cell-to-cell communication system, often referred to as quorum sensing. This phenomenon permits bacteria to sense their cell density by producing, releasing, and detecting signal molecules, also known as autoinducers. With reference to the bioluminescent bacterium Vibrio fischeri, this species employs quorum sensing to regulate light emission in a cell density-dependent manner. Each V. fischeri cell releases autoinducers into their milieu that diffuse freely across cell membranes. At low cell density, the concentration of these autoinducers remains low and does not significantly affect cellular activity. However, as the bacterial population grows, the autoinducer concentration increases. When a threshold concentration is achieved, the autoinducers bind to transcriptional regulators within the bacteria, triggering the transcription of luminescence genes and consequently, light emission. This high-density, coordinated activity benefits the bacterial community by economizing the use of energy required for light production and is only effective when performed collectively. Thus, the interplay unfolding, in this case, study exemplifies the crucial role of cell communication in aligning individual cells to function as a cohesive community, accurately process environmental cues, and efficiently modulate their behaviors for improved survival and adaptation. In figure 27, we see an example of a rudimentary signal transduction pathway. This is a prime example of the functionality and effects of cell communication. A key component of that process is cell signaling.

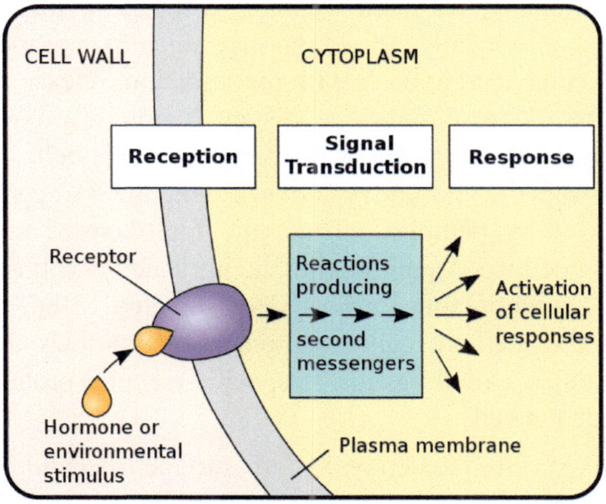

Figure 27. Example of a signal transduction pathway.

Source: By Original: Yaneeporn Vector: Pixelsquid - Own work based on: 1Signal Transduction Pathways Model.jpg:, CC BY-SA 4.0, https://commons.wikimedia.org/w/index.php?curid=97419672

Cell signaling can occur through various mechanisms, including direct cell-cell contact, local diffusion of signaling molecules, and long-range signaling through the circulatory system (Yasuhiko et al., 2021; Ye et al., 2021; Yla-Anttila, 2021; Yoshimatsu et al., 2022). Cell signaling, a core process governing cellular responses and inter-cellular interactions, can persevere through an ensemble of mechanisms, embracing direct cell-cell contact, local diffusion of signaling molecules, and long-range signaling augmented by the circulatory system. Noteworthy is the immune response, which encapsulates all these signaling mechanisms cohesively to execute a comprehensive defensive operation. The initial stage in an immune response, typically following a microbial invasion, engages direct cell-cell contact: Antigen-presenting cells, like dendritic cells, internalize and process foreign antigens, subsequently presenting these on their surface using major histocompatibility complex (MHC) proteins. Immune cells such as T-cells recognize these complexes through their T-cell receptors, instigating a cascade of intracellular signals that lead to T-cell activation. This direct interaction is supplemented with local diffusion of signaling molecules or cytokines, secreted by dendritic cells, which create a chemical gradient to guide other immune cells towards the infection site, exemplifying paracrine signaling. Additionally, activated T-cells also release cytokines to amplify the immune response locally. The long-range signaling component surfaces when immune cells, having dealt with the infection, release cytokines into the bloodstream. These circulate to distant lymphoid organs, prompting the production and mobilization of additional immune cells. This systemic response empowers the immune system to defend the host comprehensively against the invading pathogen. Thus, the multilayered communication within the immune response highlights the intricate yet robust mechanisms of cell signaling essential for coordinated and effective cellular functionality. Regardless of the method, cell communication involves the binding of signaling molecules, called ligands, to specific receptor proteins on the cell surface or inside the cell.

1. Cell Surface Receptors: Cell surface receptors are proteins located on the cell membrane that bind to extracellular signaling molecules (Quintero-Carmona, 2021; Radulovic et al., 2022; Rao et al., 2022). These receptors typically have an extracellular ligand-binding domain, a transmembrane domain, and an intracellular domain responsible for transmitting signals to the cell's interior. Cell surface receptors, proteinaceous complexes anchored in the cell membrane, act as a primary line of communication, modulating cellular activity through binding with extracellular signaling molecules. A glaring exemplar emanates from one of the most abundant and versatile classes of cell surface receptors, the G-protein coupled receptors (GPCRs). These are characterized by their seven-transmembrane-segment architecture and their ability

to bind to guanine-nucleotide-binding proteins (G proteins). A classic model representing the functionality of GPCRs is the β2-adrenergic receptor, vital in the fight-or-flight response. Upon the binding of adrenaline, an extracellular signaling molecule, to the β2-adrenergic receptor, a conformational change instigates GDP-GTP exchange on the G protein. This facilitates the dissociation of the G protein into its α and βγ subunits. The activated α subunit then interacts with adenylate cyclase, stimulating it to convert ATP into cyclic AMP (cAMP), a second messenger that initiates a cascade of intracellular events leading to physiological responses such as glycogen breakdown or smooth muscle relaxation. This illustration underscores the intricate role of cell surface receptors in transducing external signals into specific, cellular responses, thereby highlighting the interface of chemistry and biology in cell communication.

One important class of cell surface receptors is G-protein coupled receptors (GPCRs). When a ligand binds to a GPCR, it activates an associated G-protein, triggering a cascade of intracellular events (Soon et al., 2021; Smole et al., 2022; Spangler et al., 2022). These events often involve the activation of intracellular signaling molecules like cyclic AMP (cAMP) or inositol triphosphate (IP3), which ultimately lead to cellular responses such as gene expression regulation or changes in cell metabolism. In the context of cell signaling, G-protein-coupled receptors (GPCRs) function as critical intermediaries, transducing extracellular chemical cues into a cascade of intracellular events. A case in point is the activation of the visual phototransduction pathway, which kicks off when a photon of light strikes the eye. The photon is absorbed by the 11-cis-retinal chromophore of the GPCR known as rhodopsin, transforming it into all-trans-retinal and causing the receptor to change shape. This conformational shift propagates to the G-protein, termed transducin, which is associated with the receptor. A GDP molecule on the alpha subunit of transducin is replaced by a molecule of GTP, thereby activating the G-protein. The Gα-GTP then dissociates from the Gβγ subunits and rhodopsin but moves within the plane of the membrane to the effector enzyme phosphodiesterase (PDE). This stimulation of PDE facilitates the hydrolysis of cyclic guanosine monophosphate (cGMP), reducing its concentration within the photoreceptor cells. The decline in cGMP levels causes the cGMP-gated ion channels to close, resulting in hyperpolarization of the cell and a decreased rate of neurotransmitter release, which is sensed as light. Once the alpha subunit hydrolyzes GTP to GDP, the G-protein deactivates, and the alpha and beta-gamma subunits reassemble, marking a full cycle of the GPCR signaling process in vision. Thereby, the study of these processes reveals the profound orchestration and synchronization brought about by complex biochemical events in conveying extracellular signals, demonstrating the magnificence of cellular machinery.

Another type of cell surface receptor is receptor tyrosine kinases (RTKs). RTKs are involved in growth factor signaling and cell proliferation (Das et al., 2021; de Albuquerque et al., 2021; De Camilli, 2021; de Torres-Jurado et al., 2022). Ligand binding to an RTK leads to the activation of its intrinsic kinase activity, resulting in the phosphorylation of specific tyrosine residues on the receptor. This triggers downstream signaling pathways, such as the MAPK/ERK pathway, that regulate cell growth and differentiation. Receptor tyrosine kinases (RTKs), a specialized class of cell-surface receptors, play a predominant role in growth factor signaling and cell proliferation, with their malfunction often implicated in various forms of cancer. Epidermal Growth Factor Receptor (EGFR), a representative member of the RTK family, furnishes a compelling case for understanding the mechanics of RTKs in cellular growth signal transduction. The process commences when a growth factor ligand, such as Epidermal Growth Factor (EGF), binds to the extracellular domain of two EGFR monomers. This ligand-induced dimerization triggers the intrinsic kinase activity of the cytoplasmic domains of EGFR and the trans-phosphorylation of specific tyrosine residues in the kinase domain of the receptor, activating the receptor. These phosphorylated residues serve as docking sites for various intracellular signaling proteins containing Src Homology 2 (SH2) or phosphotyrosine binding (PTB) domains. One such protein is Growth Factor Receptor-bound protein 2 (Grb2), which, upon binding to phosphorylated EGFR, recruits the protein SOS, a guanine nucleotide exchange factor (GEF). This Grb2-SOS complex facilitates the conversion of the inactive GDP-bound form of the small G-protein Ras to its active GTP-bound form, earmarking the activation of the cell proliferation pathway known as the Ras-Raf-MEK-ERK pathway. ERK, being the terminal kinase of the pathway, translocates to the nucleus upon activation, where it phosphorylates numerous proteins, including transcription factors such as Myc that regulate the expression of genes necessary for the progression of the cell cycle and thus, cellular proliferation. Therefore, the fine-tuned regulation and activity of RTKs like EGFR illustrate the intricate control exerted on cell growth and proliferation, laying bare the sharp complexity and precision of cellular signaling systems.

2. Intracellular Receptors: Intracellular receptors are found within the cytoplasm or nucleus of the cell and primarily bind to hydrophobic ligands, such as steroid hormones (Sudhof, 2021; Summers et al., 2022; Sundaram & Chen, 2022). Upon ligand binding, these receptors undergo a conformational change and translocate to the nucleus, where they directly regulate gene expression. For example, the binding of cortisol to its intracellular receptor modulates the expression of genes involved in stress response and metabolism. Intracellular receptors, as opposed to their cell-surface counterparts, reside domestically within the cell, either in the cytoplasm or the nucleus, and predominantly cater to the hydrophobic ligands such as steroid hormones. Figure 28 illustrates receptors and ligands in cells.

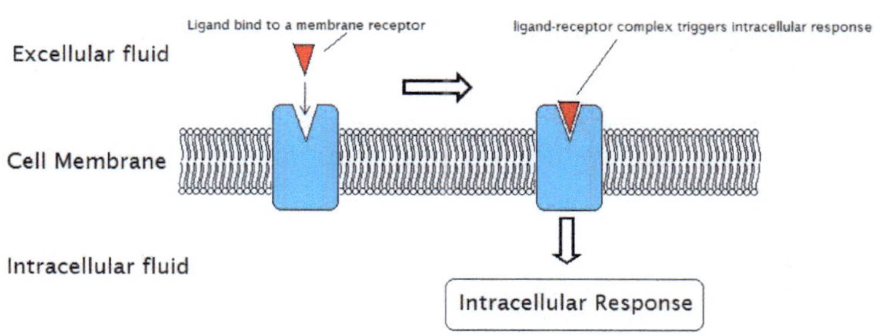

Figure 28. Illustration of ligands and receptors in cells.

Source: By Laozhengzz - Own work, Public Domain, https://commons.wikimedia.org/w/index.php?curid=6605970

Detailed comprehension of its modus operandi can be found in the glucocorticoid receptor (GR), a classic representative of this receptor family, that responds to corticosteroids, a class of steroid hormones. In its inactive state, GR is sequestered in the cytoplasm, bound to heat shock proteins (HSPs) that prevent it from entering the nucleus. The arrival of a ligand, specifically glucocorticoids, such as cortisol, dissipates this limitation. Being hydrophobic, cortisol can freely diffuse across the plasma membrane into the cytoplasm where it binds to GR. Hormone binding triggers the release of HSPs, causing GR to undergo a conformational change that exposes its nuclear localization signal (NLS). The NLS-bearing activated GR then translocates to the nucleus, where it acts as a transcription factor. It regulates gene expression by binding to glucocorticoid response elements (GREs) located in the promoter region of target genes, either enhancing or repressing their transcription. As a result, it modulates the expression of an array of genes encoding proteins involved in immune response, metabolism, and stress response. Therefore, the hormone-induced intracellular signaling coordinated by receptors like GR illuminates the survival and adaptive strategies of cells in response to their endogenous and exogenous environments.

3. Autocrine and Paracrine Signaling: In autocrine signaling, cells produce and release signaling molecules that bind to receptors on their own cell surface, initiating cellular responses within the same cell (Kroll et al., 2022; Kim et al., 2023; Klingemann, 2023; Laiman et al., 2023). This type of signaling is essential for self-regulation and maintenance of cell function. Autocrine signaling, a mode of cell-to-cell communication, is characterized by cells synthesizing, releasing, and responding to their own signals, which engenders self-regulation and provision of a context-specific response to environmental cues, using the same cell as both the source and recipient of signaling molecules. An important model that encapsulates this phenomenon is the autocrine signaling pathway associated with the Transforming Growth Factor-Beta (TGF-β) seen in several types of cells including fibroblasts, epithelial cells, and various cancer cells. TGF-β, synthesized as a large precursor molecule in the endoplasmic reticulum, is cleaved into a smaller active molecule and a large latency-associated protein. The latter non-covalently attaches to the mature TGF-β, forming a latent complex. Once synthesized, this complex is secreted out of the cell, but remains anchored to the cell surface by a

transmembrane proteoglycan. The activation of TGF-β embedded in its latency complex, residing on the cell surface, initiates the autocrine signaling pathway. Upon activation, TGF-β binds to the TGF-β receptors present on the same cell surface, initiating a signal transduction cascade. Importantly, two types of TGF-β receptors are involved - the TGF-β type I receptor (TGFBR1) and type II receptor (TGFBR2), the binding of TGF-β with TGFBR2 recruits and activates TGFBR1 that subsequently phosphorylates the receptor-regulated SMADs. These activated SMADs then form a complex with the SMAD4 protein and translocate into the nucleus, where they bind with other co-regulators facilitating the transcription of target genes. This intricate process highlights the high degree of precision and sophistication inherent in autocrine signaling, wherein the coordinated interactions of cellular elements enable the functioning of the mechanism within the same cell that manufactures the signaling molecules, demonstrating the cell's inherent capacity for self-regulation and dynamic response to varying environmental signals.

Paracrine signaling involves cells releasing signaling molecules into the extracellular fluid, which then act on neighboring cells in the vicinity (Hynds et al., 2021; Jardin et al., 2022; Iachettini et al., 2023). This type of communication is crucial for local communication and coordination of cellular activities within tissues. Paracrine signaling proposes a critical regulatory system through which cells communicate with their immediate neighboring cells in a spatially precise manner by releasing signaling agents into the extracellular milieu. An archetypal illustration of paracrine signaling is the function of neurotransmitters in the synapses of neurons. Neurons are rarely in physical contact with each other; instead, they rely on this highly efficient paracrine signaling system to transmit electrical signals across the synaptic cleft that separates one neuron from the next. When an action potential reaches the synaptic terminal of a neuron, voltage-gated calcium channels open, allowing an influx of calcium ions. This influx triggers the relocation and fusion of neurotransmitter-filled vesicles with the presynaptic membrane leading to the exocytosis of the neurotransmitters into the synaptic cleft. The neurotransmitters, acting as the signaling molecules, diffuse across this cleft to interact with specific receptors, usually ion channels, on the membrane of the receiving, or postsynaptic, neuron. A critical next step is the conversion of this chemical signal back into an electrical one. The binding of these neurotransmitters to their receptors changes the permeability of the membrane, leading to the influx or efflux of ions that changes the membrane potential. If the change in potential is significant enough to pass a certain threshold, it triggers the propagation of another action potential further down the neuron. Thus, this finely tuned paracrine signaling mechanism efficiently facilitates the crucial process of rapid, precise communication between neighbouring neurons and underlines the importance of paracrine signaling in maintaining physiological homeostasis.

4. Endocrine Signaling: Endocrine signaling involves the release of hormones into the bloodstream by endocrine glands (MacArthur & Dawlaty, 2021; Ma et al., 2022; Ma & Wu, 2022). These hormones travel through the circulatory system and bind to specific receptors on distant target cells, initiating cellular responses. The endocrine system is responsible for coordinating physiological processes and maintaining homeostasis throughout the body. Endocrine signaling can be understood as a mode of long-distance intercellular communication that is facilitated by the systemic circulation of signaling molecules or hormones, which are secreted by specialized endocrine glands. This type of signaling is of profound importance in the coordination and regulation of physiological processes across the body. A principal example demonstrating the finesses of endocrine signaling can be studied in the context of the hypothalamic-pituitary-adrenocortical (HPA) axis, which is instrumental in managing the body's response to stress. The initiation of the response emerges from the hypothalamus, a part of the brain that acts as a primary endocrine gland. In response to stress, the paraventricular nucleus of the hypothalamus secretes corticotropin-releasing hormone (CRH) into the hypophyseal portal system - a localized blood circulation pathway that directly connects the hypothalamus and the anterior pituitary gland. Acting in an endocrine manner, CRH in the bloodstream travels to the anterior pituitary and binds to specific receptors on the corticotrope cells, stimulating them to synthesize and release adrenocorticotropic hormone (ACTH). ACTH, once secreted, enters the systemic circulation and is transported to the

adrenal glands. Adrenal cortex cells bear receptors for ACTH, and once bound, these cells are triggered to produce and release glucocorticoids, primarily cortisol, into the bloodstream. These hormones regulate a plethora of physiological processes, such as glucose metabolism and immune response, to manage the overall stress response. In summary, the highly regulated sequence of hormone release, systemic transport, target recognition, and signal transduction exemplifies endocrine signaling's integral role in coordinating multiorgan responses to maintain systemic homeostasis.

Overall, cell communication mediated by signaling pathways and receptor proteins plays a central role in coordinating cellular activities, maintaining tissue homeostasis, and regulating various physiological processes in multicellular organisms. Dysregulation of cell communication can lead to numerous diseases, highlighting the importance of understanding these intricate mechanisms in cell biology. The final section of this chapter will consider cell adhesion and extracellular matrix with reference to the maintenance of tissue integrity.

3.5 CELL ADHESION AND EXTRACELLULAR MATRIX: MAINTAINING TISSUE INTEGRITY

Cell adhesion and the extracellular matrix (ECM) are critical components in maintaining tissue integrity and ensuring the proper functioning of multicellular organisms (Shen et al., 2021; Siu et al., 2021; Shi et al., 2022; Shoup & Priola, 2023). These processes play a pivotal role in various cellular functions, including tissue development, maintenance, and repair. Cell adhesion and the extracellular matrix (ECM) serve as crucial facilitators of tissue structure and multi-cellular functions. An integral representation of this is the behavior of epithelial cells in the formation and maintenance of skin, a front-line protective barrier essential for all higher organisms. Epithelial cells align to establish the epithelial layer of the skin, adhering to each other with utmost precision through specialized structures known as tight junctions, adherens junctions, and desmosomes. These cell junctions are formed through interacting transmembrane proteins like cadherins, claudins, and occludins, forming an adhesive belt around the cell and creating an impermeable, protective seal. Contrarily, the side of the epithelial cells facing away from the tissue interior anchors to the ECM - a complex network of proteins and polysaccharides that provides support and separation between tissues - through another type of adhesive structure, the hemidesmosome. Intracellularly, hemidesmosomes connect to keratin filaments, a kind of intermediate filament present in epithelial cells, and extracellularly, they bind to a type of ECM protein - laminin. This protein, in turn, associates with additional ECM components like collagen and fibronectin, providing a matrix for cell adhesion. This intricate interplay between cell adhesion and the ECM ensures the formation of a robust and dynamic structure - the skin. From a broader perspective, similar mechanisms of cell adhesion and interactions with the ECM guide the development, maintenance, and repair of different tissues, reinforcing their fundamental importance in sustaining the structural and functional integrity of multicellular organisms.

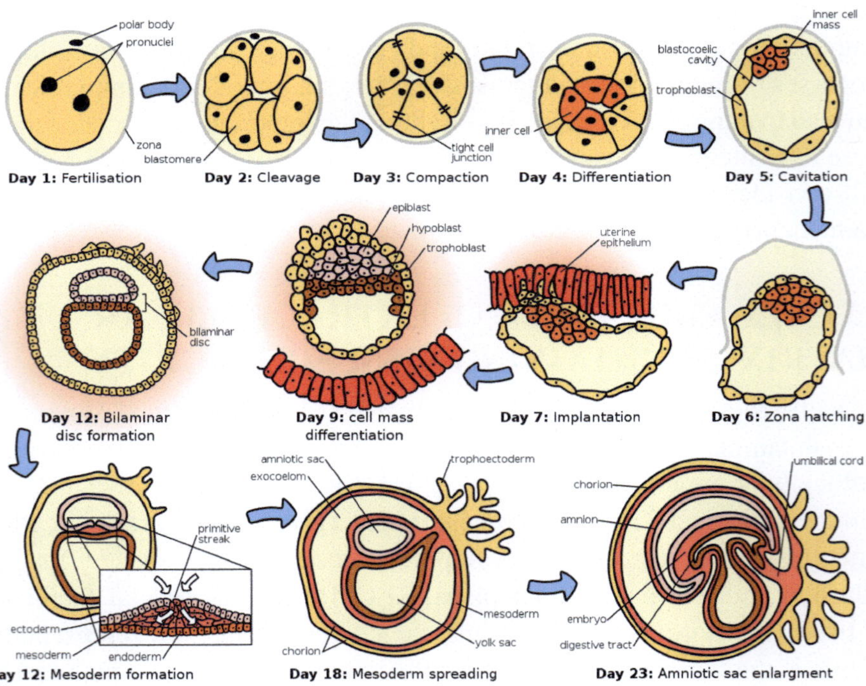

Figure 29. Illustration of human embryonic development.

Source: Wikipedia (https://en.wikipedia.org/wiki/Human_embryonic_development)

Cell adhesion refers to the process by which cells bind to neighboring cells or the ECM (D. Wang et al., 2021; J. J. Wang et al., 2021; Wang et al., 2022). It is essential for cell-to-cell communication, tissue organization, and overall tissue stability. Cell adhesion molecules (CAMs) are proteins found on the cell surface that mediate these interactions. CAMs can be classified into several families, including cadherins, integrins, and selectins. Cell adhesion, the process where cells attach to each other or to the extracellular matrix (ECM), is a fundamental biological phenomenon of paramount significance that critically orchestrates the architecture and functionality of multicellular organisms.

A prime example elucidating this process can be found in the context of embryogenesis (see figure 29), where cells aggregate to form the embryo's structure. During early embryogenesis, specifically in the process named gastrulation, an intricate dance of cell adhesion and de-adhesion governs the transformation of the single-layered blastula into a three-layered gastric structure comprising the ectoderm, mesoderm and endoderm. Initially, cells of the blastula bind tightly together via E-cadherin, a calcium-dependent adhesion molecule found on their cell surfaces. The binding of E-cadherin molecules on respective neighboring cells mediates this adhesion. Integral to gastrulation is the formation of the primitive streak, which is characterized by downregulation of E-cadherin, allowing cells to de-adhere momentarily. Concurrently, these cells acquire other adhesion molecules like N-cadherin and fibronectin receptors, enabling their binding to the ECM. Once adhered to the ECM, cells migrate along it to their new positions in the developing embryo. Subsequently, these migrated cells upregulate E-cadherin again to re-establish cell-to-cell contacts, which results in forming distinct embryonal layers. Hence, the regulated fluctuation of cell adhesion and de-adhesion, mediated by the temporal and spatial expression of different cell adhesion molecules, meticulously directs the complex choreography of embryogenesis. It underscores the importance of cell adhesion in sculpting not only the physical blueprint of multicellular organisms but also ensuring the orderly progression of development.

1. Cadherins: Cadherins are calcium-dependent transmembrane proteins that mediate homophilic cell-cell adhesion (Sudhof, 2021; Summers et al., 2022; Sundaram & Chen, 2022). They bind to identical cadherins on adjacent cells, promoting strong adhesion and forming adherens junctions. Adherens junctions are critical for maintaining tissue integrity, as they provide mechanical strength and resistance to forces applied to the tissue. Cadherins, a group of calcium-dependent transmembrane proteins, play an integral role in mediating homophilic cell-cell adhesion, a fundamental mechanism maintaining the structural integrity of tissues within multicellular organisms. Homophilic

interaction refers to the specific binding of identical molecules on two different cells, a key feature of cadherins that facilitates selective and structured cell adhesion. One of the most studied examples of cadherins at work is the role of E-cadherin in the formation and preservation of epithelial tissue, one of the four primary tissue types in the body. Found densely populated on the surfaces of epithelial cells, E-cadherin molecules interact to create a 'zipper-like' structure, facilitating tight cell-cell adhesion and forming a protective layer known as an epithelium. This layer executes several key functions, including barrier formation, filtration, and absorption, across various organ systems such as the skin, gastrointestinal tract, and kidneys. The binding of each E-cadherin protein to a counterpart on an adjacent cell occurs through calcium-dependent adhesion sites present in their extracellular domain. Importantly, the cytoplasmic tail of E-cadherin builds a bridge to the cell's structural framework or cytoskeleton by binding with p120-catenin and beta-catenin. The latter then interacts with alpha-catenin, which forms a link to actin filaments, thus integrating cell adhesion with the cytoskeletal dynamics. These interactions also form a nexus for intracellular signaling, impacting cellular processes such as proliferation, differentiation, and migration. Consequently, abnormal cadherin expression patterns, or mutations in their gene, can disrupt tissue architecture and function, contributing to various pathological conditions, including cancer. Through this study of E-cadherin, the significance of cadherins, the calcium-dependent gatekeepers of cell-cell adhesion, in preserving tissue integrity and regulating cellular behavior, can be appreciated.

E-cadherin is one of the most well-known cadherins and is essential for the integrity of epithelial tissues (McFall & Stites, 2021; Melo et al., 2022; Mendes, 2023; Mierke, 2023). Loss of E-cadherin expression is associated with the loss of tissue organization and epithelial-mesenchymal transition, which is a crucial process during embryonic development and tumor metastasis. E-cadherin, recognized as a cardinal member of the cadherin family, plays a pivotal role in maintaining the integrity of epithelial tissues. This protein, as a calcium-dependent cell-cell adhesion molecule, executes the function of adhering epithelial cells to each other, thereby forming stable epithelial cell sheets and ensuring the robustness of the tissue structure. A quintessential demonstration of the role E-cadherin fulfills can be understood by observing its functionality in the skin, the body's largest organ and primary epithelial barrier. In the skin, E-cadherin-mediated junctions, commonly termed adherens junctions, are formed between neighboring keratinocytes, which constitute the majority of epidermal cells. Each E-cadherin molecule, spanning the cell membrane, binds to a counterpart E-cadherin on another cell through its extracellular domain in a calcium-dependent manner, thereby ensuring a firm adhesive interaction. Notably, its cytoplasmic domain connects to the actin cytoskeleton within the cell via intermediary proteins - alpha, beta, and gamma-catenin - creating a link between cell adhesion and the cell's internal skeleton, thus providing structural stability and helping maintain cell shape. It further provides a platform for multiple signaling cascades that regulate critical cellular processes such as growth and differentiation. The importance of E-cadherin becomes evident when its function is lost or compromised, as seen in certain skin disorders and cancers. Its loss leads to a breakdown in cell-cell contact, tissue disorganization, and impaired barrier function, underlining the integral function E-cadherin fulfills in preserving tissue integrity and modulating cellular behavior within epithelial contexts. Consequently, the study of its function and regulation serves as a valuable sentinel in understanding epithelial physiology and pathophysiology.

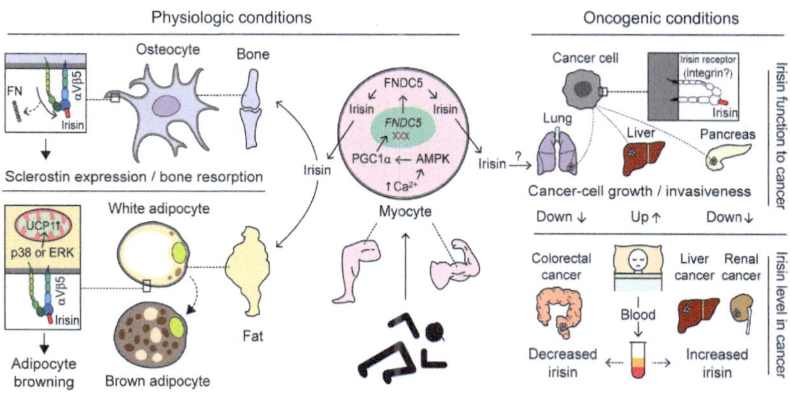

Figure 30. Biological interactions mediated by integrins with irisin.

Source: Park, E. J., Myint, P. K., Ito, A., Appiah, M. G., Darkwah, S., Kawamoto, E., &Shimaoka, M. (2020). Integrin-ligand interactions in inflammation, cancer, and metabolic disease: Insights into the multifaceted roles of an emerging ligand irisin. Frontiers in Cell and Developmental Biology, 8, 588066.

2. Integrins: Integrins are transmembrane receptors that mediate cell-ECM adhesion (Arber et al., 2021; Ardalan et al., 2022; Arhontoulis et al., 2022). They bind to specific ECM components, such as fibronectin, laminin, and collagen, enabling cells to anchor to the ECM and providing mechanical support (see figure 30). Integrins are essential for cell migration, tissue organization, and wound healing. Integrins, a large family of transmembrane receptors, are pivotal for mediating cell adhesion to the extracellular matrix (ECM), a specialized network of proteins and carbohydrates that provides support and organization for cells within tissues. A characteristic example of integrin function can be elucidated by studying its role in the process of wound healing, a highly coordinated biological event, which requires cellular migration, a process inextricably linked to cell-ECM adhesion. Upon injury, fibrin and fibronectin, among other components, aggregate to form a provisional ECM that acts as a substrate for cell migration to the wound site. It is here the role of integrins becomes critical. Integrins, located on the surface of migrating cells such as fibroblasts, bind to specific motifs present on molecules such as fibronectin in the provisional ECM. Particularly, the integrin subtypes α5β1 and αvβ3 recognize and bind to a conserved arginine-glycine-aspartic acid (RGD) sequence in fibronectin. The binding of integrins to the ECM triggers intracellular signaling pathways that lead to the rearrangement of the cytoskeleton, driven primarily by actin polymerization, facilitating cell migration towards the wound. Additionally, integrins function in a bidirectional manner; while they transmit information about the ECM composition to the cell (outside-in signaling), they also communicate cellular status to the ECM (inside-out signaling), thereby constantly modulating cell attachment and detachment. The disturbance in integrin-mediated cell-ECM interactions can have significant implications, including impaired wound healing and fibrosis. Through this window into the role of integrins in wound healing, their fundamental significance in mediating cell-ECM adhesion, influencing cell behavior, and thereby contributing to the maintenance of tissue homeostasis and response to injury is evident.

For example, during inflammation, leukocytes use integrins to adhere to the endothelium, facilitating their migration from the bloodstream into tissues (Soon et al., 2021; Smole et al., 2022; Spangler et al., 2022). Additionally, integrins are critical for cell signaling and regulating cellular processes like proliferation and differentiation. The process of inflammation is marked by a cascade of cellular responses that facilitate the recruitment of leukocytes to sites of infection or tissue injury. Integrins, key transmembrane receptors, play an indispensable role in this event, mediating the adhesion of leukocytes to the vascular endothelium and subsequently assisting their navigational trajectory from the bloodstream into underlying tissues. A detailed enquiry of the integrin function in inflammation can be seen in the case of neutrophil extravasation, a crucial onset stage of acute inflammatory response. Upon initialization by injury or infection, upregulation of selectin adhesion molecules on endothelial cells allows rolling and transient tethering of circulating neutrophils along the endothelial layer. Concomitantly, inflammatory signals from the site of injury activate G-protein-coupled receptors (GPCRs) on neutrophils, provoking changes in the conformation of integrins - notably, LFA-1 (lymphocyte function-associated antigen-1) and Mac-1 (macrophage-1 antigen), making them "high-affinity" for their ligands. These ligands, primarily ICAM-1 (intercellular adhesion molecule-1), reside on the vascular endothelium. Consequently, the interaction between high-affinity integrins on neutrophils and ICAM-1 molecules induces firm adhesion, effectively halting the neutrophil rolling. The engagement also instigates intracellular signaling events - known as outside-in signaling - triggering the reorganization of the neutrophil cytoskeleton to allow for spreading and crawling on the endothelium. Finally, neutrophils migrate through the gaps between endothelial cells (the process of diapedesis) and into the injured tissue, guided by chemical cues (chemotaxis). Disruptions to integrin function can impair inflammatory responses, emphasizing herein their paramount role as orchestrators of leukocyte adhesion and migration during inflammation.

3. Selectins: Selectins are cell adhesion molecules involved in cell-cell interactions, especially during inflammation and immune responses (Arber et al., 2021; Ardalan et al., 2022; Arhontoulis

et al., 2022). They mediate the initial tethering and rolling of leukocytes on the endothelium of blood vessels, allowing them to adhere to the site of injury or infection. Selectins constitute a significant category of cell adhesion molecules that play a central part in mediating cell-cell interactions, particularly during immune responses and the process of inflammation. A well-documented illustration of selectin activity is observed in the early stages of the leukocyte extravasation cascade, a key procedure during the inflammatory response that enables leukocytes to exit the bloodstream and infiltrate the site of inflammation. This cascade commences with the rolling of leukocytes along the endothelial cells that line blood vessels, a step enabled predominantly by the interaction of selectins with their specific ligands. E-selectin and P-selectin, expressed on activated endothelial cells, interact with P-selectin glycoprotein ligand-1 (PSGL-1) expressed on leukocytes. Specifically, the highly glycosylated extracellular domain of P-selectin and E-selectin binds to sialylLewisx (sLeX) carbohydrates present on PSGL-1, facilitating the tethering and rolling of leukocytes on the endothelium. This interaction, although relatively weak, slows down the high-velocity leukocytes in the bloodstream and brings them into close contact with the endothelium. Such selectin-mediated rolling is a precondition for subsequent firm adhesion and diapedesis of leukocytes, steps enabled by integrins as described previously. Importantly, dysregulation of selectin-mediated interactions could lead to diseases such as chronic inflammation and autoimmune disorders, underscoring their vital role in immune responses. Through the lens of this case study, the indispensable role of selectins as facilitators of cell-cell interactions, especially in coordinating intricate immune responses and inflammation events, can be aptly appreciated.

The extracellular matrix (ECM) is a complex network of proteins and carbohydrates that surrounds cells in tissues (Ma & Hoffmann, 2021; et al., 2022; Lynes et al., 2022). It provides structural support, mechanical strength, and biochemical cues necessary for cell function and tissue integrity. The ECM is composed of various components, including fibrous proteins like collagen and elastin, proteoglycans, and glycoproteins. The extracellular matrix (ECM) is a diverse, intricate assembly of proteins and carbohydrates that envelop cells within tissues, imparting both structural stability and functional regulation to cellular communities. One pivotal representation of the ECM's role can be observed in the context of connective tissue, particularly in cartilage, which serves as a linchpin of the skeletal system, providing flexible yet firm structures. Cartilage ECM is a non-cellular three-dimensional network built predominantly of collagen, particularly type II collagen, and proteoglycans, a class of glycoproteins, with the principal member being aggrecan. Collagen fibers confer tensile strength and resilience to deformation, while proteoglycan aggregates imbue the property of compressive resistance, primarily due to their hydration capacity. This intricate network forms a medium that hosts and physically supports chondrocytes, the cells responsible for the synthesis and maintenance of the ECM of cartilage. Beyond providing structural support, the ECM modulates a myriad of cellular behaviors such as differentiation, migration, proliferation, and shape, thus guiding tissue development, healing, and homeostasis. For instance, cartilage's avascular nature, due to specific components of its ECM, limits chondrocyte proliferation and entrains these cells in a low metabolic state. This low metabolic state in turn decreases requirements for nutrients, meeting the condition imposed by cartilage's lack of blood supply. Consequently, dysregulation in ECM composition or organization can induce disorders such as osteoarthritis, known to be instigated by degradation of cartilage ECM, further emphasizing the significance of the ECM in tissue function and health. Taken together, the study of cartilage illustrates the vital role the ECM plays in forming the architectural skeleton of tissue and in regulating cellular functions, both crucial for the survival and functionality of multicellular organisms.

1. Collagen: Collagen is the most abundant protein in the ECM and provides tensile strength to tissues (Altulea et al., 2021; Alvord et al., 2022; Aniento et al., 2022; Aprile et al., 2023). It forms fibrils that resist stretching and contribute to tissue integrity. Different types of collagen are found in various tissues, contributing to the unique properties of each tissue type. Collagen, acknowledged as the most abundant protein in the extracellular matrix (ECM) among a plethora of biomolecules, lends a substantial tensile strength to tissues, thereby proving its criticality in maintaining the structural integrity of various organs and tissues. An exemplary illustration of the functional importance of collagen lies within the dermis layer of the skin, a multicellular organ that serves as the primary outer protective layer of the body. The dermis is characterized by a 3-dimensional collagen network

primarily composed of type I collagen, which constitutes approximately 90% of the skin's collagen content. This collagen network follows a basket-weave-like pattern, with interwoven collagen fibers arranged in an intricately organized manner to establish a resilient ECM scaffold. The mechanical strength of collagen fibers is derived from its triple-helix structure formed by three polypeptide chains, each bearing a repeating Gly-X-Y motif, where X and Y are frequently Proline and Hydroxyproline, respectively. This structure imparts stability through a host of intermolecular hydrogen bonds, facilitating resistance against physical forces. Additionally, collagen fibers intertwine to form fibrils, and these fibrils aggregate into fibers that collectively provide the tensile strength that gives the skin its characteristic durability and elasticity. Furthermore, the collagen scaffold anchors other dermal cells such as fibroblasts, which produce and maintain the ECM, and immune cells. Thus, through the fibrous collagen network, the dermis is structurally fortified to withstand pressure and tension, reinforcing the skin's function as a protective barrier. Permutations in collagen production or assembly can result in various skin disorders, including wrinkles and certain forms of dermatitis, indicating the pivotal role of collagen in underpinning tissue architecture and strength.

2. Proteoglycans: Proteoglycans are large molecules composed of a core protein and long chains of carbohydrates called glycosaminoglycans (GAGs). They contribute to the hydration and compressibility of tissues, as well as act as reservoirs for growth factors and other signaling molecules (Granados-Riveron & Aquino-Jarquin, 2021; Graziadei & Rappsilber, 2022). Proteoglycans, expansive molecules constituted of a core protein allied with lengthy chains of polysaccharides known as glycosaminoglycans (GAGs), play critical roles in the extracellular matrix (ECM) of various tissues. A notable example illustrating the function of proteoglycans is observed in the cornea, the anterior most transparent structure of the eye, arranged in an organized lamellar form to ensure sufficient passage of light for optimal vision. It is rich in proteoglycans, particularly keratan sulfate and chondroitin/dermatan sulfate proteoglycans. The core protein of these proteoglycans associates covalently with linear GAG chains, forming bottlebrush-like structures. Particularly

important is the proteoglycan known as lumican, a keratan sulfate proteoglycan, abundant in the corneal stroma - the thick, middle layer of the cornea. Lumican interacts with collagen fibrils, facilitating the regulation of the fibrilar diameter and alignment to maintain cornea transparency. Importantly, the highly negatively charged GAG side chains attached to lumican and other corneal proteoglycans attract and bind water, thereby contributing to the maintenance of the cornea's hydration state and transparency. The regulation of corneal clarity and mechanical resilience orchestrated by proteoglycans is thus essential for appropriate visual function, and disruptions to their presence or organization can underlie corneal diseases including corneal dystrophies. Consequently, the study of proteoglycans, with their unique composition and elaborate structure, epitomizes their widespread significance in maintaining tissue architecture and contributing to various biological processes.

3. Fibronectin: Fibronectin is a glycoprotein that plays a crucial role in cell adhesion and migration (Doostmohammadi & Ladoux, 2022; Drozdz et al., 2022; Du et al., 2023). It binds to integrins on the cell surface and to other ECM components, facilitating cell attachment and movement. Fibronectin is essential for wound healing and tissue remodeling. Fibronectin, a high-molecular weight, multifunctional glycoprotein, plays a fundamentally critical role in cellular processes such as adhesion and migration, facets that are pivotal in tissue repair, development, and homeostasis. A model case demonstrating fibronectin's function is observed during the complex process of embryogenesis. In the early embryonic stage, mesenchymal cells – the precursors of various tissues – migrate and differentiate to shape the embryo. This controlled cell migration is guided heavily by fibronectin present in the extracellular matrix (ECM). As an integral component of the ECM, fibronectin forms a multi-dimensional lattice that provides a structural framework for cell anchorage, primarily mediated via integrins, a class of transmembrane cell adhesion receptors present on the cell surface. Twin functional domains in fibronectin - the RGD sequence and the synergy site - bind with the cell surface integrins, particularly integrin $\alpha 5\beta 1$. This interaction prompts a cascade of signal transduction events inside the cell and stimulates the reorganization of the actin cytoskeleton.

It also regulates the formation and disassembly of focal adhesions, dynamic structures connecting the actin cytoskeleton to the ECM via integrins. Consequently, this regulates the cell's adhesive and migratory behavior, driving cell movement during embryonic development. Besides, fibronectin is also implicated in various pathological conditions, including cancer metastasis and fibrosis, where aberrant cell adhesion and migration play a significant role. Accordingly, the examination of fibronectin's role in cellular dynamics underscores this glycoprotein's foundational role in cell adhesion, migration, ECM organization, and consequently in maintaining tissue function and architecture.

The interactions between cells and the ECM are dynamic and regulated by numerous signaling pathways. These interactions influence cellular behavior, including cell proliferation, differentiation, and migration. Proper cell adhesion and ECM organization are essential for tissue development, maintenance, and repair, as well as for normal physiological processes in multicellular organisms. Dysfunction in these processes can lead to various diseases, including cancer and tissue degeneration. Understanding the mechanisms of cell adhesion and ECM biology is crucial for advancing cell biology research and developing potential therapeutic strategies. We will now summarize the third chapter and look ahead to the fourth chapter.

CHAPTER 3 SUMMARY

Chapter 3 of the book "Introduction to Living Cells: The Building Blocks of Life" delves into the intricacies of cellular transport mechanisms, crucial for maintaining cell function and homeostasis. This chapter encompasses five significant topics, each exploring different aspects of cellular transport.

Section 3.1, Passive Transport, elucidates fundamental processes such as diffusion, osmosis, and facilitated diffusion. These mechanisms enable molecules to move across the cell membrane without the expenditure of cellular energy. Diffusion involves the spontaneous movement of molecules from regions of high concentration to low concentration, ensuring an even distribution within the cell. Osmosis, on the other hand, is the movement of water molecules across a semipermeable membrane, balancing solute concentrations on either side. Facilitated diffusion, in turn, relies on specific transport proteins to move molecules across the membrane efficiently.

Section 3.2, Active Transport, investigates the ATP-driven movement of molecules, occurring against their concentration gradient and requiring cellular energy. Active transport mechanisms, such as the sodium-potassium pump, play a crucial role in maintaining ion gradients and cellular electrical potentials, essential for nerve impulses and muscle contractions.

Section 3.3 delves into endocytosis and exocytosis, which are bulk transport processes responsible for moving large substances into and out of the cell, respectively. Endocytosis involves the cell's invagination of the plasma membrane to engulf extracellular substances into vesicles, while exocytosis releases intracellular substances by fusing vesicles with the plasma membrane. These processes are vital for the uptake of nutrients and the secretion of cellular products.

Section 3.4 focuses on cell communication, exploring signaling pathways and receptor proteins that enable cells to sense and respond to their environment. Cell surface receptors, such as G-protein coupled receptors and receptor tyrosine kinases, mediate the interactions with extracellular signaling molecules, triggering intracellular events and cellular responses.

Section 3.5 addresses cell adhesion and the extracellular matrix, fundamental for maintaining tissue integrity and cellular organization. Cell adhesion molecules, including cadherins and integrins, mediate cell-cell and cell-ECM interactions, providing mechanical strength and support to tissues. The extracellular matrix, composed of proteins like collagen and proteoglycans, surrounds cells and offers structural support, as well as biochemical signals necessary for cell function.

This chapter offers a comprehensive exploration of cellular transport mechanisms, cell communication, and tissue organization. Understanding these complex processes is crucial to comprehend the dynamic and interconnected nature of living cells and their role in maintaining the overall integrity and functionality of multicellular organisms. The fourth chapter will explore the cell cycle and cell division.

Chapter

4

CELL CYCLE AND CELL DIVISION

CONTENTS

In this chapter, we delve into the intricate processes governing the cell cycle and cell division, fundamental aspects of cellular life. The cell cycle represents a dynamic sequence of events that enables cells to grow, duplicate their genetic material, and ultimately divide into two daughter cells. This orchestrated series of events plays a pivotal role in a wide array of biological processes, such as growth, development, tissue repair, and reproduction.

The chapter commences with a comprehensive exploration of Interphase, a crucial preparatory stage preceding cell division. During Interphase, the cell diligently carries out various essential functions, such as synthesizing proteins, replicating DNA, and increasing its organelle count, all in preparation for the ensuing division process.

Next, we delve into mitosis, a pivotal process responsible for somatic cell division. Mitosis enables the accurate distribution of duplicated genetic material to the daughter cells, ensuring genetic fidelity and the maintenance of proper cell function. By dissecting the stages of Mitosis, we gain insight into the intricate mechanisms that guide chromosomal segregation and cytokinesis, culminating in the formation of two genetically identical daughter cells.

The cell cycle regulation, a topic of utmost significance, follows in our exploration. The cell diligently monitors its progression through the cell cycle, ensuring that each phase is completed accurately and efficiently. Key regulatory checkpoints play a vital role in detecting any errors or DNA damage, allowing the cell to halt the cycle or trigger repair mechanisms when necessary. The meticulous orchestration of these control mechanisms ensures genomic stability and minimizes the risk of cellular dysfunction or transformation.

Diverging from the somatic cell division, we then plunge into meiosis, a specialized form of cell division dedicated to sexual reproduction. During meiosis, specialized cells known as gametes undergo two rounds of division, generating genetically diverse offspring. The process of crossing over during meiosis contributes to the genetic variability of offspring, leading to the diversity observed in sexually reproducing species.

Throughout this chapter, we emphasize the paramount importance of cell division in the context of growth, development, and repair. The ability of cells to divide and replicate is fundamental to the growth of multicellular organisms, the renewal of tissues, and the repair of injuries or damages. This process forms the foundation of life's continuity and complexity, underpinning the vast diversity and resilience observed in the living world.

In conclusion, the chapter on cell cycle and cell division provides a profound understanding of the intricate machinery regulating cell growth, duplication, and division. The remarkable coordination of these processes ensures the perpetuation of life, driving the development and functioning of complex organisms. As we delve into the details of interphase, mitosis, cell cycle regulation, meiosis, and the significance of cell division, we gain invaluable insights into the fundamental principles governing cellular life. To open the chapter, we will explore the interphase preparation for cell division.

4.1 INTERPHASE: PREPARATION FOR CELL DIVISION

Interphase is a pivotal phase in the cell cycle, encompassing the period before cell division when the cell actively prepares for the forthcoming division process (Featherston & Cotruvo, 2021; Ferrari et al., 2022; Ferrari et al., 2023). During Interphase, the cell undertakes crucial activities that are essential for successful division, including DNA replication, protein synthesis, and the expansion of cellular organelles. It is important to note that Interphase is not a passive stage but an active and dynamic period of cellular activity. Interphase, a critical stage of the cell cycle, encapsulates the precursory period preceding cellular division where the cell undergoes intensive preparations for the imminent division. One such observation of this fundamental phase can be elucidated through a detailed examination of Escherichia coli, a common bacterium and prominent subject in the field of cell microbiology. This Gram-negative bacterium, frequently found in the human gut, exhibits a unique period of interphase called the B-period, marked by cellular enlargement. During this period, the singular chromosome of E. coli is duplicated, thus preparing the bacteria for cell division. Additionally, critical biochemical transformations occur, including the synthesis of various proteins and DNA following the singular, circular chromosome's replication. This chromosome replication results from the process of binary fission, where the cell divides, producing two genetically identical daughter cells. Moreover, the B-period's conclusion, the chromosome replication cessation, leads to the cell's inauguration into the D-period, involving a meticulous segregational process of these duplicated chromosomes into the nascent daughter cells. This D-period is characterized by the synthesis of septal peptidoglycan and the onset of cytokinesis, thus marking the conclusion of interphase. Deeper analyses of the E. coli interphase phase have improved our broader comprehension of cellular microbiology, offering pivotal insights into microbial growth, DNA replication, and cell division mechanics. Further high-throughput investigations and advanced microscopy techniques, such as single-molecule tracking and fluorescent imaging, have echoed the vital significance of the labile interphase period in the dynamic and complex nature of the cellular life cycle.

The first stage of Interphase is known as the G1 (Gap 1) phase (Pendharkar et al., 2021; Perera, 2021; Peng et al., 2022). During this stage, the cell increases in size, synthesizes new proteins, and accumulates the required resources and energy for the ensuing stages. The G1 phase serves as a checkpoint to assess if the cell is ready to progress into the next phase. Factors such as nutrient availability, growth signals, and DNA integrity are assessed to ensure that the cell is in optimal condition for cell division. In cell biology, the first stage of Interphase, known as the G1 (Gap 1) phase, represents a critical period during which the cell engages in essential preparatory processes before proceeding to the subsequent stages of the cell cycle. An exemplary case study that underscores the significance of the G1 phase can be found in cancer research, where abnormal cell cycle regulation contributes to the uncontrolled proliferation of cancer cells. For instance, in the context of breast cancer, studies have revealed dysregulation of cyclin-dependent kinases (CDKs) and their associated cyclins, pivotal players in cell cycle progression, during the G1 phase. Specifically, the G1 phase is tightly controlled by the CDK4/6-cyclin D complex, which drives the cell from G1 into the S phase by phosphorylating the retinoblastoma protein (Rb). In normal cells, Rb acts as a tumor suppressor, inhibiting the progression of the cell cycle by binding to transcription factors, thus preventing the expression of genes required for S phase entry. In cancer cells, however, the CDK4/6-cyclin D complex is overactivated, leading to hyperphosphorylation of Rb and its inactivation, thereby promoting unchecked cell cycle progression. Consequently, cancer cells bypass G1 phase checkpoints and continue to divide uncontrollably (see figure 31), contributing to tumor growth and metastasis. Understanding the intricate regulatory mechanisms governing the G1 phase and their dysregulation in cancer offers crucial insights into the development of targeted therapies that aim to restore cell cycle control and halt tumor proliferation. By unraveling the complexities of the G1 phase, researchers can advance our knowledge of cell biology and devise novel strategies to combat cancer and other diseases characterized by aberrant cell cycle regulation.

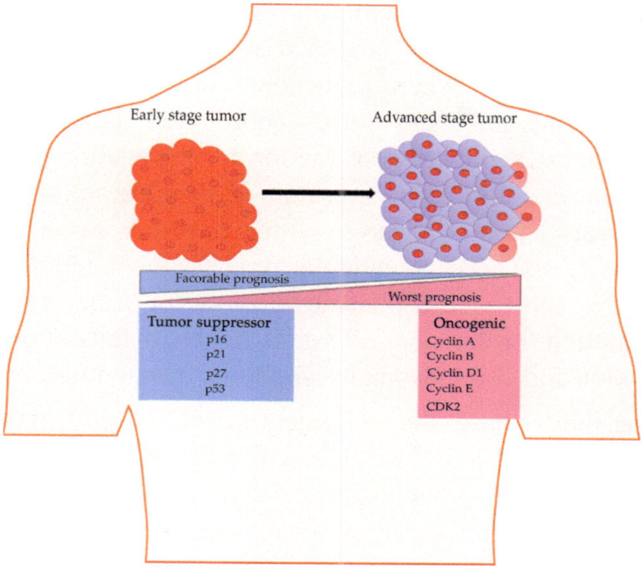

Figure 31. Cell cycle control in cancer.

Source: Kashyap D, Garg VK, Sandberg EN, Goel N, Bishayee A. Oncogenic and Tumor Suppressive Components of the Cell Cycle in Breast Cancer Progression and Prognosis. Pharmaceutics. 2021; 13(4):569. https://doi.org/10.3390/pharmaceutics13040569

Upon successful completion of the G1 phase, the cell enters the S (Synthesis) phase, the central event of interphase (Westman & Grinstein, 2021; Wiedner & Giudice, 2021; Whiteley et al., 2022). Here, the cell undergoes DNA replication, resulting in the duplication of its entire genome. DNA replication is a precise and complex process orchestrated by various enzymes and regulatory factors to ensure the accurate copying of genetic material. This duplication of DNA ensures that each daughter cell receives a complete set of genetic information during division. Following the successful navigation of the G1 phase, the cell embarks on the S phase, oftentimes considered the primary event of Interphase. One robust case study pertinent to this phasic progression is the nascent modeling of yeast Saccharomyces cerevisiae. Known for its rapid growth rate and unicellular nature, S. cerevisiae operates as a significant model organism within the cellular microbiology canon. At the onset of the S phase, the DNA within the yeast cell's nucleus undergoes a meticulous replication process. This process, facilitated by DNA polymerases and multiple replication factors, ensures the creation of two complete sets of genetic material, both of which are identical to the parent cell's original genetic blueprint. The DNA replication in the S phase is guided by origins of replication, defined sequences in the genome where replication commences. It is also during the S phase that the cell meticulously coordinates the duplication of its centrosomes, the organelles responsible for the segregation of DNA during cell division. Despite a transient cessation in tangible cellular growth, the yeast cell actively engages in DNA synthesis and repair, a marker of its metabolic activity and vital preparation for subsequent cell division. Thus, studies of the S. cerevisiae S phase exhibit the intricate machinations of cellular preparation for division, offering valuable details into processes of DNA duplication and chromosomal segregation during Interphase.

Following the S phase, the cell enters the G2 (Gap 2) phase (Khanna & Villa, 2022; Kim & Choi, 2022; Kikuchi et al., 2023). During this stage, the cell further prepares for division by synthesizing necessary proteins, especially those required for cell division, such as microtubules and other components of the mitotic apparatus. Additionally, the cell continues to grow and undergoes quality control checks to ensure that the replicated DNA is free of errors and mutations. These surveillance mechanisms are vital in maintaining genomic stability and preventing the propagation of damaged DNA. Subsequent to the S phase is the G2 phase, the endmost stage of Interphase within the cellular replication cycle. This phase can be expounded through an examination of the mitosis process in HeLa cells, a human cell line used extensively in scientific research. During the G2 phase in HeLa cells, cell growth continues unabated, with the synthesis of proteins necessary for cell division occurring concurrently. Importantly, double-strand breakdown repair, a process ensuring the integrity of the newly replicated genetic material, is heightened during the G2 phase. The result is two intact sets of chromosomes, each one comprised of two genetically identical sister chromatids. It is during this critical 'gap' period that the cell undertakes stock-taking measures, meticulously checking the fully replicated DNA for errors or damage and repairing it where necessary. Concurrently, the cellular machinery necessary for mitosis is assembled, under the guidance of various cyclin-dependent kinases. This includes the construction of the mitotic spindle, a structure vital for sister chromatid separation in the following mitosis phase. Cyclin B, a particular protein, accumulates during the G2 phase, instigating the transition into the M phase, or mitosis. Thus, the G2 phase indicative of HeLa cells serves to underscore the vital preparatory processes necessary for successful, error-free cell division. The assessments of the replicated DNA and the concurrent protein synthesis, as evidenced in HeLa cells (see figure 32), traces the intricate workings of cellular proliferation, revealing the significance of the G2 phase (Smole et al., 2022; Soon et al., 2021; Spangler et al., 2022).

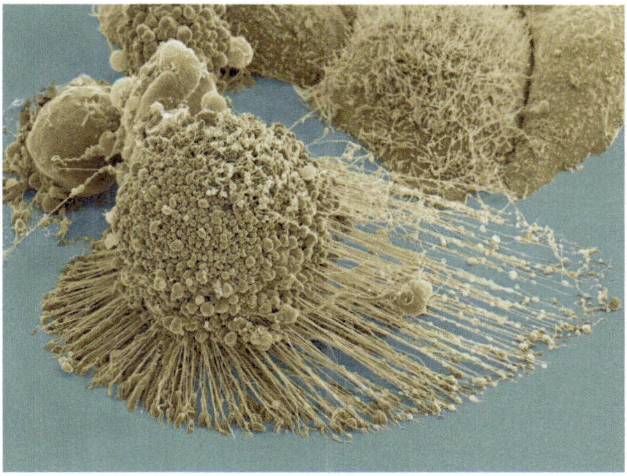

Figure 32. Visual of HeLa cells.

Source: NCI Visuals Online - National Cancer Institute https://visualsonline.cancer.gov/details.cfm?imageid=11870

Interphase culminates with a final checkpoint, the G2/M checkpoint, which assesses whether the cell is ready to proceed with mitosis, the actual process of cell division (MacArthur & Dawlaty, 2021; Ma et al., 2022; Ma & Wu, 2022). At this point, the cell undergoes an additional round of checks to ensure that all components necessary for cell division are in place and that the replicated DNA is intact and accurately duplicated. The culmination of Interphase is marked by a crucial juncture referred to as the G2/M checkpoint. This critical checkpoint assesses whether the cell is adequately prepared to initiate the subsequent phase of mitosis, a process comprising cell division in its entirety. To elucidate this pivotal checkpoint's function, we can closely inspect the G2/M phase transition in the HeLa cell line, an immortal cell line derived from cervical cancer cells. Primarily, this G2/M checkpoint operates as a surveilling mechanism to ensure that both DNA replication from the preceding S phase is completed without error and that any potential DNA damage is sufficiently repaired. The assessment is founded on the cell's capability to activate a specific enzyme, namely cyclin-dependent kinase 1 (CDK1). The activation of CDK1, in collaboration with cyclin B1, forms the maturation-promoting factor (MPF), a crucial determinant promoting progression into the mitotic phase. At this stage, the checkpoint assessment takes place, regulated by crucial proteins such as the tumor suppressor protein p53 and the DNA repair protein BRCA1. If the assessment reveals any abnormalities in DNA replication or presents indications of DNA damage, the cell cycle progression halts, allowing the cell to employ repair mechanisms or, if necessary, instigate programmed cell death, also known as apoptosis. Therefore, through an in-depth exploration of the HeLa cell line, it can be substantiated that the G2/M checkpoint, the climax of Interphase, embodies a vital control measure underpinning accurate cell division, thereby maintaining genomic integrity in cell microbiology.

Overall, Interphase plays a critical role in preparing the cell for division. It is a highly regulated and coordinated process that ensures that each daughter cell receives the necessary genetic material and cellular machinery for proper functioning. The intricate interplay of regulatory factors, signaling pathways, and surveillance mechanisms during Interphase underscores the complexity and precision of cellular activities. Understanding Interphase is essential in deciphering the fundamental mechanisms that govern cell biology and the dynamic process of cell division, which lies at the heart of life's continuity and diversity. The next section of this chapter will focus on mitosis which is the process of somatic cell division.

4.2 MITOSIS: THE PROCESS OF SOMATIC CELL DIVISION

Mitosis is a fundamental process of somatic cell division in cell biology, wherein a single diploid cell gives rise to two genetically identical daughter cells (Lamb et al., 2021; Lamptey et al., 2021; Lannagan et al., 2021; Langford, 2022). This highly orchestrated and tightly regulated process is critical for the growth, development, and maintenance of multicellular organisms. Mitosis is divided into distinct phases, each characterized by specific events that ensure accurate segregation of genetic material and the equal distribution of cellular components. Mitosis, the process of somatic cell division in cell microbiology, is a fundamental biological mechanism whereby a single diploid cell proliferates to produce two genetically identical daughter cells. A comprehensive understanding of this process can be gleaned from embryonic stem cells and their in-depth studiy reports. Embryonic stem cells, capable of differentiating into any cell type, underscore the mitotic process's criticality by their capacity to segregate and distribute chromosomes equally to daughter cells, thereby ensuring genetic constancy. Anchored in the phases of prophase, metaphase, anaphase and telophase, mitosis launches with the condensation of chromosomal material into identifiable structures in prophase, followed by the lining up of these chromosomes along the metaphase plate. Crucially, in anaphase, the protein complex cohesin, which initially clasped the sister chromatids together, undergoes cleavage, regulating sister chromatid separation adequately. Henceforth, the genetically identical sister chromatids, now individual chromosomes, traverse to opposing cell poles, aided by the mitotic spindle – a biphasic structure comprised of microtubules. Finally, in telophase, nuclear membranes reincorporate around chromosomes at each cell end, and the chromosomes relax into a more dispersed form of chromatin. The mitotic phase subsequently concludes with cytokinesis, where the cellular cytoplasm is divided equally, resulting in two identical daughter cells. Thus, through thorough studies of embryonic stem cell mitosis, it is evident that this process is a meticulously regulated cellular event that maintains the genetic continuity crucial to the cellular life cycle, cell growth, tissue repair, and development in cell biology.

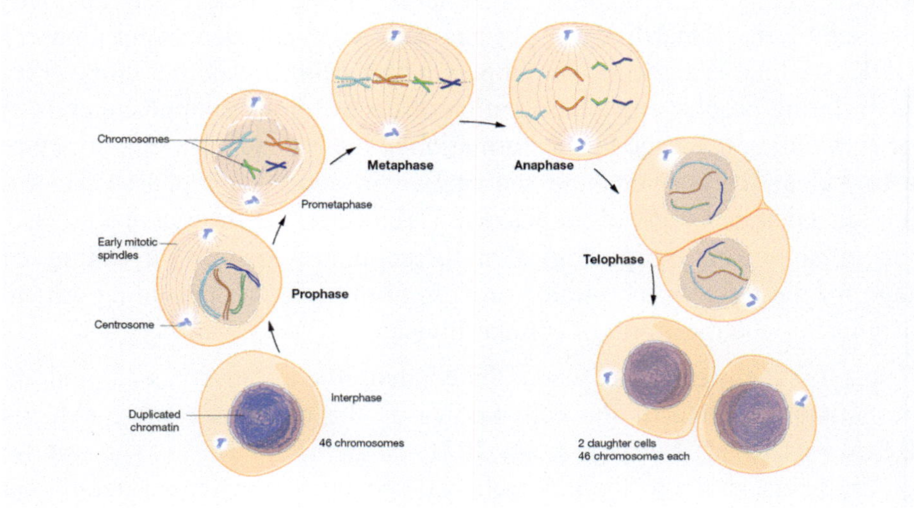

Figure 33. Illustration of mitosis.

Source: National Human Genome Research Institute https://www.genome.gov/genetics-glossary/Mitosis

The first phase of mitosis is prophase (see figure 33), during which chromatin condenses into visible chromosomes, and the nuclear envelope disassembles (Westman & Grinstein, 2021; Wiedner & Giudice, 2021; Whiteley et al., 2022). The centrosomes, containing centrioles, move to opposite poles of the cell, and microtubules begin to assemble, forming the mitotic spindle, a crucial structure that aids in chromosome movement. Prophase, the inaugural phase of mitosis, is characterized by the condensation of chromatin into identifiable chromosomes and the disassembly of the nuclear envelope. Detailed investigations of these phenomena have been undertaken using the human HeLa cell line, a cornerstone of cell biology research. As prophase initiates, the chromatin within the cell nucleus, normally diffused and transcriptionally active, undergoes progressive condensation. This concatenation process is driven by condensin and cohesin complexes which work in close coordination with topoisomerase II enzymes to manipulate the three-dimensional structure of chromatin. The outcome is the formation of visible, highly compacted chromosomes, each composed of two genetically identical sister chromatids. Concurrently, the nuclear envelope, the cell nucleus's protective barrier, starts to disassemble, a procedure necessitating the phosphorylation of nuclear lamins, fibrous proteins acting as structural components of the nuclear lamina. This disassembly is primarily driven by the enzyme cyclin-dependent kinase 1 (CDK1), which is activated in complex with Cyclin B. Another noteworthy event during prophase is the migration of centrosomes to opposite ends of the cell, setting the stage for the formation of the mitotic spindle, a structure instrumental for chromosome segregation in subsequent phases. Hence, through a detailed study of prophase in HeLa cells, we observe that this is a stage of significant morphological and structural reorganization, setting the stage for the tightly orchestrated process of mitosis, and providing critical insights into the dynamics of cellular division.

Next is the metaphase, where the condensed chromosomes align along the equatorial plane of the cell, known as the metaphase plate (Morris & Edwards, 2021; Murphy et al., 2021; Nagle et al., 2021). The mitotic spindle fibers attach to the kinetochores, specialized protein structures on the centromeres of chromosomes, ensuring proper alignment and tension across the chromatids. Metaphase, the subsequent phase following prophase in the chronology of mitosis, is typified by the alignment of condensed chromosomes along the cell's equatorial plane, appropriately termed the metaphase plate. Detailed illumination of this event can be achieved through the investigation of Chinese Hamster Ovary (CHO) cells, renowned in biological research for their robust growth and genetic stability. As metaphase commences, the centrosome-driven mitotic spindle, established during prophase, capitalizes on the combination of motor proteins and dynamic microtubules to guide the maneuvering of chromosomes to the metaphase plate. It is precisely along this imaginary equatorial line, bisecting the cell, where every chromosome, now visibly composed of two sister chromatids, aligns in anticipation of subsequent separation. Integral to the alignment process is the kinetochore, a protein complex assembled on the centromere region of each sister chromatid, serving as an attachment site for microtubule fibers of the spindle. The tension generated across these sister chromatids, now biorientated because of their attachment to microtubules from opposite cell poles, facilitates their alignment at the metaphase plate, facilitating the subsequent separation of sister chromatids into two daughter cells with faithful genome replication. Therefore, through rigorous study of metaphase in CHO cells, we perceive the sophisticated coordination and alignment of chromosomes, underscoring the intricate molecular choreography underlying cell division.

Anaphase follows, marked by the separation of sister chromatids as the mitotic spindle fibers contract (Cahan et al., 2021; Caldas et al., 2021; Burke et al., 2023). The chromatids are pulled towards opposite poles of the cell, ensuring that each daughter cell receives an identical set of chromosomes. Anaphase, the subsequent phase following metaphase, is characterized by the separation of sister chromatids triggered by the contraction of mitotic spindle fibers. To elucidate this process, let us consider the mitotic divisions in the blastula-stage zebrafish embryo, a model organism widely used in developmental biology and genetics due to its external fertilization and transparent embryonic stages. During anaphase, each pair of sister chromatids, previously aligned at the metaphase plate, are drawn apart due to the shortening of kinetochore microtubules, which are attached to the kinetochores at their centromeres. This simultaneous, coordinated movement ensures that each daughter cell receives one complete set of chromosomes. The commencement of anaphase is signaled by the activation of separase, an enzyme that cleaves the protein glue known as cohesin, binding the sister chromatids together. The breakdown of cohesin permits the chromatids to separate, which are then guided toward opposing cell poles by the shrinking microtubules of the mitotic spindle. In parallel, the non-kinetochore microtubules continue to lengthen, thus further elongating the cell. It is noteworthy that during anaphase, the sister chromatids, upon their separation, are referred to as daughter chromosomes. The insights from anaphase in zebrafish blastulae highlight the highly dynamic, orchestrated nature of cell division, offering an intriguing glimpse into the principles of chromosomal segregation and the preservation of genetic fidelity across generations.

Lastly, during telophase, the separated chromatids reach the opposite poles, and a new nuclear envelope forms around each set of chromosomes (LeGuennec et al., 2021; Leidal & Debnath, 2021; Lendahl, 2022). The chromosomes begin to decondense back into chromatin, and the mitotic spindle disassembles. Telophase is followed by cytokinesis, where the cytoplasm divides, and two distinct daughter cells are formed. Telophase, the final stage in the mitosis process, signifies the juncture wherein the segregated chromatids reach the cell's opposing poles and the formation of new nuclear envelopes around each chromosome set ensues. For illustration, a deep examination of this process in onion (Allium cepa) root tip cells, commonly utilized for studying cell division due to their rapid proliferation, reveals essential aspects of this phase. As telophase initiates, the now distinct daughter chromosomes complete their migration to the cell's opposite poles, propelled by the retracting spindle fibers attached to the kinetochores. A crucial characteristic of this phase is the reformation of the nuclear envelope around each chromosome cluster, a process facilitated primarily by the dephosphorylation of nuclear lamins. Concurrently, the chromosomes begin de-condensing, resuming their extended and diffuse state seen as chromatin in Interphase. This transformation eases access to the DNA for gene transcription in the subsequent interphase. Furthermore, nucleoli, which disappeared during the early stages of mitosis, reappear, signifying the resumption of ribosome production, critical for protein synthesis. Additionally, the mitotic spindle disassembles, marking its functional conclusion post-chromosomal migration. Therefore, observations from telophase developments in Allium cepa root tip cells exemplify this phase's unique orchestration of events facilitating the cellular transition from a mitotic state back to interphase while ensuring the new daughter cells possess accurate genetic information.

Cell biology research has elucidated key regulatory mechanisms that control mitosis, ensuring the accurate execution of each phase (Ajayi-Smith et al., 2021; Al-Marsoummi et al., 2021; Alao et al., 2021; Albrecht et al., 2021). Cyclin-dependent kinases (CDKs) and cyclins play crucial roles in regulating the cell cycle and initiating mitosis. Cyclin B combines with CDK1 to form the active maturation promoting factor (MPF), which triggers the entry of the cell into mitosis. MPF phosphorylates various substrates involved in mitotic processes, leading to the precise coordination of mitotic events. Investigation into cell microbiology has unveiled critical regulatory mechanisms that govern the seamless execution of mitosis, facilitating a meticulous progression through each phase. A comprehensive analysis of this regulation can be obtained from breakthrough research on the fission yeast Schizosaccharomyces pombe, a model organism, offering a straightforward representation of eukaryotic cell cycle regulation. At the heart of these regulatory mechanisms are the cyclically active kinases, particularly the cyclin-dependent kinases (CDKs). The oscillating activity levels of CDKs, driven by their associating proteins, the cyclins, dictate the progression from one mitotic phase to the next. These kinases exert control by phosphorylating target proteins involved in different mitotic events, subsequently altering their activity or stability. For instance, the initiation of DNA replication during the S phase is tightly controlled by a surge in S-CDK activity, while the commencement of anaphase is heralded by the activation of separase, under the regulation of metazoan M-CDK. In contrast, the transition from metaphase to anaphase is governed by the Anaphase-Promoting Complex/Cyclosome (APC/C), a significant ubiquitin ligase that targets securin and mitotic cyclins for degradation, thus freeing separase for activation. Additionally, the G2/M transition (see figure 34) and the early mitotic events are managed by the activity of M-CDK, with rising M-CDK levels triggering nuclear envelope breakdown during prophase and chromatid separation during anaphase. Further checkpoints ensure DNA integrity and replication completion, thus preventing erroneous cell division. Accordingly, comprehensive studies of the fission yeast S. pombe have clarified the cogent interplay of various regulatory mechanisms in ensuring the faithful execution of mitosis, thereby elucidating the intricate choreography of cellular division.

Disruptions in mitotic regulation can result in cell cycle abnormalities and diseases, such as cancer (Patriarca et al., 2021; Pauletto et al., 2021; Passanha et al., 2022). For instance, mutations in genes encoding mitotic proteins can lead to defective chromosome segregation and aneuploidy, a condition characterized by an abnormal number of chromosomes in the daughter cells. Disruptions in the finely tuned regulation of mitosis can yield cell cycle abnormalities, giving rise to diseases such as cancer. An exemplary case in point is observed in the aberrant cell mechanisms occurring in breast cancer, one of the most prevalent and thoroughly studied types of malignancies on a global scale. In breast cancer cells, there is often a deregulation of the cyclin-dependent kinase inhibitors (CKIs). These proteins ordinarily function by modulating the CDK-cyclin activity, maintaining stringent control of the progression through different cell cycle phases. However, alterations in the genes encoding CKIs, most notably in the p16INK4a, p21CIP1 and p27KIP1 inhibitors, have been implicated in the pathogenesis of various breast cancer subtypes. These alterations can lead to unchecked cell proliferation due to the unchecked activity of CDK-cyclin complexes, thereby promoting tumor development. Furthermore, dysregulation of proteins involved in the G2/M checkpoint, such as the tumor suppressor p53 and the kinase Chk1, can impede the cell's ability to repair damaged DNA before entering mitosis, ultimately resulting in genetic instability and an increased likelihood of malignant transformation. Such findings in breast cancer cells underscore the need for maintaining the correct balance in mitotic regulation for preserving cellular fidelity and preventing disease. Thus, the profound understanding of these aberrant mechanisms opens avenues for targeted therapeutic strategies, underscoring the invaluable contribution and potential of cell microbiology in the sphere of disease management.

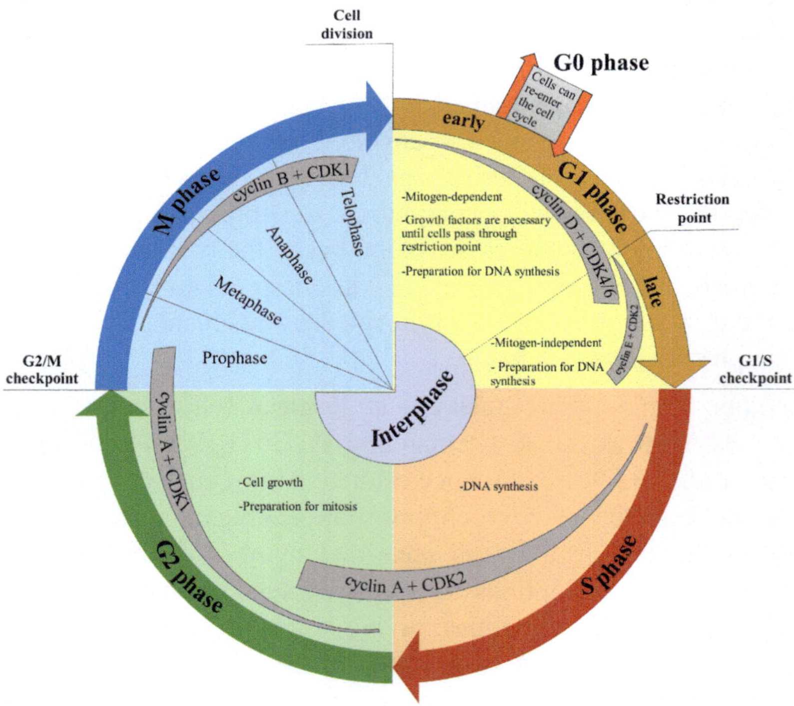

Figure 34. Model of cell cycle checkpoint activation.

Source: Ligasová A, Frydrych I, Koberna K. Basic Methods of Cell Cycle Analysis. International Journal of Molecular Sciences. 2023; 24(4):3674. https://doi.org/10.3390/ijms24043674

Mitosis is a remarkably intricate and highly regulated process that underpins the growth and maintenance of multicellular organisms. Understanding the molecular mechanisms governing mitosis is essential for comprehending cell biology and the development of therapeutic interventions targeting cell cycle-related disorders. The precise orchestration of mitosis ensures the faithful transmission of genetic material, safeguarding genomic stability and supporting the integrity of complex organisms. By delving into the intricacies of mitosis, researchers can uncover novel insights into cell biology and pave the way for advancements in medicine and biotechnology. In the next section, we will consider the regulation of the cell cycle through the use of checkpoints and controls.

4.3 REGULATION OF THE CELL CYCLE: CHECKPOINTS AND CELL CYCLE CONTROL

Regulation of the cell cycle is a tightly orchestrated process in cell biology, ensuring the accurate progression and coordination of events throughout the cell cycle (Khanna & Villa, 2022; Kim & Choi, 2022; Kikuchi et al., 2023). To prevent aberrant cell division and maintain genomic stability, cells employ various checkpoints and control mechanisms that monitor critical stages of the cell cycle. Regulation of the cell cycle; a series of successive events from cell birth to cell division; is a meticulously coordinated process in the realm of cell microbiology, warranting precise progress and orchestration of events throughout its course. Engaging HeLa cells, a human cell line employed extensively in biomedical research, provides an apt case study for elucidating this sophisticated regulatory procedure. Central to the cell cycle regulation are cyclin-dependent kinases (CDKs), a group of protein kinases whose activating cyclin subunits, the eponymous cyclins, oscillate in their concentrations at specific phases of the cell cycle. The fluctuating activity of these CDKs, in conjunction with their cyclin partners, functions as a molecular clock that regulates cell cycle transitions. Additional layers of regulation are provided by CDK inhibitors, which can halt the cell cycle, and checkpoint mechanisms, which ensure that preceding events have been executed correctly before progression into the subsequent phase. For instance, the G1/S checkpoint halts the cell cycle until any DNA damage incurred during the G1 phase is repaired, thereby preventing the replication of damaged DNA. Similarly, the G2/M checkpoint ensures complete and accurate DNA replication (see figure 35) before entering mitosis. Parallelly, the spindle assembly checkpoint prevents the onset of anaphase until all chromosomes are correctly attached to the spindle. Thus, HeLa cell studies highlight that regulation of the cell cycle is an intricately choreographed process which, through a system of interconnected regulatory proteins and checkpoints, ensures precise coordination and execution of events throughout the cell cycle. Not only does this understanding offer insights into fundamental cellular processes, but it also forms the basis for discerning various pathologies associated with cell cycle dysregulation.

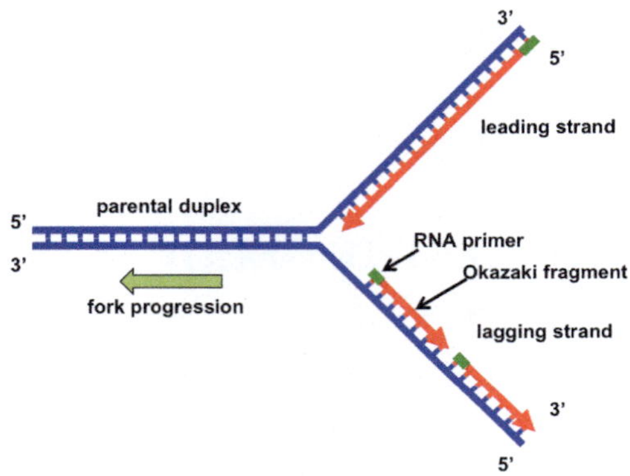

Figure 35. Illustration of a DNA replication fork.

Source: Leman AR, Noguchi E. The Replication Fork: Understanding the Eukaryotic Replication Machinery and the Challenges to Genome Duplication. Genes. 2013; 4(1):1-32. https://doi.org/10.3390/genes4010001

One of the essential regulatory elements is the cyclin-dependent kinases (CDKs), which are a family of serine/threonine kinases (Skoniecka et al., 2021; Smith et al., 2022; Smith et al., 2023). CDKs form complexes with specific regulatory proteins called cyclins, and their activity is cyclically regulated during the cell cycle. These CDK-cyclin complexes serve as key drivers of cell cycle progression, phosphorylating various substrates to activate or inactivate them, leading to the orderly transition from one cell cycle phase to the next. One of the pivotal regulatory components in the biological orchestra of the cell cycle is the cyclin-dependent kinases (CDKs), a family of serine/threonine kinases. For a detailed examination of this paramount regulation, we can refer to the yeast Saccharomyces cerevisiae, a favored model organism in cell cycle studies owing to its rapid growth and simplified, yet representative, cell cycle machinery. CDKs, as their name suggests, are dependent on associating proteins known as cyclins for their appropriate activation. The key feature of cyclins is their periodically fluctuating concentration within the cell cycle, which impacts the activation and deactivation of their associated CDKs. As a result, various cell cycle phases are subject to control by distinct CDK-cyclin complexes. In S. cerevisiae, for example, the onset of the S phase, or the DNA synthesis phase, is regulated chiefly by the S-CDK, a complex of CDK and S cyclin. Additionally, the CDK-cyclin complex also mitigates the transition from the G2 to the M phase via M-CDK, a combination of CDK and mitotic cyclins. Importantly, CDKs do not indiscriminately phosphorylate all substrates; instead, they showcase substrate specificity courtesy of the regulatory subunits that are cyclins. Hence, by studying the CDK kinetics in S. cerevisiae, it becomes evident that these serine/threonine kinases play an instrumental role in synchronizing and driving the cell cycle's sequential progression, an understanding indispensable to grasping eukaryotic cell biology's inherent complexity.

The cell cycle is divided into several checkpoints, where the cell undergoes stringent assessments to ensure proper cellular conditions before advancing to the subsequent phase (Yuan et al., 2021; Yu et al., 2022; X. J. Zhang et al., 2023; X. Y. Zhang et al., 2023). The G1/S checkpoint, also known as the restriction point, is a critical decision point during the G1 phase. At this checkpoint, the cell assesses external growth signals and the integrity of its genetic material. If conditions are favorable, the cell commits to DNA synthesis and initiates the S phase. However, if there are unfavorable conditions, the cell may enter a non-dividing state called the G0 phase or undergo apoptosis. The G1/S checkpoint, oftentimes referred to as the restriction point, embodies a crucial decision point within the G1 phase of the cell cycle. For a comprehensive understanding of this decision-making juncture, the analysis of fibroblasts, and connective tissue cells involved in wound healing, presents an optimal case study. The G1/S checkpoint essentially operates as a surveillance mechanism that assesses whether conditions are favorable for cell division to proceed. Two key aspects considered by the cell at this checkpoint are the availability of cellular nutrients and the integrity of the genomic DNA. The regulation of this checkpoint primarily involves the Retinoblastoma protein (Rb), a tumor suppressor protein, which hinders cell cycle progression by inhibiting the E2F transcription factor responsible for activating genes needed for the S phase. Upon receiving appropriate growth signals, cyclin D-CDK4/6 complexes begin to phosphorylate Rb during the late G1 phase, prompting Rb to release E2F. This allows the cell to cross the G1/S restriction point and prepares it for DNA replication by activating S phase genes. However, if DNA damage is detected, the p53 protein pathway comes into play, preventing the transition to S phase until the damage is repaired. Thus, the case of fibroblasts effectively elucidates how the G1/S checkpoint acts as a regulatory interface, ensuring the conditions are conducive for cell division, underscoring the profound intricacy of cellular dynamics.

Another crucial checkpoint is the G2/M checkpoint, which occurs at the end of the G2 phase (Hynds et al., 2021; Jardin et al., 2022; Iachettini et al., 2023). At this stage, the cell verifies whether DNA replication is complete and whether any DNA damage has been repaired. Additionally, the proper assembly of the mitotic spindle is evaluated. If the cell passes this checkpoint, it proceeds to mitosis and cell division. However, if any issues are detected, the cell cycle is arrested to allow for DNA repair or other necessary corrections. Another integral checkpoint in the coordination of cell cycle events is the G2/M checkpoint, occurring at the conclusion of the G2 phase. Elucidation of this checkpoint's function can be achieved through the examination of melanoma cells, a type of skin cancer cell common in oncological studies. The G2/M checkpoint functions fundamentally as a cell's secondary inspection point, where a final assessment is contrived towards verifying the completion of all DNA replication and the repair of any potential DNA damage that occurred during the S phase.

If the inspection is deemed successful, the cell is then permitted to commence mitosis. Critical to the functioning of the checkpoint is the cyclin B-CDK1 complex, the activating of which represents the primary driver of cells into mitosis. The pathway in which the protein kinases ATM and ATR, in response to DNA damage, block CDK1 activation by stimulating the activity of the kinase Wee1 and suppressing Cdc25 phosphatases, plays a crucial role in halting the cell cycle at this checkpoint. This delay permits adequate time for the repair of DNA damage, protecting genomic integrity. If the inflicted damage is considered irreparable, cell cycle arrest may be induced, leading the cell to fatefully enter senescence or apoptosis (see figure 36), hence forestalling the division of cells comprising damaged DNA. Consequently, through melanoma cells, the G2/M checkpoint's role is starkly highlighted as a crucial safeguard attached to the labyrinthine machinery of the cell cycle, restricting the propagation of potential genetic errors, thereby fortifying our understanding of the intricacies involved in cell microbiology.

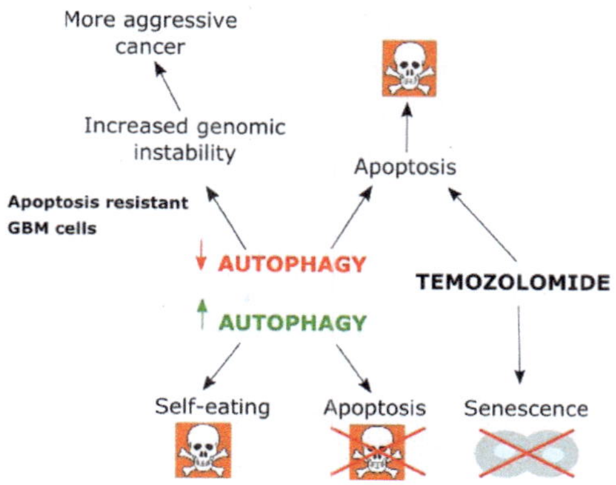

Figure 36. The interplay between senescence and apoptosis is cell type specific.

Source: Pawlowska E, Szczepanska J, Szatkowska M, Blasiak J. An Interplay between Senescence, Apoptosis and Autophagy in Glioblastoma Multiforme—Role in Pathogenesis and Therapeutic Perspective. International Journal of Molecular Sciences. 2018; 19(3):889. https://doi.org/10.3390/ijms19030889

The last checkpoint is the spindle checkpoint, also known as the mitotic checkpoint, which operates during the metaphase phase of mitosis (Altulea et al., 2021; Alvord et al., 2022; Aniento et al., 2022; Aprile et al., 2023). This checkpoint ensures that all chromosomes are correctly aligned on the metaphase plate and attached to the mitotic spindle via kinetochores. Only when all chromosomes are properly aligned and tension is evenly distributed across the chromatids, the cell is allowed to progress into anaphase, ensuring the accurate segregation of chromosomes. The terminal checkpoint in the cell cycle is the spindle checkpoint, also recognized as the mitotic checkpoint, operating during the metaphase of mitosis. For a comprehensive understanding of this vital checkpoint, the exploration of human osteosarcoma U2OS cells, a commonly used cell line in the study of spindle checkpoint activities, proves enlightening. The spindle checkpoint serves to ensure that all the chromosomes are correctly attached to the spindle fibers before progressing to anaphase, thereby preventing chromatid segregation errors. Central to this checkpoint's function is the Anaphase-Promoting Complex/Cyclosome (APC/C), a ubiquitin ligase whose activity marks the end of metaphase and onset of anaphase. However, the APC/C remains inactive until every kinetochore, a protein structure located on chromatids, is stably attached to spindle fibers. The regulation of APC/C depends on multiple spindle checkpoint proteins including, Mad1, Mad2,

Mad3, Bub1, Bub3, and BubR1. These proteins impede the APC/C complex's activation, thereby delaying the onset of anaphase, providing the cell ample time to rectify improperly attached chromosomes. Thus, the essential functioning of the spindle checkpoint, as investigated in U2OS cells, highlights its role as a guardian of genomic integrity during mitosis, ensuring the correct segregation of chromosomes, thereby maintaining genetic continuity in subsequent cell generations.

To regulate the cell cycle, several tumor suppressor proteins and cyclin-dependent kinase inhibitors (CKIs) act as gatekeepers (LeGuennec et al., 2021; Leidal & Debnath, 2021; Lendahl, 2022). For example, the retinoblastoma protein (Rb) is a key tumor suppressor that controls the G1/S transition by inhibiting the activity of E2F transcription factors. Phosphorylation of Rb by CDK-cyclin complexes during the G1 phase releases its inhibition, allowing the transcription of genes essential for DNA synthesis. Regulation of the cell cycle is vitally governed by various tumor suppressor proteins and cyclin-dependent kinase inhibitors (CKIs), acting as vigilant gatekeepers to ensure flawless transition between stages. A profound understanding of their role in cell cycle modulation can be attained by exploring the widely used HeLa cell line, a human cervical cancer cell line renowned for its consistency and exceptional suitability for experimentation. Tumor suppressor proteins, such as p53 and Rb, play a critical role in modulating the cell cycle. p53, often termed the "guardian of the genome," gets activated in response to DNA damage or genomic stress, leading to cell cycle arrest or programmed cell death, apoptosis, to prevent the propagation of damaged cells. Similarly, Rb protein works collaboratively with E2F transcription factors to control cell cycle progression, particularly at the G1/S checkpoint. On the other hand, CKIs control the cell cycle by inhibiting cyclin-dependent kinases. Proteins like p21Cip1 and p27Kip1 belong to the Cip/Kip family of CKIs that inhibit a broad spectrum of cyclin-CDK complexes, whereas proteins like p15Ink4b, p16Ink4a, p18Ink4c, and p19Ink4d belong to the INK4 family that specifically inhibit CDK4 and CDK6. Functioning collectively, these regulators maintain the precision and orderliness of the cell cycle. Hence, through the analysis of HeLa cells, the paradoxically complex and stringent regulation of the cell cycle becomes lucid, shedding light on the multitude of molecular events that ensure appropriate cellular division and genomic stability.

Cell biology research has illuminated the intricate signaling pathways and regulatory networks involved in cell cycle control. Dysregulation of these checkpoints and regulatory mechanisms can lead to uncontrolled cell division and the development of cancer. Understanding the complexities of cell cycle regulation is vital in advancing our knowledge of cell biology and identifying potential therapeutic targets for the treatment of various diseases related to cell cycle dysfunction. By maintaining the precise balance of cell cycle checkpoints, cells ensure the faithful transmission of genetic material, safeguard genomic integrity, and support the homeostasis of multicellular organisms. Through further investigations in cell biology, scientists continue to unravel the intricate web of cell cycle regulation and its implications for health and disease. The fourth section of this chapter explores meiosis as a form of cell division for sexual reproduction.

4.4 MEIOSIS: CELL DIVISION FOR SEXUAL REPRODUCTION

Meiosis is a highly specialized form of cell division in cell biology, essential for sexual reproduction in eukaryotic organisms (Stancliffe & Patti, 2022; Stratman et al., 2022; Stowers, 2023). Unlike mitosis, which produces genetically identical daughter cells, meiosis generates gametes (sperm and egg cells) with half the number of chromosomes as the parent cell. This reduction in chromosome number is crucial for maintaining the correct chromosome number in sexually reproducing species and ensuring genetic diversity in offspring. Meiosis, a specialized cell division variant, is indispensable for sexual reproduction in eukaryotic organisms. This process can be aptly demonstrated through observing Saccharomyces cerevisiae (baker's yeast), a user-friendly, unicellular organism extensively employed in molecular biology studies. In contrast to mitosis, meiosis constitutes a reduction division, wherein a diploid cell undergoes two consecutive divisions, namely meiosis I and meiosis II, to produce four haploid cells or spores. These divisions comprise largely analogous phases to those observed in mitosis but exhibit distinct differences principally in meiosis I. During prophase I of meiosis, homologous chromosomes pair and exchange genetic material through a process known as crossing over, resulting in recombined chromosomes which are the carriers of genetic variance. Subsequent to crossing over, individual homologs, each comprised of two sister chromatids, are segregated into separate cells during anaphase I. Meiosis II, relatively akin to mitosis, subsequently follows, with the separated sister chromatids of each homolog further segregating into separate cells. Thus, observations of S. cerevisiae during meiosis manifest that this highly specialized form of cell division is not merely aimed at cell multiplication, but significantly generates genetic diversity, a critical aspect of evolution. Its importance becomes more pronounced in higher eukaryotes, where the meiotic division of specific cells in reproductive organs leads to the production of sex cells or gametes, paving the pathway for sexual reproduction. Therefore, meiosis remains a critically fundamental aspect of study in cell biology. Figure 37 is an illustration of meiosis.

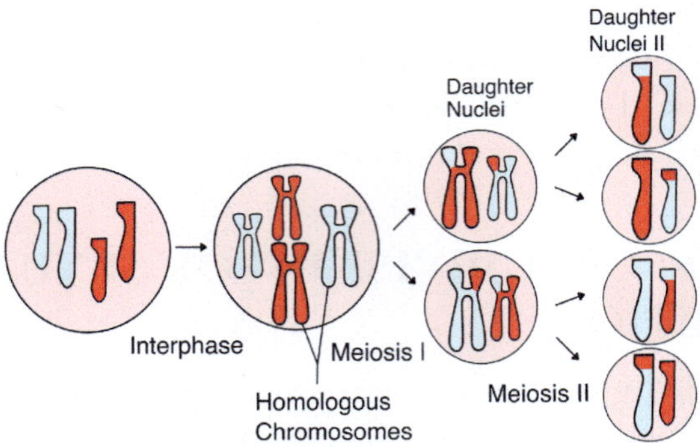

Figure 37. Illustration of meiosis.

Source: Wikipedia (https://en.wikipedia.org/wiki/Meiosis)

The process of meiosis occurs in two consecutive stages: meiosis I and meiosis II (McFall & Stites, 2021; Melo et al., 2022; Mendes, 2023; Mierke, 2023). Meiosis I is marked by the pairing of homologous chromosomes, known as synapsis, followed by crossing over between non-sister chromatids. This genetic exchange between homologous chromosomes results in the reshuffling of genetic material, contributing to genetic variability in offspring. During anaphase I, homologous chromosomes separate and migrate to opposite poles of the cell. As a result, each daughter cell produced at the end of meiosis I contains one set of chromosomes, with each chromosome consisting of two sister chromatids. Meiosis I is a distinctive phase in the cell cycle characterized by the pairing of homologous chromosomes, a process commonly termed synapsis, which is eventually succeeded by an event of crossing over between non-sister chromatids. For a detailed exploration of this phenomenon, let us consider the process as it unfolds within Drosophila melanogaster, the fruit fly, a popular model organism extensively utilized in genetics research. During prophase I of Drosophila meiosis, each chromosome initially replicates to yield two identical sister chromatids, resulting in a pair of homologous chromosomes. These homologous pairs align lengthwise in close proximity, an event facilitated by a complex web of proteins forming the synaptonemal complex. This intimate alignment permits the occurrence of crossing over, or recombination, wherein non-sister chromatids from each homologous pair undergo a precise breakage and subsequent rejoining. Specifically, segments of genetic material are effectively swapped between these chromatids, creating novel combinations of alleles within each chromosome, a source of genetic variation in offspring. Ensuing the recombination event, the homologous pairs desynapse, though remain connected at the site of the crossover, often called chiasmata, which ensure proper segregation of homologs during anaphase I. Thus, by delving into the details of meiosis I in Drosophila melanogaster, we grasp the ability of this cell division process to generate genetic diversity, a cornerstone of evolutive progression, and observe the complex mechanisms that enable the precise execution of these critical genetic reshuffling events.

Meiosis II, resembling a mitotic division, involves the separation of sister chromatids, ensuring that each daughter cell receives a haploid set of chromosomes (Collins et al., 2021; Costa & Sousa, 2022; Colin et al., 2023). At the end of meiosis II, four haploid daughter cells are produced, each with a unique combination of genetic material. These daughter cells are gametes and serve as the vehicles for transmitting genetic information from one generation to the next during sexual reproduction. Meiosis II, bearing resemblance to a mitotic division, facilitates the segregation of sister chromatids, thereby guaranteeing that every daughter cell is endowed with a haploid set of chromosomes. This statement finds verification through the comprehensive examination of a case study focusing on Saccharomyces cerevisiae, a type of budding yeast commonly adopted as a model organism in cell biology. Embarking upon meiosis after encountering conditions of nutrient scarcity, this microorganism sets in motion a sequence of events strikingly similar to mitosis during Meiosis II. In anaphase II, the kinetochore microtubules, originating from opposite poles, pull apart the sister chromatids via depolymerization, a process catalyzed by the anti-poleward forces exerted at kinetochores, coupled with the poleward forces applied at centromeres. The orchestrated movement of these microtubules ensures orderly chromosome segregation. Pursuing the anaphase, telophase II witnesses the decondensation of chromosomes, followed by the formation of a nuclear envelope around each set of chromosomes, thereby formally marking the segregation process's completion. Cytokinesis then seals the process with cellular division. Consequently, each of the four daughter cells crafted from the parent yeast cell is given a haploid set of chromosomes.

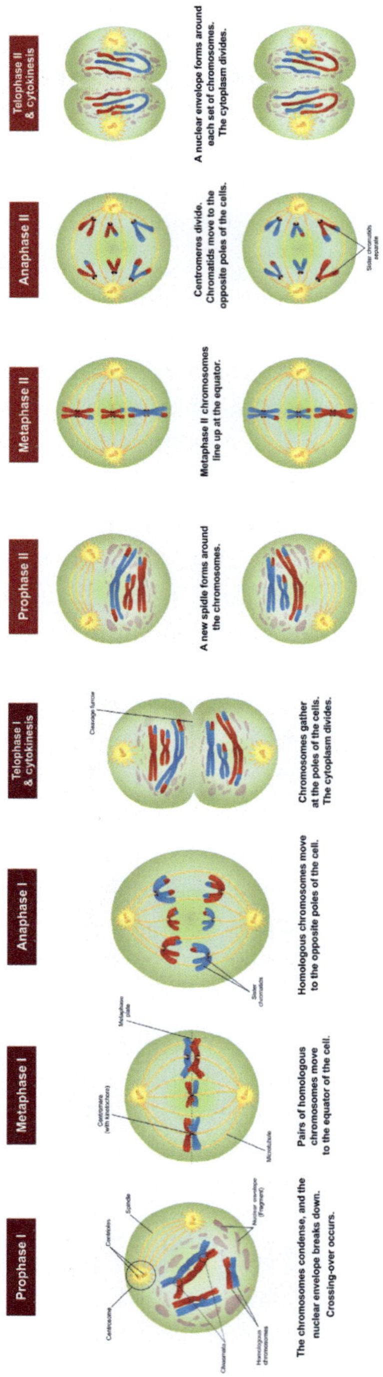

Figure 38. Illustration of sex cell division or meiosis.

Source: By Ali Zifan - Own work using: information from Campbell Biology (10th Edition) by: Jane B. Reece & Steven A. Wasserman., CC BY-SA 4.0, https://commons.wikimedia.org/w/index.php?curid=49630204

Through such an effective course of action, Saccharomyces cerevisiae successfully positions Meiosis II as a mitosis-like cell division phase pivotal in bestowing daughter cells with genetic uniformity, ergo bolstering biodiversity whilst maintaining the consistency of their genetic material.

The process of meiosis is intricately regulated to ensure the proper alignment and segregation of chromosomes (Lamb et al., 2021; Lamptey et al., 2021; Lannagan et al., 2021; Langford, 2022). Various checkpoints, similar to those in mitosis, monitor the integrity of DNA and the accurate separation of homologous chromosomes and sister chromatids. Errors in meiotic regulation can lead to aneuploidy, a condition characterized by an abnormal number of chromosomes in gametes, which may result in developmental disorders or infertility in offspring. The process of meiosis, perpetrated to ensure the apt alignment and segregation of chromosomes, hinges upon a complex network of regulatory mechanisms. A compelling illustration of this regulatory complexity is underpinned by an investigation of the oocytes of Mus musculus, or the common house mouse. Initiated during embryogenesis, the primary oocytes of this mammal arrest development in prophase of the first meiotic division, lingering in this dormant state until the initiation of an adult reproductive cycle.

This strand of suspended animation serves as the inaugural platform for meiotic regulation. A sophisticated assortment of molecular signals regulates the maintenance of this prophase I arrest including cyclic AMP, which inhibits phosphodiesterases, induces high intracellular concentration of cAMP and thus propagates the arrest. Subsequent to the onset of the reproductive cycle, the drop in luteinizing hormones and follicle-stimulating hormones stimulate the resumption of the first meiotic division. Furthermore, this hormonal cascade also regulates the metaphase I to metaphase II transition, underpinned by a surge in anaphase-promoting complex activity which precipitates the degradation of securin, catalyzing the separation of homologous chromosomes. The exquisite coordination of these steps is evident in the alignment along the metaphase plate and the equational division associated with anaphase II, ensuring proper segregation of chromatids into divided daughter cells (see figure 38). Through the examination of this precise, multilayered regulation of meiosis in the primary oocytes of Mus musculus, the intrinsic nuances and the vital importance of regulatory processes in ensuring the proper alignment and segregation of chromosomes in meiosis are vividly evoked.

Meiosis plays a vital role in creating genetic diversity within populations (Spiers et al., 2021; Stacey & Healy, 2021; Stallaert et al., 2022). By promoting genetic recombination through crossing over and independent assortment of chromosomes, meiosis contributes to the variability observed in sexually reproducing species. This genetic diversity is essential for the adaptation and survival of species in changing environments. In the biological theatre, meiosis assumes a pivotal role in engendering genetic diversity within populations. This statement gains traction through a comprehensive study of meiotic recombination in Arabidopsis thaliana, a model plant species often entrusted to unveil eukaryotic biology. Meiosis impacts genetic diversity predominantly through two processes: independent assortment of chromosomes and recombination. This latter phenomenon, studied in-depth in Arabidopsis thaliana, involves the exchange of genetic material between homologous chromosomes, creating novel combinations of genes on each chromosome. Throughout prophase I of meiosis in Arabidopsis thaliana, double-strand breaks are introduced into the DNA, mediated by an enzyme called SPO11. The formation of these breaks instigates genetic recombination, wherein homologous chromatids 'cross over', exchanging segments of DNA. Sequentially, these recombinant chromosomes are segregated into haploid gametes during the meiotic divisions, thereby transmitting the novel combinations of genetic material into progeny. Consequently, a population of Arabidopsis thaliana, even one derived from a single individual, exhibits extensive genetic diversity, attributed to the recombination events facilitated through meiosis. Ergo, the study of meiotic processes in Arabidopsis thaliana unravels the essential role of meiosis in fostering genetic diversity within populations, simultaneously underscoring ecology's principle of adaptability conferring survival advantage to diverse populations.

In cell biology, meiosis research provides valuable insights into the molecular mechanisms governing chromosome pairing, synapsis, and segregation (Tolkin & Hubbard, 2021; Tomoda & Kime, 2021; Tuieng et al., 2021). Understanding the complexities of meiosis has significant implications for human health, agriculture, and conservation biology. In humans, errors in meiosis can lead to chromosomal abnormalities, such as Down syndrome, highlighting the critical role of meiotic regulation in human development. Indeed, the study of meiosis in cell biology furnishes significant insights into the molecular mechanisms orchestrating chromosome pairing, synapsis, and segregation. C. elegans, a non-parasitic nematode and well-recognized model organism renowned for its cell biological advantages, provides an illustrative case for comprehending these mechanisms. The meiotic processes in its germline cells provide valuable glimpses into the molecular interactions facilitating the pairing of homologous chromosomes, a prerequisite for their accurate segregation later. The chromosomes in C. elegans at the beginning of meiosis commence with a 'polarized' configuration. This structural asymmetry, maintained by a balance of forces exerted by microtubules and motor proteins, is fundamental to the pairing process. Chromosomes orient themselves such that their pairing centers, sequences of DNA necessitating auspicious pairing interactions, face the cellular periphery where cytoplasmic microtubules reside. Driven by forces originating from dynein motors, the chromosomes actively move along microtubule tracks, enabling homologous chromosomes to search, encounter and pair up with each other, thus achieving synapsis. This dynamic nature of chromosome mobility continues through the pachytene stage, allowing the chromosomes to correct mispairing, and allowing their homologue bias pairing. Subsequently, synaptonemal complex proteins connect these paired chromosomes, yielding a fully synapsed tetrad. The synapsed chromosomes then establish chiasmata, allowing their biphasic segregation, initially at metaphase I segregating homologous chromosomes followed by separation of sister chromatids at metaphase II in anaphase. The investigation of meiotic processes in C. elegans thus culminates in a multifaceted understanding of the interplay between molecular and cellular mechanisms steering chromosome pairing, synapsis and segregation.

In summary, meiosis is a crucial process in cell biology, enabling sexual reproduction and generating genetically diverse gametes. The accurate execution of meiosis ensures the transmission of genetic material and contributes to the remarkable biodiversity observed in nature. Continued research in cell biology will undoubtedly shed further light on the intricacies of meiosis, advancing our understanding of reproductive biology and its implications for the diversity and survival of life on Earth. This fourth chapter will close by highlighting the importance of cell division in growth, development, and repair.

4.5 IMPORTANCE OF CELL DIVISION IN GROWTH, DEVELOPMENT, AND REPAIR

Cell division plays a pivotal role in cell biology and is of paramount importance in various biological processes, including growth, development, and tissue repair (Prazak et al., 2021; Pressly et al., 2022; Qazi et al., 2022). These fundamental functions ensure the continuity and functionality of multicellular organisms, enabling them to adapt and thrive in dynamic environments. Cell division undeniably plays a crucial role in diverse aspects of cell biology, with its expansive influence permeating growth, development, to repair in biological organisms. Reviewing cell division in the context of Saccharomyces cerevisiae, a single-celled yeast, elucidates its profound role in these contexts. Commencing with growth, Saccharomyces cerevisiae responds to nutrient availability by regulating cell division rates. When nutrients are plentiful, this yeast undergoes rapid cell divisions, increasing its population size and hence its colony growth. Conversely, in an environment with scarce nutrients, the yeast adopts a growth arrest state, halting cell divisions until conditions improve. As a unicellular organism, this growth alteration in S.cerevisiae is akin, in essence, to developmental changes experienced by multicellular organisms.

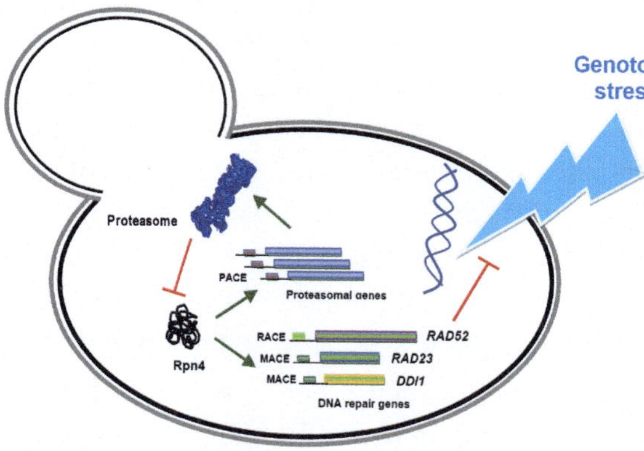

Figure 39. Example of Yeast Rpn4 Links the Proteasome and DNA Repair via RAD52 Regulation.

Source: Spasskaya, D. S., Nadolinskaia, N. I., Tutyaeva, V. V., Lysov, Y. P., Karpov, V. L., &Karpov, D. S. (2020). Yeast Rpn4 links the proteasome and DNA repair via RAD52 regulation. International Journal of Molecular Sciences, 21(21), 8097.

Transitioning now to tissue repair, although Saccharomyces cerevisiae does not possess tissue in a conventional multicellular sense, it does employ cell division in an interesting repair mechanism - the formation of 'bud scars'. These scars are the remnants of previous daughter cells after division, and if the mother cell's wall undergoes damage, it uses existing bud scars to segregate damaged cell wall components into a future bud, thereby limiting the damage and marking a form of tissue repair. Thus, whether viewed through the lens of yeast colony growth or its unique 'tissue repair' mechanism, cell division dictates the core processes of growth, development, and repair in Saccharomyces cerevisiae, reinforcing cell division's overarching importance in cell biology. The figure below is an example of Yeast Rpn4 Links the Proteasome and DNA Repair via RAD52 Regulation which involves some of the processes that have been described above and will be detailed more in this section.

Growth: Cell division is the driving force behind the growth and expansion of organisms (Russo et al., 2021; Roy et al., 2022; Roth & Taatjes, 2023). During development and throughout an organism's life, cells divide to increase their number, allowing tissues and organs to grow in size and complexity. In embryonic development, rapid and controlled cell division leads to the formation of the three germ layers and the subsequent differentiation into various cell types, tissues, and organs. In adults, cell division is essential for tissue renewal and replacement, maintaining the homeostasis of organs, and ensuring the functionality of vital systems. The balance between cell division and cell death (apoptosis) is crucial for maintaining tissue size and function. The grounding assertion that cell division is the driving force behind the growth and expansion of organisms vividly depicts a seminal precept of cell biology, a precept that finds its expansive elucidation in the unicellular algae Chlamydomonas reinhardtii. Coveted as a model organism, the life cycle of Chlamydomonas reinhardtii provides an equitably detailed platform for observing the modulatory impact of cell division on growth and organismal expansion. The single-celled green algae, existing predominantly in haploid form, undergo cellular divisions during vegetative growth when subjected to optimal growing conditions. The process involves the replication of the cell's DNA and essential organelles, followed by mitotic division, resulting in a twofold increase in the cell count. Each subsequent round of cell division effectively doubles the population size, thus facilitating the expansion and growth of the algal colony. Mitotic cell division also critically underlies the organism's adaptive response to variations in nutrient availability. Under conditions of nutrient scarcity, the alga undergoes gametic differentiation, characterized by a cessation of mitotic divisions. Upon the restoration of favorable nutrient conditions, the fusion of these differentiated gametes initiates a round of zygotic meiotic divisions, creating a surge of genetically unique, haploid daughter cells. Through each round of cell division, whether mitotic or meiotic, Chlamydomonas reinhardtii synthesizes its survival strategy, regulating its growth, expansion, and adaptation to changing external conditions. Thus, this cellular journey of Chlamydomonas reinhardtii vividly affirms the underlying centrality of cell division in driving the growth and expansion of organisms.

Development: The precise regulation of cell division is essential for orchestrating the complex processes of development (Pinson & Huttner, 2021; Postler, 2021; Pratt et al., 2021). During embryogenesis, cell division is tightly controlled to ensure that cells differentiate into specialized cell types in the correct spatial and temporal patterns. The orderly progression of cell division is critical in shaping the body plan and establishing functional tissues and organs. Errors in cell division during development can lead to severe congenital disorders and developmental abnormalities. The precision in the regulation of cell division is indeed essential for orchestrating the intricate processes that underlie development. This assertion gravitates toward certitude upon examining the embryogenesis of Caenorhabditis elegans, a nematode worm well-established in developmental biology research. The embryo of Caenorhabditis elegans commences development as a single cell, the fertilized egg, which embarks upon a stereotypical and exquisitely timed series of cell divisions. These divisions engender an assemblage of cells that progressively differentiate and give rise to distinct tissues and organs, adhering to a rigorously consistent pattern among individuals. This rigorous consistency is owed to the remarkable precision in regulating the cell division processes. For instance, the orientation of cell division, dictated by the mitotic spindle, is under intricate control. Several polarizing cues inscribed within the egg, including the PAR proteins and the Wnt signaling pathway, coordinate to define the axis of the first cell division. Post embryonic cell divisions carry forward this meticulous legacy with cells synchronizing their divisions in a lineage and temporal-specific manner, contingent on growth signals and environmental cues. Such precision in the regulation of these divisions ensures that each cell assumes its exact position and fate, directing it to contribute to a specific part of the organism. The developmental patterns of Caenorhabditis elegans thus vividly exemplifies the indispensability of the precise regulation of cell division in orchestrating the complex developmental processes, ultimately scripting the genesis of an organism from a single cell.

Tissue Repair: Cell division is vital for tissue repair and regeneration after injuries or damage (Kroll et al., 2022; Kim et al., 2023; Klingemann, 2023; Laiman et al., 2023). When tissues are injured, neighboring cells can proliferate and divide to replace the damaged or lost cells. In the case of minor injuries, cell division enables tissue healing and restoration of normal function. For example, during wound healing, skin cells at the wound site rapidly divide and migrate to close the wound. In more severe injuries, cell division may be complemented by stem cell activation and differentiation to replace specialized cell types that cannot undergo cell division.

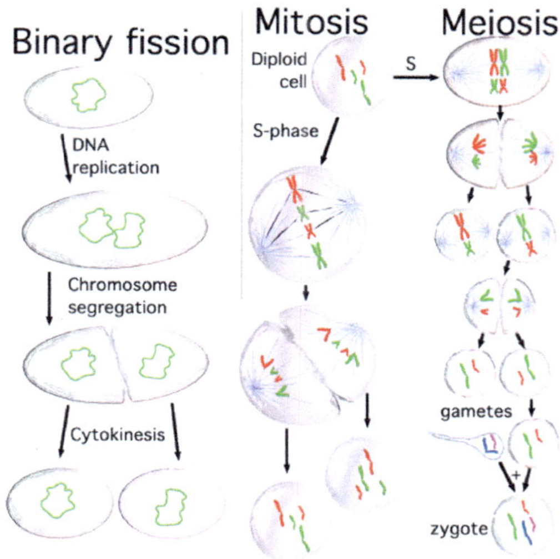

Figure 40. The main functions of cell division.

Source: CC BY-SA 3.0, https://commons.wikimedia.org/w/index.php?curid=327384

The foundational role of cell division in the processes of tissue repair and regeneration after injuries or damage is decisively demonstrated through extensive research on the regenerative abilities of planarian flatworms. Planarians, characterized by an exceptional regenerative capability, owe this trait to the presence of pluripotent stem cells, called neoblasts, that initiate cell division in response to tissue damage or injury. Upon injury, signals from the wound site instigate neoblasts to proliferate, navigating them toward the damaged site. These neoblasts undergo a series of regulated cell divisions, producing a blastema, a mass of cells capable of growth and regeneration into new organs or body parts. Crucially, these neoblasts are also diverse, with different sub-populations contributing to specific cell lineages, mirroring the controlled differentiation seen in embryonic stem cells. The orchestrated cell division and differentiation of these neoblasts underlie the planarian's astonishing ability to regenerate lost or damaged tissues and even entire body parts. Ascertaining the types of cells derived from individual neoblasts and understanding the signals that control their proliferation, migration, and differentiation, therefore, holds the promise of deciphering the regulatory processes that drive tissue repair and regeneration. In conclusion, the case study of planarian flatworms unequivocally underscores the imperative of cell division in the repair, regeneration, and maintenance of tissues upon encountering injuries or damage. Figure 40 summaries the functions of cell division.

Cell biology research has elucidated the complex molecular and genetic mechanisms governing cell division and its regulation (Lamb et al., 2021; Lamptey et al., 2021; Lannagan et al., 2021; Langford, 2022). The cell cycle, checkpoints, and various signaling pathways tightly control the process, ensuring accurate replication and distribution of genetic material. Defects in cell division regulation can lead to pathological conditions, such as cancer, where cells divide uncontrollably and invade surrounding tissues. Moreover, the understanding of cell division has significant implications in medicine and biotechnology. In cancer treatment, therapies are designed to target rapidly dividing cancer cells, exploiting the differences in cell division rates between normal and cancerous cells. Similarly, research on cell division is vital for regenerative medicine, where scientists aim to harness the potential of stem cells and cell division to repair and replace damaged tissues and organs.

In conclusion, the importance of cell division in growth, development, and repair cannot be overstated in cell biology. It is a fundamental process that shapes the destiny of cells, tissues, and organisms, allowing them to adapt, grow, repair, and reproduce. The intricate control of cell division ensures the orderly progression of life, laying the foundation for the remarkable complexity and diversity observed in living organisms. Through continued research in cell biology, we can further uncover the mysteries of cell division and unlock its potential for medical advancements and our understanding of life itself.

CHAPTER 4 SUMMARY

Chapter 4 of the textbook delves into the intricate processes of the cell cycle and cell division, which are fundamental aspects of cell biology. The chapter begins with an exploration of interphase (4.1), a critical stage that precedes cell division. Interphase is a dynamic period during which the cell undergoes various preparatory activities, such as DNA replication, protein synthesis, and organelle expansion, to ensure successful division.

The chapter then moves on to discuss mitosis (4.2), a highly regulated process responsible for somatic cell division. Mitosis ensures the accurate distribution of genetic material to two genetically identical daughter cells. It involves distinct phases, including prophase, metaphase, anaphase, and telophase, each characterized by specific events that ensure precise chromosome segregation and cytokinesis.

Next, the regulation of the cell cycle (4.3) is explored, which includes checkpoints and control mechanisms that govern the orderly progression of the cell cycle. These checkpoints monitor the integrity of DNA, cell size, and other critical factors, ensuring that the cell cycle proceeds accurately. Dysregulation of these checkpoints can lead to cell cycle abnormalities and diseases, such as cancer.

The chapter proceeds with an in-depth discussion of meiosis (4.4), a specialized form of cell division dedicated to sexual reproduction. Meiosis involves two rounds of division, resulting in the generation of haploid gametes with unique genetic combinations. The process of crossing over and genetic recombination contributes to genetic diversity in sexually reproducing species.

Finally, the chapter highlights the importance of cell Division in growth, development, and repair (4.5). Cell division is fundamental to the growth and development of multicellular organisms, as it allows for tissue expansion, renewal, and maintenance. Additionally, cell division plays a crucial role in tissue repair and regeneration after injuries, promoting the healing process.

Throughout the chapter, the intricate mechanisms of cell cycle regulation and division are emphasized, with a focus on their significance in various biological processes. The understanding of these processes has profound implications for medical research, particularly in cancer treatment and regenerative medicine. By unraveling the complexities of the cell cycle and cell division, researchers gain crucial insights into cell biology, contributing to advancements in various fields of science and medicine. The fifth chapter will expand some of these lessons into the wider topic of genetics and inheritance.

Chapter

5 GENETICS AND INHERITANCE

CONTENTS

In this pivotal chapter, we delve into the captivating realm of Genetics and Inheritance, uncovering the fundamental principles that underpin the inheritance and transmission of genetic traits across generations. Within these pages, we embark on a journey through the intricate molecular mechanisms that govern life's blueprint – the DNA structure and the genetic code it encapsulates.

Section 5.1 unveils the enigmatic DNA structure, an intricate double-helix configuration that safeguards the hereditary information within living organisms. This section delves into the genetic code, deciphering the language written in nucleotide sequences that determine an organism's unique traits and characteristics.

Moving forward, in Section 5.2, we explore DNA replication – a remarkable process that ensures the perpetuation of genetic information with remarkable fidelity. Unraveling the intricacies of this fundamental biological process, we examine how cells meticulously copy their genetic material during cell division, safeguarding the continuity of life itself.

Transitioning from the realm of DNA, Section 5.3 sheds light on the intricacies of transcription and translation – the remarkable transformations that bridge the gap between DNA and proteins. Here, we embark on a captivating journey that leads us from the genetic code's transcription into RNA molecules, ultimately culminating in the complex process of translation, wherein proteins are synthesized based on the genetic instructions encoded in RNA.

In Section 5.4, the focus shifts to Mendelian Genetics – a pivotal realm that expounds the timeless laws of inheritance discovered by Gregor Mendel. We delve into the principles of dominant and recessive traits, examining how alleles and genetic variation contribute to the diversity of traits observed in living organisms.

Finally, Section 5.5 brings to the fore the intriguing world of chromosomal aberrations and genetic disorders. Here, we uncover the intricate interplay of genetics and disease, exploring the various mutations and chromosomal irregularities that give rise to genetic disorders and their far-reaching implications for human health.

With a comprehensive exploration of these five interrelated sections, this chapter endeavors to equip readers with a profound understanding of Genetics and Inheritance. As we delve into the wonders of the genetic world, we discover the threads that weave through the fabric of life, providing us with unparalleled insight into the fascinating complexities of heredity and genetic inheritance. Let us begin this chapter by exploring DNA structure and genetic code.

5.1 DNA STRUCTURE: THE GENETIC CODE

At the heart of the fascinating world of cell biology lies the fundamental cornerstone of life – DNA (Deoxyribonucleic Acid). The elucidation of DNA's intricate structure and the deciphering of its genetic code have revolutionized our understanding of biology, genetics, and inheritance (Featherston & Cotruvo, 2021; Ferrari et al., 2022; Ferrari et al., 2023). The transformative revolution in the field of biology, genetics, and inheritance has primarily emerged due to the elucidation of deoxyribonucleic acid's elaborate structure and the successful deciphering of its genetic code. To delve into this epoch-making event, a detailed case study of cell microbiology serves as the bedrock. An analysis of the seminal work done by Dr. Rosalind Franklin in the early 1950s in X-ray crystallography offers a vivid demonstration of these ground-breaking strides in cell biology.

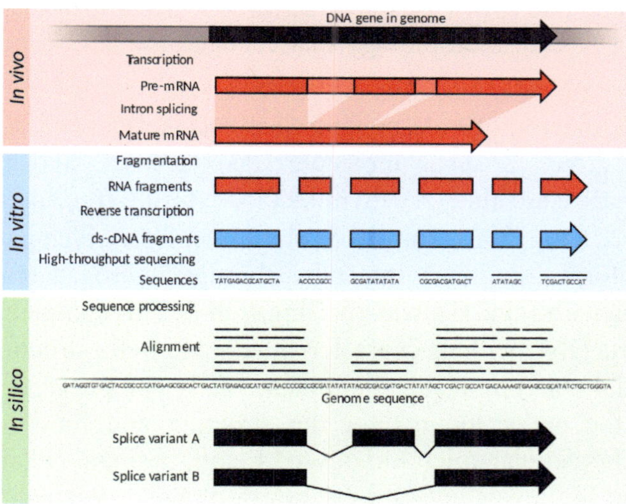

Figure 41. Illustration of RNA sequencing.

Source: Wikipedia (https://en.wikipedia.org/wiki/RNA-Seq)

Franklin's iconic Photo 51, an X-ray diffraction image of a DNA molecule, rendered visible the molecule's double-helix structure for the first time. This optical milestone formed the cornerstone of further inquiries into the physical representation of this molecule, which would drill down to the atomic level, splitting open the compact nature of DNA's hereditary material (Hummert et al., 2021; Husnik et al., 2021; Hudson & Wieland, 2023). The double-helix structure of DNA, as an embodiment of two antiparallel strands spiraling around a shared longitudinal axis with an external sugar-phosphate backbone and internalized pairs of adenine-thymine and guanine-cytosine, offered the platform from which the human understanding of genetics grew (see figure 41). The elucidation of this structure also simplified the interpretation of DNA's replication process, thereby emphasizing its role in creating progeny cells with the same genetic information. In parallel, the decoding of the genetic code formed the gateway to understanding gene expression and regulation. As a tertiary structure composed of nucleotide sequences, DNA's coding sequences, or exons, produce RNA sequences in a process termed transcription. The RNA sequences then supplied the template for the synthesis of proteins, commonly referred to as translation. This uniform genetic code, with specific triplets dictating specific amino acids, thus established the central dogma of molecular biology—DNA to RNA to protein. Consequently, the correlation between the DNA structure and the genetic code showcases how these two facets fostered our holistic understanding of cell microbiology that changed the landscape of biology and genetics.

DNA is a macromolecule composed of nucleotides, each comprising a phosphate group, a sugar molecule (deoxyribose), and one of four nitrogenous bases: adenine (A), thymine (T), cytosine (C), and guanine (G) (Charpentier & King, 2021; Carroll & Marqusee, 2022; Chang et al., 2023). These bases form complementary pairs through hydrogen bonds, with A always pairing with T and C always pairing with G. This complementary base pairing imparts DNA with its characteristic double-helix structure. DNA, as a quintessential macromolecule, is fundamentally composed of nucleotides that anchor the underpinnings for the structure and function of living organisms. The cellular microbiology of E.coli, a single-celled bacterium that resides in the lower intestines of warm-blooded organisms, serves as an instructive case study to depict the intricate architecture of DNA. The inherent simplicity of its genome harbors a rife template for the illustration of DNA's structure and the corresponding roles of its constituent building blocks. Each nucleotide in the DNA macromolecule comprises a phosphate group, a deoxyribose sugar molecule, and one of the four nitrogenous bases—adenine (A), thymine (T), cytosine (C), or guanine (G)—these three components are arranged along a line, from phosphate to deoxyribose to base. The combined segments of deoxyribose and phosphate present the "backbone" structure of the DNA molecule's double helix, with the bases extending inward. The chemical bonding rules ensure that adenine pairs with thymine through two hydrogen bonds, whilst guanine pairs with cytosine via three hydrogen bonds, solidifying the bases' strategic positions and their respective partner molecules. Consider, for example, the LacZ gene in E.coli that codes for an enzyme essential in lactose metabolism. The codons in this gene that designate specific amino acids in the protein consist of sequences of the aforementioned base pairs, demonstrating the critical role that the formation and selection of the A-T and C-G base pairs play in directing protein synthesis. Consequently, the detailed understanding of its DNA at the nucleotide level has not only led to the understanding of its behavior and mutations but also offered invaluable insight into eukaryotic and other prokaryotic DNA structures and functions, underscoring the constitutive and biological significance of the components of DNA nucleotides.

The DNA double helix is like a twisted ladder, with the sugar-phosphate backbone forming the sides and the paired nitrogenous bases forming the rungs (X. Wang & X. W. Chen, 2023; X. C. Wang et al., 2021; Y. F. Wang & H. R. Chen, 2023). The helix has a right-handed twist, and the base pairs are held together by hydrogen bonds, which ensure the stability of the molecule while allowing for the unwinding and replication processes. In the realm of cell microbiology, describing the DNA structure as a twisted ladder provides a comprehensible model for the understanding of its components and interaction mechanisms. This metaphorical depiction of deoxyribonucleic acid finds vindication in the groundbreaking research performed on the prokaryote Escherichia coli, renowned for its simplicity and straightforward genome. The DNA double helix of E. coli is a microscopic manifestation of this 'ladder', with the sugar-phosphate backbone, composed of alternating deoxyribose sugar and phosphate groups, coiling to constitute the 'sides' of the ladder. This stable backbone architecture owes its stability to the negatively charged phosphate groups that confer resilience and integrity to the covalent bonds between the sugar and phosphate molecules. Ascending vertically, interconnected by the 'sides', lie the 'rungs', composed of paired nitrogenous bases; adenine binds to thymine, and guanine couples with cytosine, thereby establishing an additional layer of stability via the hydrogen bonds that section the DNA molecule into consistent, predictable pairs. These base pairs permit the smooth relay of genetic information during replication, ensuring the orchestrated transfer of genetic material from the parent to the child cells. An example is the E. coli's lac operon, a sequence of genes enabling lactose metabolism. Its smooth functioning is mediated by precise interactions of the nitrogenous base pairs, solidifying the comprehension of the DNA's twisted ladder model. Therefore, E.coli's apparent simplicity masks a world of biochemical complexity that elucidates DNA workings, bringing into the scientific lexicon the visualization of DNA as a double helix resembling a twisted ladder.

The linear sequence of nucleotides along the DNA molecule holds the key to the genetic code (Pendharkar et al., 2021; Perera, 2021; Peng et al., 2022). This code is written in the order of the nitrogenous bases, with each set of three bases, called a codon, corresponding to a specific amino acid. Amino acids are the building blocks of proteins, and proteins play crucial roles in nearly all cellular processes, making them essential for the functioning of living organisms. The recognition of the DNA molecule's linear nucleotide sequence as the vault to the genetic code revolutionized our grasp on molecular biology and subsequently, the understanding of life itself. A detailed inspection of the Severe Acute Respiratory Syndrome (SARS) virus provides an exemplar for the exploration of this concept. As a single-stranded RNA virus, SARS utilizes the host replication machinery, translating its RNA genome into functional and structural proteins requisite for its replication.

The crucial aspect here remains the virus's RNA sequence, effectively the mirror image of a DNA sequence, that directly dictates the type and order of proteins synthesized. The sequence, delineated as linear, reflects the order in which nucleotides are arranged along the virus's genome. A single nucleotide comprises a phosphate group, ribose sugar, and one of four nitrogenous bases: adenine, uracil, cytosine, or guanine. Each trio of these bases, termed codons, acts as signals for specific amino acids during protein synthesis. As an example, within the SARS RNA sequence, the codon GAU codes for aspartic acid, while CCA codes for proline, integral for the structure and function of the SARS spike protein. Thus, the virus's structural functionality and pathological potency are inherently linked to the specific linear sequence of nucleotides within its RNA. By elucidating and manipulating these sequences, molecular biologists were able to gain ample insights into the virus's biology, facilitate vaccine development, and enhance therapeutic strategies, thus showcasing the crucial role played by the linear sequence in genetic coding.

The genetic code is degenerate, meaning that multiple codons can code for the same amino acid (see figure 42). For example, the amino acid phenylalanine can be specified by two different codons: UUU and UUC.

However, the code is not ambiguous, as each codon only codes for one specific amino acid. Moreover, the genetic code is nearly universal, with a few minor exceptions across different organisms.

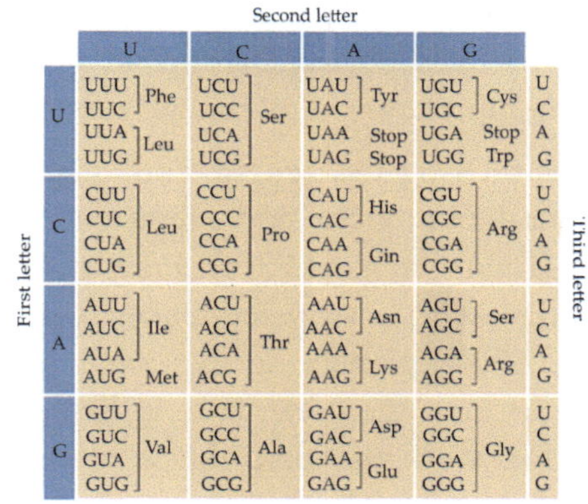

Figure 42. Illustration of how the genetic code is degenerate.

Source: By NIH - http://history.nih.gov/exhibits/nirenberg/HS5_cracked.htm, Public Domain, https://commons.wikimedia.org/w/index.php?curid=33555245

The genetic code's degenerate nature signifies that multiple codons can encode the same amino acid, a function underlining the redundancy and robustness that characterize the genetic makeup of living organisms. This critical aspect in the field of molecular biology is demonstrably seen through the case study of Saccharomyces cerevisiae, commonly known as baker's yeast.

This unicellular eukaryote possesses about 6000 genes, offering a comprehensive platform for investigating the degeneracy of the genetic code. Precisely, each protein-coding gene in S. cerevisiae consists of a unique sequence of codons, where a codon represents a trio of nucleotides. However, despite the myriad of unique codon sequences, only 20 amino acids form proteins, encapsulating the essence of the genetic code's degeneracy. For instance, the amino acid Leucine is coded by six different codons—UUA, UUG, CUU, CUC, CUA, and CUG. The critical question arises as to why nature has adopted such a seemingly redundant code.

One element of the answer lies in the buffering of the effects of mutations. The degeneracy of the genetic code means that a single mutation in a DNA sequence does not necessarily lead to a change in the amino acid sequence of a protein, consequentially minimizing potential deleterious effects. Additionally, multiple codons for a single amino acid may also factor into the regulation of protein expression, with certain codons preferentially selected for optimal function, demonstrating the nuanced advantage of codon degeneracy. Hence, decoding the intricacies of the this organism's genome has refined our understanding of degeneracy in genetic coding, thereby enriching the broader comprehension of genetic robustness and complexity.

The process of deciphering the genetic code began with the landmark experiments of Marshall Nirenberg and Heinrich Matthaei in the early 1960s (Li et al., 2023Liang et al., 2021; Liang et al., 2022). They used synthetic RNA molecules with known sequences to decipher which amino acids they coded for. These pioneering experiments laid the foundation for our understanding of how the genetic code is read and translated into proteins. The pioneering work that laid the foundation for decoding the genetic code can be attributed to the instrumental experiments of Marshall Nirenberg and Heinrich Matthaei in the early 1960s. These experiments set the stage for understanding the process by which cells manufacture proteins from the information etched into the DNA's structure. Employing a comprehensive case study of cell biology, the proposition to unravel the genetic language hinged on decoding the cryptic sequences of nucleotides in DNA to their corresponding sequences of amino acids in proteins. The bacterium Escherichia coli served as their experimental model, providing the bacterial lysate to enforce the process of protein synthesis. Nirenberg and Matthaei were the first to decipher a specific genetic codon by synthesizing an artificial piece of RNA in a laboratory containing only uracil nucleotides. They observed that when this newly synthesized RNA was immersed into the bacterial lysate, it produced a peptide consisting solely of phenylalanine amino acid. This seminal observation illuminated that the RNA triplet (UUU) coded for the amino acid phenylalanine, thus deciphering the first letter of the genetic code. This experiment ultimately established that the language of DNA is written in a linear manner, not in a two-dimensional or three-dimensional space. The discovery laid a path for future scientists to attract the remaining 61 codons to their precise amino acids, thereby revealing that genetic instructions are stored within DNA. Thus, the process that started with the precipitous observations of Nirenberg and Matthaei underwent radical advancements, forging a deeper understanding of how genetic information is translated into functional proteins.

DNA's genetic code is not only responsible for protein synthesis but also plays a crucial role in regulating gene expression (Quintero-Carmona, 2021; Radulovic et al., 2022; Rao et al., 2022). Various sequences within

DNA, such as promoters and enhancers, interact with specific proteins to control the initiation and regulation of transcription. This elaborate regulatory network ensures that genes are expressed at the right time and in the right cell type, thereby orchestrating the complex processes of growth, development, and adaptation in multicellular organisms. The eminent role of DNA's genetic code transcends the simplistic conception of being solely responsible for protein synthesis, for the code also plays a decisive role in orchestrating gene expression. The case study of cell biology involving the regulation of lactose metabolism in Escherichia coli - widely regarded in scientific discourse as the lactose or lac operon - provides a quintessential example of this dual role. The lac operon, a set of genes that includes lacZ, lacY, and lacA, gets turned 'on' only in the presence of lactose and in the absence of glucose.

This metabolic adaptation is made feasible through the selective expression of genes based on the prevailing biochemical milieu within the cells, testifying to the influence of DNA on gene regulation. The lac operon's activation initiates with the absorption of lactose into E.coli, which induces the transformation of some lactose molecules into allolactose. Allolactose then binds to the lac repressor, a protein that inhibits the transcription by blocking RNA polymerase's access to the promoter site. Upon binding to allolactose, the repressor protein changes shape and releases from the DNA, thus facilitating transcription. In essence, the capacity of E. coli to turn "on" and "off" the genes for lactose metabolism, relies not merely on the genetic code's instructions for protein synthesis. Instead, it requires the application of the genetic code as a regulator of gene expression. This instance encapsulates the definitive role of the genetic code in coordinating genomes and delivering organized responses to environmental changes, thereby aligning the activities of the cell with its environment. Consequently, the intimate link between the genetic code and gene regulation underscores the elegant yet fundamental versatility of the DNA molecule.

In summary, DNA's remarkable structure and the intricate genetic code it harbors form the backbone of cellular life. Understanding the genetic code has far-reaching implications, from medical applications such as gene therapy to evolutionary studies that shed light on the diversity of life. As our knowledge of DNA structure and the genetic code continues to expand, so does our appreciation of the wonders and complexities of the microscopic world that governs all living beings. In the next section, we consider how DNA replication is essential for ensuring genetic continuity.

5.2 DNA REPLICATION: ENSURING GENETIC CONTINUITY

DNA replication is a pivotal process in cell biology that ensures the faithful transmission of genetic information from one generation to the next (Kim et al., 2023; Klingemann, 2023; Laiman et al., 2023). This fundamental process is vital for maintaining the integrity and continuity of life's blueprint, as errors during replication can lead to genetic mutations and potential disruptions in cellular function. The impetus given to DNA replication as a critical process in cell microbiology stems from its role to ensure the faithful propagation of genetic information from one cellular generation to the next. Delving into the specific case of the bacterium Escherichia coli deepens our understanding of this process. The circular DNA of E. coli offers an uncomplicated and streamlined framework to study the precise mechanisms of DNA replication. Integral to this process is the DNA polymerase enzyme, which catalyzes the synthesis of new DNA by adding nucleotides to a preexisting DNA strand. The initiation site, termed as the origin of replication or ori, marks the beginning of the replication process where DNA strands are locally unwound. This unwinding forms a structure called a replication fork, which resembles a Y-shape and facilitates the antiparallel replication of DNA. On the leading strand, DNA polymerase III synthesizes a new complementary DNA strand continuously, while on the lagging strand, it generates a new strand in a discontinuous and piecemeal fashion known as Okazaki fragments due to the 5' to 3' direction of DNA synthesis. DNA polymerase I then replaces the RNA primers used to initiate replication in the lagging strand with DNA, and the ligase enzyme connects the fragments into a continuous chain. Thus, E. coli genome is replicated exactly before cell division to ensure the equal partitioning of a full complement of genes into each of the two daughter cells. The meticulous nature of this process underscores its criticality in preserving the genetic integrity and biological continuity across generations, imposing DNA replication as an essential and fundamental cellular process in all forms of life.

The process of DNA replication (see figure 43) occurs during the cell cycle in the synthesis (S) phase, wherein the cell duplicates its genetic material in preparation for cell division (Ricci & Srivastava, 2021; Reiche et al., 2022; Reuschl et al., 2022). The replication process is semiconservative, meaning that each newly synthesized DNA molecule consists of one original (parental) strand and one newly synthesized (daughter) strand. The synthesis (S) phase is when a cell duplicates its genetic material, known as DNA replication, gearing up for cell division - a process underscored through an examination of the human keratinocyte, a predominant cell type in the epidermis. These cells undergo a well-delineated cycle, including a distinct S phase, pivotal for the maintenance and renewal of the skin barrier function.

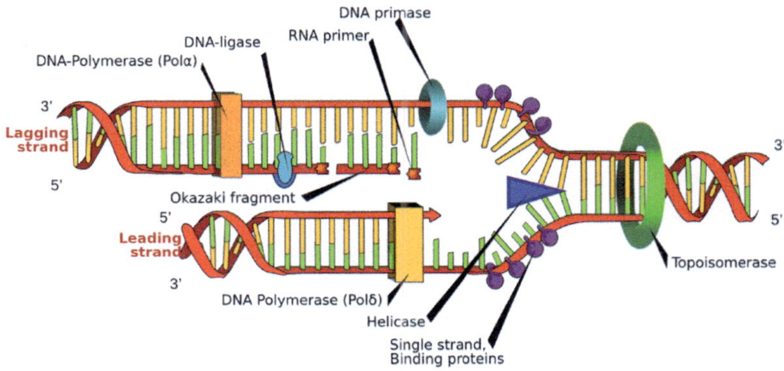

Figure 43. The process of DNA replication.

Source: Wikimedia Commons (https://commons.wikimedia.org/wiki/File:DNA_replication_en.svg)

The initiation of the S phase marks the period of DNA replication, the process that ensures each daughter cell receives a complete set of genetic instructions. This phase's commencement calls into play a consortium of enzymes that collectively undertake the job of unzipping the DNA double helix, with the enzyme DNA helicase breaking the hydrogen bonds between the base pairs. Concurrently, single-strand binding proteins attach to the separated strands, preventing them from rejoining and preserving the 'replication fork,' where new nucleotides are added. DNA polymerase extends the primer, adding nucleotides in the 5' to 3' direction, assisted by the enzyme DNA primase, which synthesizes short RNA sequences that act as templates for DNA polymerase. It is noteworthy that replication is bi-directional, with a leading strand synthesized continuously toward the replication fork while the lagging strand undergoes a discontinuous form of replication away from the replication fork. Once the replication fork has traversed the entire length of the DNA, two identical copies of the genetic material are ready, which, in keratinocytes, is finally used to produce new layers of skin cells. The preciseness of this process, instrumental in preserving genetic constancy, testifies to the DNA replication's magnificent role during the S phase of the cell cycle.

The replication machinery is a complex ensemble of enzymes and proteins that work together with remarkable precision (Li et al., 2023Liang et al., 2021; Liang et al., 2022;). It begins with the unwinding of the DNA double helix, catalyzed by helicase enzymes. This creates two single-stranded DNA templates that serve as templates for the synthesis of new strands. Unfurling the intricacies of DNA replication necessitates an in-depth understanding of the replication machinery, a formidable ensemble of enzymes and proteins that cooperate to ensure exceptional precision during replication. To illustrate this, an excellent case study is the bacterium Escherichia coli, a widely known prokaryotic model in cell microbiology studies. The primary replicate enzyme at the heart of this DNA replication process, DNA polymerase III, is a dimeric enzyme complex that synthesizes both the leading and lagging strands simultaneously at the replication fork. However, the precise replication of the parental DNA strand involves not just DNA polymerase III but also a retinue of other proteins and enzymes. For instance, DNA helicase unwinds the parental double helix at the replication fork, producing two template strands ready for replication. Single-strand DNA-binding proteins then attach to the unpaired DNA strands, preventing them from reannealing and keeping them ready for copying. The primase enzyme synthesizes a short RNA sequence, the primer, that is complementary to the DNA template, thus serving as the starting point for DNA polymerase III action. Topoisomerase prevents the over-winding of DNA ahead of the replication fork, whereas ligase catalyzes the formation of phosphodiester bonds between Okazaki fragments in the lagging strand, ensuring that the new strand is continuous. Together, these enzymes coordinate to accomplish the penultimate goal: the faithful duplication of the DNA sequence. Thus, the replication machinery of E.coli demonstrates the remarkable concert of multiple proteins and enzymes working together with unparalleled precision to sustain life's continuity and evolution through accurate DNA replication.

At each single-stranded template, an enzyme called DNA polymerase binds to initiate the replication process (Han et al., 2023; Haraoka et al., 2023; Hartmann & Mayor, 2023). DNA polymerases are responsible for adding nucleotides to the growing daughter strands, following the rules of complementary base pairing. The original DNA strand's 3'-OH end serves as the starting point for DNA synthesis, while DNA polymerases add nucleotides to the new strand in the 5'-to-3' direction, creating a continuous daughter strand (leading strand). The initiation of the DNA replication process hinges on the action of an enzyme called DNA polymerase that espouses each single-stranded template, which is astutely exemplified in the study of Escherichia coli bacterium. In the context of E. coli, DNA polymerase III is the primary enzyme involved in DNA replication, which operates in conjunction with a series of proteins and enzymes to ensure the accurate relay of genetic information. Prior to DNA polymerase III attachment, the double-stranded DNA molecule undergoes unwinding, mediated by DNA helicase. This results in the formation of a replication fork comprising two single-stranded templates.

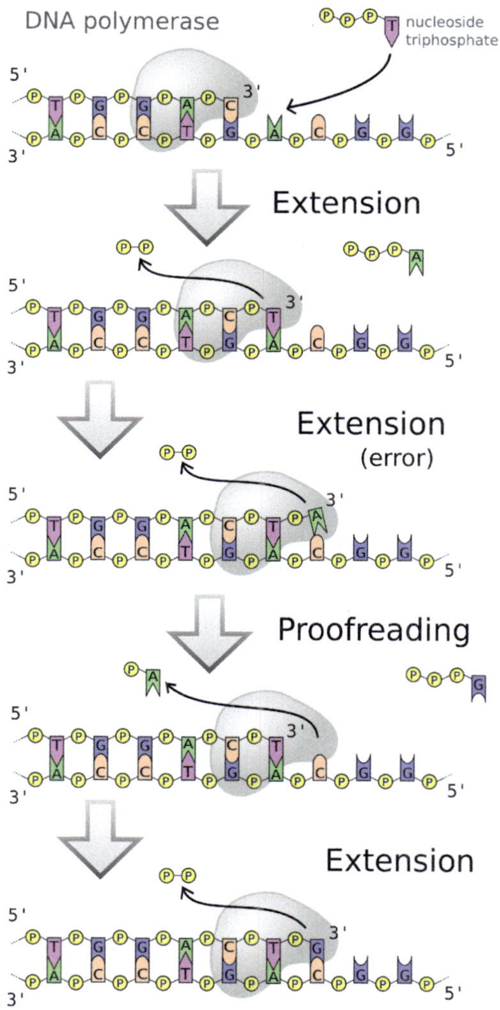

Figure 44. Illustration of DNA polymerase.

Source: By I, Madprime, CC BY-SA 3.0, https://commons.wikimedia.org/w/index.php?curid=2527732

Notably, DNA cannot initiate the replication process without a primer. DNA primase synthesizes a short ribonucleic acid (RNA) primer complementary to the DNA strand, providing a location for DNA polymerase III to anchor. Binding to the primed location, DNA polymerase III begins to synthesize a new DNA strand along the template, adding complimentary nucleotides in the direction of 5' to 3'. For the leading strand, replication is continuous, but for the lagging strand, formed discontinuously as Okazaki fragments, the process requires multiple RNA primers. Eventually, these RNA primers are replaced by DNA nucleotides by the action of DNA polymerase I, and DNA ligase then creates phosphodiester bonds to seal any gaps, thereby ensuring a continuous DNA strand. The subsequent release of DNA polymerase III denotes the termination of DNA replication. Thus, through an appreciation of DNA replication, the significant role of DNA polymerase in initiating the replication process finds profound resonance, thereby revealing the inherent plasticity and precision that governs the transfer of life's blueprints across generations. Figure 44 is an illustration of DNA polymerase.

The other single-stranded template cannot be synthesized continuously due to its 5'-to-3' orientation (Biermann & Reya, 2022; Bohnert & Johnson, 2022; Bomont, 2021). Instead, it is synthesized in short fragments called Okazaki fragments. RNA primers are initially laid down on the lagging strand by primase, and then DNA polymerases extend these primers with DNA nucleotides. The resulting fragments are later connected by another enzyme called DNA ligase, forming a complete daughter strand. The unidirectionality of DNA replication, which occurs exclusively in the 5'-to-3' direction, imposes intrinsic challenges when replicating the other template strand, or the lagging strand, comprehensively illustrated in the cellular replication of the bacterium Escherichia coli. During the replication process, DNA helicase unwinds the parent double-stranded DNA, resulting in a 'Y-shaped' region known as the replication fork. However, the antiparallel nature of DNA strands means that while one strand, termed the leading strand, aligns in the 3'-to-5' direction facing the replication fork and is synthesized continuously towards the fork by DNA polymerase III, the lagging strand, oriented in the 5'-to-3' direction, cannot be synthesized continuously. To circumnavigate this orientation problem, the replication machinery synthesizes the lagging strand discontinuously, in short segments termed Okazaki fragments. These fragments are synthesized in the direction away from the replication fork, initiated by an RNA primer that is laid down by primase. Following the synthesis of each fragment, DNA polymerase I replaces the RNA primers with DNA, while DNA ligase connects the Okazaki fragments, ensuring the completion of a continuous complementary strand. Thus, despite its 5'-to-3' orientation imposing an initial hurdle, the lagging strand's replication exhibits the exemplification of biological ingenuity in maintaining the fidelity of genetic information transfer.

The replication process is highly coordinated to ensure accuracy and minimize errors (Riley et al., 2021; Robaszkiewicz et al., 2021; Roeder et al., 2022). DNA polymerases have proofreading capabilities, allowing them to detect and correct errors that may occur during replication. Additionally, various repair mechanisms are in place to fix any mistakes that escape the polymerase's proofreading activity. The process of replication is a critically coordinated event that prioritizes accuracy and minimizes errors, which is aptly elucidated through an in-depth study of the yeast Saccharomyces cerevisiae. In this organism, DNA replication commences at multiple points along the genome, called origins of replication, thus optimizing the time taken for the complete copying of genetic material. To maintain the orderliness of this process, various proteins and complexes are employed. At the outset, the Origin Recognition Complex (ORC) identifies and binds to the origins, facilitating the assembly of additional proteins. The subsequent binding of cell division cycle 6 (Cdc6) and Cdt1 proteins facilitates the loading of the Mini-Chromosome Maintenance (MCM) complex, a helicase that unwinds the double-stranded DNA. This pre-replication complex is then activated by additional proteins, leading to the separation of DNA strands and the enlistment of DNA polymerases. Notably, the process has inherent safeguards to ensure accuracy. DNA polymerase has a proofreading activity, allowing it to recognize and correct errors during replication. The DNA mismatch repair system further minimizes errors by fixing any mispaired bases that escaped the notice of DNA polymerase. Moreover, repeated replication from the same origin within the same cell cycle is prevented by geminin, a protein that inhibits the re-loading of MCM helicase. Therefore, the mechanisms in S. cerevisiae exemplify the high degree of coordination in the replication process that safeguards genetic fidelity, reinforcing DNA's role as the stable repository of genetic information.

Ensuring the fidelity of DNA replication is of utmost importance, as errors can lead to genetic mutations that have profound consequences for cellular function and the organism as a whole (Skoniecka et al., 2021; Smith et al., 2022; Smith et al., 2023). Mutations can alter protein structure and function, affecting essential cellular processes such as metabolism, growth, and response to the environment. While some mutations may be harmless, others can be detrimental, leading to genetic disorders and diseases. Maintaining the fidelity of DNA replication is a cardinal principle in cell biology, as inaccuracies can engender genetic mutations with far-reaching effects on cellular function and potentially severe repercussions for the host organism. A noteworthy example of this is evident in the study of cancer cells, particularly the case of malignant melanoma, a highly aggressive form of skin cancer often linked to DNA damage induced by ultraviolet (UV) radiation. The details surrounding this condition offer an informative demonstration of the detrimental impact of DNA replication errors. DNA replication is a highly accurate process involving multiple cellular machinery working in unison to ensure that the genetic material is precisely copied. However, in the presence of UV radiation, there can be the formation of cyclobutane pyrimidine dimers (CPDs), a photoproduct that causes distortion in the DNA double helix, leading to errors during DNA replication. If the DNA repair machinery, such as the Nucleotide Excision Repair pathway, does not rectify these lesions, the replication machinery may misread the CPDs, leading to the incorporation of incorrect nucleotides and thus causing mutations. In melanoma, a common mutation caused by UV radiation is the C to T transition mutation noticed primarily in the BRAF gene, resulting in a valine to glutamic acid change at position 600 (V600E). This mutation triggers uncontrolled cellular proliferation leading to the development of cancer. Hence, the replication errors in melanoma cells underscore the urgency of maintaining accurate DNA replication, the failure of which can escalate into serious pathological conditions.

Moreover, DNA replication is not without challenges (Tolkin & Hubbard, 2021; Tomoda & Kime, 2021; Tuieng et al., 2021). The DNA double helix is remarkably long, and the replication process is rapid, occurring at rates of up to 1,000 nucleotides per second in human cells. As such, the replication machinery must work efficiently to complete the process within a limited time frame. Indeed, the process of DNA replication, though finely orchestrated, encounters numerous obstacles that challenge the fluidity of its operation. Understanding these challenges demands close scrutiny of a detailed case study in cell microbiology. The bacteriophage T4, a virus that infests E. coli, is worth examining due to the complex nature of its DNA structure that iterates the challenges of replication. For instance, unlike the linear structure displayed by eukaryotic DNA, T4 DNA comprises long, concatemeric strings, which consist of numerous genome replicas linking head to tail. Such a configuration significantly complicates the replication process, given the inability to use a single replication fork due to both topological constraints and the risk of structurally compromising the DNA. Consequently, T4 has had to evolve a rather innovative replication mechanism, involving intricate recombinational processes. The T4 virus overcomes the challenge by forming a loop structure, causing the replication to occur from both end-points of the loop towards the junction in between, eventually leading to two complete genome copies. This mode of replication, known as the rolling circle mechanism, offers a potent workaround to the obstacle. Interestingly, T4's replication mechanism also underlines its fight against fewer replication errors by utilizing three different DNA polymerases, each designated to a specific task, thereby reducing mutagenesis. Hence, the replication process in bacteriophage T4 serves as a compelling example of how replication utilizes intricate molecular mechanisms to address and overcome its inherent challenges.

Overall, DNA replication is a remarkable process that ensures the continuity of genetic information across generations. It showcases the intricate precision of cellular machinery and the extraordinary measures taken to safeguard the integrity of the genetic code. Understanding the intricacies of DNA replication is essential for uncovering the mechanisms underlying genetic stability and the fascinating dynamics of cell biology. The next section analyzes transcription and translation from DNA to protein.

5.3 TRANSCRIPTION AND TRANSLATION: FROM DNA TO PROTEIN

The processes of transcription and translation play crucial roles in cell biology, serving as the molecular bridge that enables the transfer of genetic information from DNA to functional proteins (X. Wang & X. W. Chen, 2023; X. C. Wang et al., 2021; Y. F. Wang & H. R. Chen, 2023). These intricate mechanisms orchestrate the synthesis of proteins – the workhorses of the cell responsible for carrying out various essential functions. The transition of genetic information from DNA to functional proteins necessitates two pivotal processes in cell microbiology—transcription and translation. Their importance can be illuminated through the scrutiny of a detailed case study involving the bacterium Escherichia coli. For instance, the β-galactosidase enzyme, encoded by the lacZ gene in E. coli, is a crucial catalyst in the metabolism of lactose sugars. The gene-to-enzyme process testifies to the indispensability of transcription and translation. During transcription, one DNA strand serves as a template for the synthesis of a single-stranded messenger RNA (mRNA) molecule via RNA polymerase.

This event commences with the binding of RNA polymerase to a promoter region on DNA, with the enzyme then assembling a corresponding RNA molecule. After transcription, the resultant mRNA harbors the genetic code necessary for the synthesis of β-galactosidase. The mRNA then migrates from the nucleus to the bacterial cytosol, where translation -- the creation of proteins -- occurs. During translation, the mRNA's nucleotide sequence, read as triplets (codons), dictates the type and sequence of amino acids in the polypeptide. Consequently, each codon aligns with a specific amino acid, transcribed through the cellular machinery's synthesis complex named ribosome, along with transfer RNA (tRNA). Ultimately, a sequence of bonded amino acids takes shape as β-galactosidase protein, now ready to catalyze the hydrolysis of lactose sugars. Thus, the intricacies of the lactose metabolism, appreciably underscore the role of transcription and translation as pivotal junctures in the seamless transformation of gene-level information into functional proteins in an Escherichia coli cell.

Transcription is the first step in this remarkable journey. It involves the conversion of the genetic information encoded in DNA into a complementary RNA molecule (Shen et al., 2021; Siu et al., 2021; Shi et al., 2022; Shoup & Priola, 2023). The process initiates when an enzyme called RNA polymerase binds to a specific region of the DNA known as the promoter. The promoter region marks the starting point for transcription and helps regulate gene expression by controlling when and how often a gene is transcribed. The salient process of transcription constitutes the first, crucial step in the journey of genetic expression, which entails the conversion of genetic information encoded in DNA into a corresponding RNA molecule. With the lens of cell microbiology focused on the bacterium Escherichia coli, the exploration of this process is more plausible. Distinct genes in the the genome code for disparate proteins instrumental for the bacteria's survival and multiplication. An exemplary instance involves the lac operon, an operon responsible for the metabolism of lactose within the bacterium. Notably, when lactose is available and glucose (a preferred carbon source for the bacterium) is scarce, the transcription of the lac operon initiates. The molecular mechanics of transcription materialize as the RNA polymerase, an enzyme responsible for synthesizing RNA, binds to a specific DNA site known as the promoter. Subsequent to binding, the DNA double helix unwinds, allowing the RNA polymerase to read the DNA template strand in the 3' to 5' direction. The enzyme then catalyzes the assembly of nucleotide triphosphates on a growing RNA molecule, pursuing the principle of base complementarity and hence transcribing the information encoded in the DNA into an RNA molecule. For example, within the lac operon, the transcription of the lacZ gene leads to the production of an mRNA molecule, which is then transported to the ribosomes for the next stage of genetic expression: translation. Therefore, through the meticulous inspection of transcription in the E.coli model system, the complex yet efficient nature of this vital molecular process garners due acknowledgement, reinforcing the appreciation of transcription's role as the foundational step in genetic information flow.

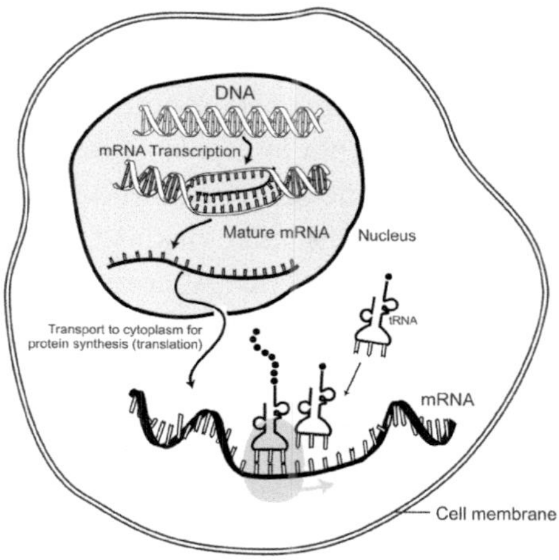

Figure 45. The nature of mRNA molecules.

Source: By National Human Genome Research Institute, NIH - mRNA illustration, NIH, Public Domain, https://commons.wikimedia.org/w/index.php?curid=126381443

Once bound to the promoter, RNA polymerase unwinds the DNA double helix, exposing the template strand (Sagwal & Bekeschus, 2021; Ryder et al., 2022; Sakamaki & Mizushima, 2023). This strand serves as a template for synthesizing a complementary RNA molecule. As RNA polymerase moves along the DNA template, it adds ribonucleotides (adenine, uracil, cytosine, and guanine) to the growing RNA strand, following the rules of complementary base pairing. However, in RNA, uracil (U) replaces thymine (T) as a complementary partner to adenine (A). The mechanism by which RNA polymerase binds to the DNA's promoter region and unwinds the double helix to expose the template strand is instrumentally illustrated through the beta-galactosidase (lacZ) gene of the Escherichia coli. The lacZ gene, a fundamental part of the lac operon system, contributes to lactose metabolism, elucidating RNA polymerase's indispensability in transcription. When the bacterium encounters lactose, a regulatory protein, the lac repressor, dissociates from the operator segment of the lac operon, freeing the promoter region for RNA polymerase binding. Once adhered to the promoter, RNA polymerase instigates the unwinding of the DNA double helix, initiating the segregation of the two DNA strands over a short stretch. This unwinding event engenders a transcription bubble, exposing one of the DNA strands, the template strand. Following strand separation, the exposed template strand aligns with the active site of RNA polymerase, primed to direct the synthesis of a corresponding mRNA strand. RNA polymerase advances along the template, synthesizing a complementary mRNA molecule by incorporating matching RNA nucleotides. This faithfully transcribes the genetic instructions encoded in the lacZ gene into the mRNA. Essential for the subsequent translation into beta-galactosidase, this protein thus ensures its survival in lactose-rich environments. Accordingly, the compelling case of the lac operon lays bare RNA polymerase's role in unwinding the DNA helix, revealing the methodological precision that underpins this gene expression's inaugural phase.

Transcription continues until RNA polymerase reaches a termination sequence on the DNA, signaling the end of the transcription process (Lewis & Wells, 2021; D. Li et al., 2021; Lerit, 2022). The newly synthesized RNA molecule, called messenger RNA (mRNA), carries a copy of the genetic information from the DNA to the ribosomes, the cellular machinery responsible for protein synthesis. Transcription, the process by which the DNA sequence is copied to a complementary RNA strand, is halted when the RNA polymerase encounters a termination sequence on the DNA. This termination sequence, often a specific collection of nucleotides, serves as a stop signal to inform the RNA polymerase that the transcription process is complete. Alluding to a specific case study involving Escherichia coli, an extensively studied prokaryote in cellular microbiology, elucidates this molecular process vividly. The involved RNA polymerase in this organism comprises a core enzyme associated with a sigma factor, which collectively initiates the transcription by promoting the polymerase's binding to the promoter sites on the DNA. As the transcription elongates along the DNA strand, the sigma factor disassociates, enabling the core enzyme to proceed. Upon reaching the termination sequence, two distinct mechanisms can be implemented: rho-dependent and rho-independent termination. The former involves a hexameric protein, rho, which disrupts the interaction between the RNA-DNA hybrid and the RNA polymerase at the termination site. Conversely, the latter entails the formation of a GC-rich hairpin loop structure in the newly synthesized RNA, whose physical tension catalyzes the disassembly of the transcription apparatus. It is noteworthy that each termination mechanism is not entirely exclusive, and the RNA polymerase's well-coordinated interplay with these elements guarantees biological efficiency and minimizes erroneous transcripts. Therefore, whether it is a common bacterium such as Escherichia coli or any cellular organism, the precise arrival at a termination sequence on the DNA designates the cessation of the transcription process, rooted in a blend of molecular mechanisms that navigate the journey from genes to transcripts.

The mRNA produced during transcription is now ready to undergo translation, the second step in the journey from DNA to protein (Gan et al., 2022; Garcia-Gutierrez & Garcia-Dominguez, 2022; Geers et al., 2022). Translation takes place in the ribosomes, which consist of large and small subunits, each containing rRNA (ribosomal RNA) and numerous proteins. The ribosomes provide the platform for the interaction between mRNA and transfer RNA (tRNA) molecules. In the cellular rigmarole of genetic expression, the process of transcription precedes and primes the mechanism of translation, wherein the mRNA synthesized is subjected to processing in order to generate a corresponding protein molecule (see figure 46).

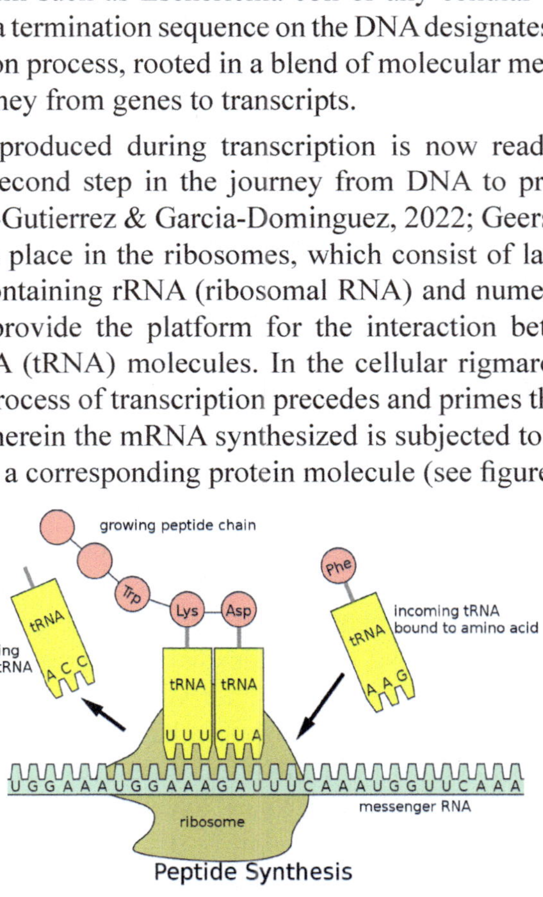

Figure 46. Scheme of translation.

Source: By Boumphreyfr vector conversion by Glrx - File:Peptide syn.png, CC BY-SA 3.0, https://commons.wikimedia.org/w/index.php?curid=101457889

These two steps embody the basic principle of central dogma in molecular biology (Sudhof, 2021; Summers et al., 2022; Sundaram & Chen,

2022). To delve deeper into the detailed mechanistic insight of these highly synchronized stages, one might consider the model organism Saccharomyces cerevisiae, arguably the most extensively studied eukaryote in the realm of cell microbiology. The mRNA transcribed in the nucleus contains a series of nucleotide bases arranged in codons, each comprising three bases capable of encoding a specific amino acid. Prior to undertaking the translation process, this mRNA undergoes a series of modifications, such as 5' capping, splicing and polyadenylation, to become mature mRNA, which then gets exported from the nucleus to the cytoplasm. In the cytoplasm, it sojourns to the protein synthesis machinery, the ribosome, where translation sets in. Each codon is then recognized by an appropriate, matching anticodon on an aminoacyl tRNA molecule during the elongation phase of translation. Concurrently, the ribosome fosters the condensation reaction between each incoming amino acid and the growing peptide chain. Once the mRNA stop codon parks in the ribosome's A site, a release factor protein triggers the process's cessation and the synthesized protein's release, followed by ribosome disassembly. Thus, in a simple, unicellular eukaryote like Saccharomyces cerevisiae, the central dogma unfurls from DNA through mRNA to protein, supported by an intricate orchestration of transcription and subsequent translation.

The tRNA molecules play a pivotal role in translation by carrying specific amino acids to the ribosome (Wu & Liu, 2021; Yang et al., 2021; Xu & Li, 2021; Xie et al., 2022; Xue et al., 2022). Each tRNA has an anticodon region that can base pair with the complementary codon on the mRNA. The ribosome facilitates the pairing of the tRNA anticodon with the mRNA codon, ensuring the correct placement of amino acids in the growing polypeptide chain. Transfer RNA (tRNA) molecules have an indispensable role in the process of translation, responsible for ferrying specific amino acids to the ribosome, the site where the protein synthesis unfolds. The importance of tRNA becomes evident by the case, for instance, of Escherichia coli, a bacterium ubiquitously employed in various cell biology studies. Each tRNA molecule, post-transcription, folds into a distinctive cloverleaf structure, with the anticodon loop and the amino acid accepting arm representing key components crucial to its function. The anticodon, a sequence of three nucleotides in a tRNA, is complementary to a corresponding codon on the messenger RNA (mRNA). Meanwhile, the amino acid accepting arm serves as the site for the covalent attachment of a specific amino acid. The attachment of an amino acid to its corresponding tRNA is a prerequisite to translation and is facilitated by a group of enzymes called aminoacyl-tRNA synthetases. Each of these synthetases, in Escherichia coli and across other species, recognizes its specific tRNA and corresponding amino acid, ensuring the accuracy of the so-called "charging" process. Once charged, the tRNA molecules are enlisted into action during translation, directed by the successive codons of an mRNA molecule being translated. As the ribosome navigates along the mRNA strand, each codon is interpreted by the anticodon of an incoming tRNA molecule bearing the appropriate amino acid. The juxtaposition of these amino acids, dictated by the codon sequence on the mRNA and facilitated by the tRNA molecules, ultimately results in the synthesis of a specific protein. This underscores the pivotal role that tRNA molecules exhibit in the dynamic process of protein synthesis via translation.

The process of translation involves three main stages: initiation, elongation, and termination (Jones et al., 2021; Karlsson & Draber, 2021; Ju et al., 2022). During initiation, the small ribosomal subunit binds to the mRNA, and the tRNA carrying the first amino acid (methionine) binds to the start codon (AUG). The large ribosomal subunit then joins, forming the functional ribosome. Elongation follows, as the ribosome moves along the mRNA, one codon at a time, and tRNA molecules bring in their corresponding amino acids. Peptide bonds form between adjacent amino acids, creating a growing polypeptide chain. This process continues until the ribosome encounters a stop codon (UAA, UAG, or UGA), signaling the termination of translation. Upon termination, the newly synthesized polypeptide is released from the ribosome, and the ribosomal subunits dissociate, ready to begin translation anew with another mRNA molecule. Translation, the molecular process of converting the information contained within messenger RNA (mRNA) into a polypeptide chain, can be succinctly described as occurring in three main stages: initiation, elongation, and termination. To elucidate these stages in detail, let us consider the case of the model organism Escherichia coli. In the initiation phase, the process begins when a ribosome assembly aligns with the start codon (AUG) on an mRNA molecule. This ribosome assembly in Escherichia coli comprises the smaller 30S subunit, where the initiation complex including initiation factors, the first aminoacyl-tRNA, and GTP, forms on the mRNA's start codon. Subsequent release of initiation factors and association with the larger 50S subunit form the complete 70S initiation complex. Here, the elongation stage commences. During elongation, an incoming aminoacyl-tRNA molecule, complementary to the next codon sequence on the mRNA and charged with its corresponding amino acid, binds at the A-site of the ribosome. A peptide bond is then formed between the newly arrived amino acid and the growing polypeptide chain, catalyzed by the large ribosomal subunit's peptidyl transferase activity. The ribosome then advances three nucleotides along the mRNA, effectively moving the uncharged tRNA into the E-site for its eventual release, and the newly positioned tRNA, now tethered to the growing peptide chain, into the P-site. The A-site is now vacant and ready for the next incoming aminoacyl-tRNA. Repeated cycles of this process add amino acids to the polypeptide chain until a stop codon (UAG, UAA, or UGA) is encountered on the mRNA, triggering the translation's termination stage. Here, protein release factors recognize the stop codon, catalyze the hydrolysis of the bond linking the polypeptide to the tRNA at the P-site, and subsequently release the completed polypeptide. The ribosome dissociates from the mRNA, heralding the end of the translation process.

In summary, the processes of transcription and translation form a tightly regulated and intricately coordinated system in cell biology, enabling the transfer of genetic information from DNA to functional proteins. These processes are essential for the proper functioning and survival of cells and organisms, as proteins play diverse roles in cellular processes such as metabolism, transport, signal transduction, and immune response. Understanding the dynamics of transcription and translation provides valuable insights into the complexities of gene expression and the fascinating world of cell biology. The next section considers the laws of inheritance under Mendelian genetics.

5.4 MENDELIAN GENETICS: LAWS OF INHERITANCE

Mendelian Genetics is the study of inheritance patterns in living organisms, particularly at the cellular level (Ajayi-Smith et al., 2021; Al-Marsoummi et al., 2021; Alao et al., 2021; Albrecht et al., 2021). It is named after Gregor Mendel, an Austrian scientist, who conducted groundbreaking experiments with pea plants in the mid-19th century. His work laid the foundation for understanding the basic principles of inheritance, which are governed by three main laws.

1. Law of Segregation: The Law of Segregation states that each individual organism carries two alleles for each trait, and these alleles separate or segregate during the formation of gametes (reproductive cells) (Cahan et al., 2021; Caldas et al., 2021; Burke et al., 2023). As a result, only one allele from each gene pair is passed on to the offspring. This process ensures genetic variation

in the population. In the exploration of cell microbiology, the Law of Segregation delivers a profound significance by elucidating the mechanism underpinning the segregation of hereditary traits. Drawing on a profound case study on pea plants conducted by Gregor Mendel, whose seminal work provided the foundations for the modern genetic theory, it is discovered that each trait within a given organism is facilitated by two alleles, potentially varying in their genetic information. These alleles are secluded during the formation of gametes, which are essentially reproductive cells. This pivotal phenomenon occurs during a stage famously known as meiosis, more specifically in the first meiotic division. Let us delve deeper into the rudimentary molecular structure of a diploid cell: It is comprised of homologous pairs of chromosomes which are essentially carriers of genetic information in the form of genes; each gene potentially carries different versions, predominately referred to as alleles. When the cell steps into the threshold of meiosis, these chromosome pairs align themselves equatorially during Metaphase I and are consistently pulled apart to opposite poles during Anaphase I. This results in the segregation or separation of each allele into separate daughter cells; each receiving one allele from the parent cell, thus validating the Law of Segregation. The marvel of this biological process upholds the evolutionary and adaptive progression in organisms, interlacing the complexity of life traits with simple, yet significant cellular mechanisms. Such exploration underscores the grandeur of genetic diversity, allowing a wide range of variations within the population subset. This meticulous process demonstrates the intricate operations of how foundational genetic principles are encoded within the diverse sequences of life.

2. Law of Independent Assortment: The Law of Independent Assortment states that different genes assort independently during gamete formation (Orr & Hynds, 2021; Ohi & Kenworthy, 2022; Okano et al., 2023). In other words, the inheritance of one gene does not influence the inheritance of another gene located on a different chromosome. This law is applicable when genes are located on different non-homologous chromosomes or are far apart on the same chromosome. The Law of Independent Assortment, put forth by Gregor Mendel, postulates that during gamete formation, different genes segregate independently, resulting in a unique

combination of alleles. A focused examination of this law can be achieved via a cell microbiology case study involving Drosophila melanogaster, or fruit flies, due to their quick generation time and easily observable phenotypes. For the purposes of this examination, consider two genes, one responsible for eye color and another for wing length. When a heterozygous parent with genes for both red and white eye color and long and short wings undergoes gamete formation, it is expected by this law that these genes will assort independently, providing four possible combinations of alleles in the offspring. This prediction flows from the fact that during meiosis I, homologous chromosomes line up along the metaphase plate randomly with respect to one another before being segregated into daughter cells. This alignment and segregation process is mediated by spindle fibers, ensuring each gamete receives only one copy from each pair of chromosomes. This segregation of homologous chromosomes and their respective genes independently is known as independent assortment. To confirm this hypothesis, an analysis of offspring from such a cross was performed. As predicted by the Law of Independent Assortment, the resulting offspring exhibited all four possible combinations of these two traits, providing empirical support for the said law. Thus, an understanding of the Law of Independent Assortment (see figure 47) becomes crucial in the field of genetics as it elegantly explains the genetic variation observed within species.

3. Law of Dominance: The Law of Dominance explains that some alleles are dominant, and others are recessive (S. Y. Li et al., 2021; Y. Li et al., 2021; Y. H. Li et al., 2021). Dominant alleles are expressed phenotypically in heterozygous individuals (organisms carrying two different alleles for a specific trait), while recessive alleles are only expressed when an individual is homozygous recessive (carrying two copies of the recessive allele). Inspecting a succinct illustration of the Law of Dominance, a fundamental principle vested within the realm of molecular genetics, one could consider the classic monohybrid cross experiments conducted meticulously by Gregor Mendel using pea plants. Mendel's experimentation with the color of pea plants, which carried alleles for both yellow and green colors, illuminates this principle profoundly. A critical feature of this law is that it dictates that some

alleles are dominant, thus expressing their traits in the phenotype even in a heterozygous configuration, while others are recessive, expressing their traits only in the absence of a dominant allele, in a homozygous condition. In Mendel's experiment, when the first filial generation, resulting from the cross between true-breeding yellow and green pea plants, consistently exhibited yellow color irrespective of their green parentage, the yellow was classified as the dominant trait. On the contrary, the green trait, though present, had not found expression in the phenotype, thus confirming its recessivity.

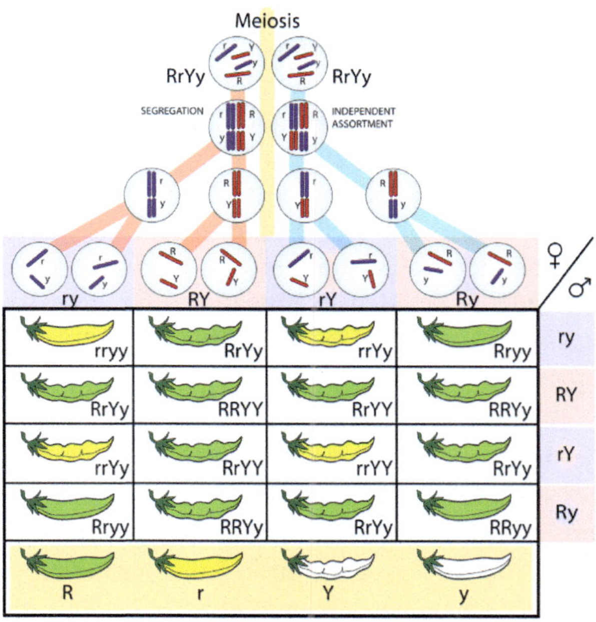

Figure 47. Illustration of Mendel's law of independent assortment

Source: By LadyofHats - Own work based on:[1], [2], [3], [4], [5], Public Domain, https://commons.wikimedia.org/w/index.php?curid=4770847

The second filial generation, resulting from the self-pollination of the first generation, exhibited both yellow and green pea plants in the inherent Mendelian ratio of 3:1. This resulted from the segregation and random fusion of dominant and recessive alleles during gamete formation and fertilization. This examination thereby unequivocally demonstrated that traits are inherited via discrete units, or genes, and that these genes can be dominant or recessive. The Law of Dominance, thus, holds the key to understanding the deterministic rules underpinning inheritance patterns and serves as a cornerstone in exploring the vast tapestry of genetic diversity.

These laws (see figure 48) have significant implications for the inheritance of traits in organisms (Granados-Riveron & Aquino-Jarquin, 2021; Graziadei & Rappsilber, 2022). For instance, when two heterozygous individuals (Aa) with a dominant and a recessive allele mate, they can produce offspring with three possible genotypes: AA (homozygous dominant), Aa (heterozygous), and aa (homozygous recessive). The resulting phenotypic ratios are 1:2:1, where one individual is homozygous dominant, two individuals are heterozygous, and one individual is homozygous recessive. Distilling the essence of Mendel's laws of inheritance, one recognizes their significant implications in the inheritance of traits in organisms. These laws, encompassing the Law of Segregation and the Law of Independent Assortment, heavily contribute to the manner in which entities propagate genetic information across generations. To survey the reach of such implications, consider a detailed examination of a case study involving maize plants, an exemplary model organism due to their large chromosomes.

The hallmark of Mendel's Law of Segregation, that each organism possesses two alleles that segregate during gamete formation, could be observed in the maize plant's production of two types of gametes during fertilization. For instance, consider a hybrid maize plant with both purple and yellow kernels, where purple is dominant. According to Mendelian inheritance, these alleles segregate during the formation of pollen and ovules ensuring that each gamete contains either the dominant purple or the recessive yellow allele. Upon fertilization, these gametes randomly combine forming a spectrum of offspring with purple or yellow kernels, in line with Mendel's prediction. Moreover, extending the case study to two traits like kernel color and kernel texture, the Law of Independent Assortment further stipulates that these traits distribute independently of each other during gamete formation. Therefore, the color of the kernels would not influence the texture, and vice versa. These combined observations underscore the effect of Mendelian inheritance in elucidating the principles of how traits are stochastically passed down to offspring, thus upholding the wide genetic variations and evolutionary adaptation evident in the biological world. Thus, Mendel's laws of inheritance furnish not merely the inheritance paradigm per se but also underpin the remarkable diversity in the biological kingdom.

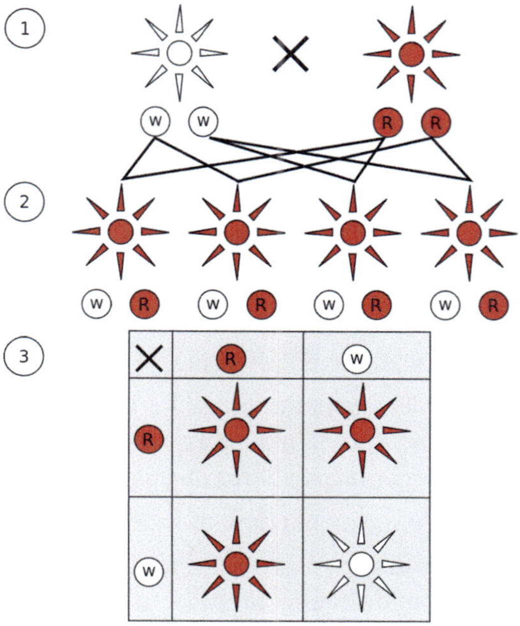

Figure 48. Mendel's laws of inheritance

Source: By Benutzer:Magnus Manske - http://de.wikipedia.org/wiki/Datei:Mendelian_inheritance_3_1.png, Public Domain, https://commons.wikimedia.org/w/index.php?curid=10396755

Provided below are guidance notes for exploring and understanding and exploring Mendel's laws of inheritance using modern laboratory equipment:

1. Preparation and Experimental Setup:
   - Select an appropriate model organism for your experiment, such as the fruit fly (Drosophila melanogaster) or the roundworm (Caenorhabditis elegans). These organisms have short generation times and are easy to handle in a laboratory setting.
   - Obtain the necessary laboratory equipment, including microscopes, Petri dishes, culture media, pipettes, and genetic tools like PCR machines and gel electrophoresis systems.
   - Obtain purebred strains of the chosen organism with well-characterized traits. For example, if studying eye color, you may need strains with red-eyed and white-eyed phenotypes.

- Ensure proper sterilization of equipment and media to maintain a contamination-free environment during the experiment.

2. Cross Breeding (Law of Segregation):
   - Set up controlled mating experiments, such as a monohybrid cross. Cross two purebred strains differing in a single trait of interest (e.g., eye color).
   - Observe the progeny and record their phenotypic ratios. In the case of eye color, you should expect a 3:1 ratio of red-eyed to white-eyed offspring.
   - Use modern genetic tools like PCR to verify the presence of specific alleles in the offspring's DNA. This will confirm the segregation of alleles during gamete formation.
3. Independent Assortment (Law of Independent Assortment):
   - Perform dihybrid crosses by choosing two traits that are located on different chromosomes (e.g., eye color and wing shape).
   - Analyze the phenotypic and genotypic ratios of the offspring. You should observe a 9:3:3:1 phenotypic ratio if the two traits assort independently.
4. Testing for Dominance (Law of Dominance):
   - Cross a purebred dominant individual (AA) with a purebred recessive individual (aa) for the same trait.
   - Examine the phenotype of the F1 generation. If the trait shows complete dominance, all offspring should display the dominant phenotype.
   - To test for incomplete dominance or co-dominance, perform a similar cross and analyze the F1 phenotypes. In cases of incomplete dominance, the offspring will exhibit an intermediate phenotype.
5. Advanced Techniques:
   - To delve deeper into the genetic basis of traits, use molecular techniques such as DNA sequencing to identify specific alleles responsible for observed phenotypes.
   - Utilize genome editing tools like CRISPR-Cas9 to introduce specific mutations and study their effects on inheritance patterns.

6. Statistical Analysis:
   - Apply statistical methods to validate the observed ratios and determine the significance of the results.
   - Use chi-square tests or other appropriate statistical tests to compare the expected and observed phenotypic ratios and assess the goodness of fit.
7. Data Interpretation and Conclusion:
   - Analyze and interpret the data collected during the experiments to draw conclusions about the inheritance patterns of the traits studied.
   - Relate the findings to Mendel's Laws of Inheritance, highlighting how the observed ratios conform to his predictions.
8. Discussion and Future Perspectives:
   - Discuss any discrepancies between the expected and observed results and propose potential reasons for these deviations.
   - Consider the limitations of the experiments and suggest improvements or alternative approaches for future studies.
9. Ethical Considerations:
   - Respect ethical guidelines and regulations regarding the use of model organisms in research.
   - Ensure the humane treatment and proper care of the organisms during the experiments.
10. Documentation and Communication:
    - Thoroughly document all experimental procedures, results, and analyses.
    - Prepare a comprehensive report or scientific paper outlining the experiment's objectives, methods, findings, and conclusions.
    - Present the results at scientific conferences or publish them in peer-reviewed journals to contribute to the broader scientific community's knowledge.

Note: Always follow established laboratory safety protocols and obtain necessary approvals from relevant authorities before conducting any genetic experiments involving living organisms.

Mendelian Genetics provides a straightforward understanding of single-gene traits, but real-life inheritance is more complex due to factors such as multiple genes, polygenic inheritance, and the interaction between genes and the environment. Nonetheless, Mendel's work remains foundational in genetics, and his laws serve as the basis for the study of inheritance patterns in cell microbiology and other fields. This chapter will end with a discussion concerning chromosomal aberrations and genetic disorders as some of the consequences of problematic inheritance.

5.5 CHROMOSOMAL ABERRATIONS AND GENETIC DISORDERS

Chromosomal aberrations refer to structural changes or numerical variations in chromosomes that can lead to genetic disorders (Hernandez-Gonzalez et al., 2021; Herranz-Iturbide et al., 2021; Hiatt et al., 2021). These abnormalities can occur during cell division, either in meiosis (gamete formation) or mitosis (somatic cell division). Chromosomal aberrations, referring to structural deviations or numerical variations in chromosomes, underpin numerous genetic disorders. To affirm the impacts and manifestations associated with such aberrations, it proves suitable to analyze Down syndrome, a genetic disorder emanating from a numerical aberration. In typical circumstances, a healthy human cell houses 23 pairs of chromosomes, each carrying genes, the molecular blueprint for individual biological traits. However, Down syndrome arises when an anomalous event called nondisjunction occurs during meiosis, specifically anaphase I or II. In anaphase I, if both homologous chromosomes in a pair travel to the same pole, or in anaphase II, if sister chromatids fail to separate, it results in one of the resulting gametes housing an extra copy of chromosome 21, a condition termed trisomy 21. Subsequent fertilization involving this aberrant gamete and a normal one leads to a zygote with an extra chromosome 21. The resultant individual exhibits the hallmarks of Down syndrome, which include distinct facial features, delayed physical and intellectual development attributed to the amplified gene dosage from the extra chromosome. This understanding of chromosomal aberrations unravels the complexities surrounding certain genetic conditions, thus pulling into focus the intricate interplay of genotype and phenotype. Hence, the detailed exploration of the genetic anomaly provides invaluable insights into genetic disorders, aiding in their identification, management, and potential therapeutic interventions. Figure 49 is an example of chromosomal aberrations.

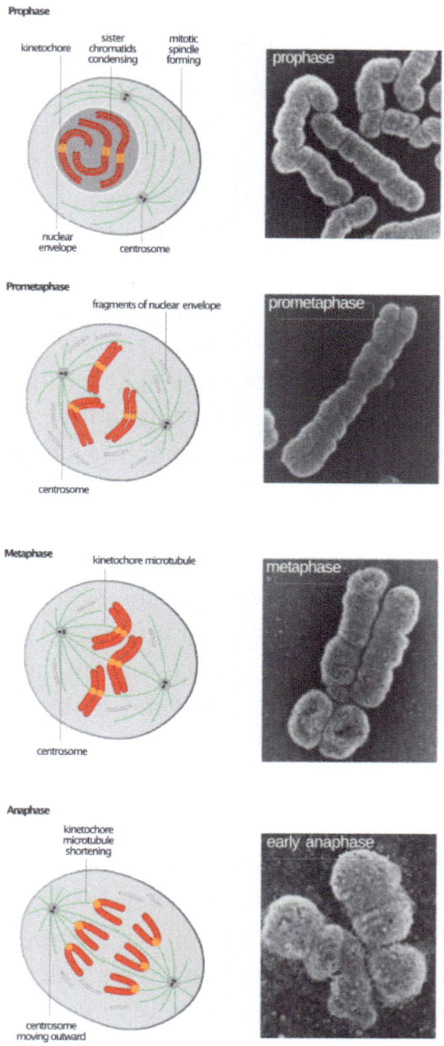

Figure 49. Chromosomes.

Source: By David O Morgan - The Cell Cycle. Principles of Control., Attribution, https://commons.wikimedia.org/w/index.php?curid=89955214

Understanding the molecular basis of these aberrations and their consequences is crucial for studying the etiology of genetic disorders in cell microbiology (Pendharkar et al., 2021 Perera, 2021; Peng et al., 2022). Illuminating the molecular dynamics underlying chromosomal aberrations and their ramifications proves indispensable to comprehend the etiology of diverse genetic disorders within the realm of cell microbiology. A probing examination of the causative aberration for Philadelphia chromosome-positive Chronic Myeloid Leukemia (CML) offers substantive insights into this aspect. The Philadelphia chromosome arises from a reciprocal translocation, a structural chromosomal aberration, between chromosomes 9 and 22, specifically transferring the ABL gene on chromosome 9 to the BCR region on chromosome 22 to form the BCR-ABL fusion gene. This fusion gene, present on the aberrant chromosome 22 or the Philadelphia chromosome, code for a dysregulated tyrosine kinse BCR-ABL protein. Under normal circumstances, the kinase activity of the ABL protein is tightly regulated; however, its fusing with BCR results in the production of a continuously active tyrosine kinase. This constitutively active tyrosine kinase drives the pathogenesis of CML by promoting uncontrolled proliferation and impeding apoptosis of myeloid cells in the bone marrow, leading to a massive increase in white blood cells. Thus, understanding the molecular blueprint of these aberrations is pivotal in discerning the pathophysiology of genetic disorders like CML. Consequently, this knowledge can be wielded to develop targeted cancer therapies such as Imatinib, which blocks the activity of the BCR-ABL tyrosine kinase, thereby highlighting the importance of understanding the molecular basis of chromosomal aberrations in the study of the etiology of genetic disorders.

1. Structural Chromosomal Aberrations: Structural chromosomal aberrations involve changes in the physical structure of one or more chromosomes (LeGuennec et al., 2021; Leidal & Debnath, 2021; Lendahl, 2022). Structural chromosomal aberrations refer to alterations in the physical structure of one or more chromosomes, and they have profound implications for cell biology and the development of genetic disorders. A striking case study of such aberrations is the Philadelphia chromosome, which is associated with chronic myeloid leukemia (CML). The Philadelphia chromosome results from a reciprocal translocation between the long arms of chromosomes 9 and 22, t(9;22)(q34;q11), leading to the fusion of the ABL1 gene on chromosome 9 with the BCR gene on chromosome 22. This translocation gives rise to a chimeric gene known as BCR-ABL1, which encodes a constitutively active tyrosine kinase that promotes uncontrolled cell proliferation. The abnormal BCR-ABL1 protein is central to the pathogenesis of CML and is responsible for the expansion

of leukemic cells in the bone marrow and peripheral blood. The Philadelphia chromosome is typically detected through cytogenetic techniques, such as karyotyping or FISH, enabling the identification of this characteristic chromosomal aberration in CML patients. The detection of the Philadelphia chromosome is not only critical for confirming the diagnosis of CML but also serves as a prognostic marker and a target for therapeutic interventions. The development of targeted therapies, such as tyrosine kinase inhibitors (TKIs) like imatinib, dasatinib, and nilotinib, has revolutionized the management of CML. These TKIs specifically inhibit the activity of the BCR-ABL1 fusion protein, leading to substantial improvements in patient outcomes and long-term survival rates. The Philadelphia chromosome and its role in CML exemplify how structural chromosomal aberrations significantly impact cell biology and contribute to the pathogenesis of genetic disorders. This case study underscores the importance of understanding the molecular basis of chromosomal aberrations, as it has not only advanced our knowledge of disease mechanisms but has also paved the way for the development of targeted therapies that have revolutionized cancer treatment. Continued research in cell biology and genetics is essential to further elucidate the complexities of chromosomal aberrations and their impact on human health, ultimately leading to improved diagnostic methods and therapeutic strategies for a wide range of genetic disorders.

Examples include:

- Deletion: A portion of the chromosome is lost due to breakage, leading to the loss of specific genes and potentially causing genetic disorders (Pendharkar et al., 2021; Perera, 2021; Peng et al., 2022). Deletion in chromosomes, pointing to the loss of a chromosomal segment due to breakage, can alter the genetic integrity of an organism by resulting in the loss of specific genes, in turn precipitating certain genetic disorders. A prime example to illuminate the effects of deletions is Cri-du-Chat Syndrome, a genetic disorder pioneered by French pediatrician and geneticist Jerome Lejeune in 1963. This disorder results from a deletion on the short arm of chromosome 5, more specifically spanning the region 5p15.2 to 5p15.3. The specific genes lost due to this deletion, including CTNND2, Semaphorin F, and delta catenin among others, consequently impact various developmental processes. CTNND2, for instance, plays a pivotal role in neural development and its deletion is predominantly implicated in the cognitive impairment seen in Cri-du-Chat syndrome. Another critical gene, Semaphorin F participates in axonal guidance during neural development and its deletion might contribute to the neurological symptoms of this disorder. Clinically, affected individuals often present a high-pitched cry, reminiscent of a cat's meowing, lending the syndrome its distinct name. Additional manifestations include intellectual disabilities, delayed development, small head size, and distinctive facial features. The case of Cri-du-Chat syndrome thus underscores how deletions in chromosomes, leading to the loss of crucial genes, constitute a key mechanism resulting in an array of genetic disorders. Therefore, the exploration of chromosomal deletions in genetic disorders fosters insights into pathogenesis, diagnosis, and potential therapeutic interventions for such conditions.

– Duplication: A segment of the chromosome is duplicated, resulting in extra copies of certain genes (Granados-Riveron & Aquino-Jarquin, 2021; Graziadei & Rappsilber, 2022). Duplication, referring to the scenario where a chromosome segment is repeated, can lead to the existence of extra copies of the genes encompassed within that segment. A concrete example that unfolds the consequences of such duplications is Charcot-Marie-Tooth disease type 1A (CMT1A). Rooted in the duplication of the peripheral myelin protein 22 (PMP22) gene located on chromosome 17, CMT1A is a hereditary neuropathic disorder. The gene PMP22 is instrumental in the formation of the myelin sheath, a protective layer ensuring efficient propagation of nerve impulses along the axon. Under normal circumstances, a diploid organism possesses two copies of the PMP22 gene; however, the duplication resulting in CMT1A leads to an individual inheriting an extra copy of this gene, thus augmenting the dosage of PMP22 by 50 percent. This overdose disrupts the normal process of myelination, causing difficulties in nerve impulse transmission. The clinical manifestation of CMT1A includes motor and sensory neuropathy, foot deformities and progressive muscle wasting, painting a vivid picture of how chromosomal duplications carrying extra copies of genes can trigger genetic disorders. The understanding derived from such observations not only unmasks the biological ramifications of chromosomal duplications but also proves vital for advances in diagnostics, disease management, and potential curative strategies.

– Inversion: A chromosomal segment breaks and reattaches in the reverse orientation, potentially disrupting gene function (Biermann & Reya, 2022; Bohnert & Johnson, 2022; Bomont, 2021). Inversion, a chromosomal aberration where a segment breaks off and rearranges itself in the reverse orientation, may potentially disrupt gene function. A critical illustration of this can be found in the case of hemophilia A, a blood clotting disorder that exhibits an association with inversions on the X chromosome. Factor VIII, a crucial component in the blood-clotting cascade, is encoded by the F8 gene, located on the X chromosome. Hemophilia A is commonly linked to an inversion affecting a segment encompassing the F8 gene, accounted for by two low copy repeat sequences flanking the gene that induces this inversion during meiosis. The inversion disrupts the F8 gene in intron 22 or intron 1, resulting in the production of functionally impaired Factor VIII or entirely preventing its synthesis. The absence or dysfunction of Factor VIII impedes the coagulation cascade, rendering the blood incapable of clotting effectively. Consequently, affected individuals experience prolonged bleeding or spontaneous hemorrhages. As this disorder is X-linked, it primarily is seen in males, while females are carriers of this inheritance. Hemophilia A elucidates how chromosomal inversions can upset critical gene functions, thereby activating a cascade of pathological processes. An in-depth comprehension of these inversion mechanisms offers constructive insights into the genesis and progression of genetic disorders, thus paving the way for enhanced diagnostic and therapeutic modalities.

– Translocation: Parts of two non-homologous chromosomes exchange places, which can lead to the fusion of unrelated genes and the formation of fusion proteins (LeGuennec et al., 2021; Leidal & Debnath, 2021; Lendahl, 2022). Translocation denotes an event whereby fragments from two non-homologous chromosomes undergo an interchange, potentially leading to the fusion of unrelated genes and the subsequent formation of fusion proteins. For a profound understanding of this aberration, we shall examine the genetic underpinnings of chronic myeloid leukemia (CML), a myeloproliferative disorder characterized by aberrant proliferation of myeloid cells in the bone marrow and peripheral blood. The pathogenic hallmark of CML is the Philadelphia chromosome, a product of a reciprocal translocation between chromosomes 9 and 22, leading to the fusion of the Abl tyrosine kinase gene from chromosome 9 with the Bcr gene on chromosome 22. This results in the creation of the BCR-ABL fusion gene, located on the diminutive chromosome 22, lending it the nomenclature, Philadelphia chromosome. The BCR-ABL fusion gene codes for a constitutively active BCR-ABL fusion protein with unregulated tyrosine kinase activity which triggers a cascade of oncogenic signaling pathways, instigating abnormal myeloid cell proliferation. Patients with CML typically present with an elevated white blood cell count, fatigue, and splenomegaly, and if left untreated, the disease progresses from the chronic phase to an accelerated phase and eventually, an acute phase or blast crisis. In conclusion, rampant translocations such as the BCR-ABL translocation in CML can underpin the initiation of malignancies, underscoring their profound implications in the pathogenesis of genetic disorders. Moreover, the elucidation of the molecular intricacies of such translocations can serve as the basis to tailor targeted therapeutic strategies, like tyrosine kinase inhibitors in the case of CML.

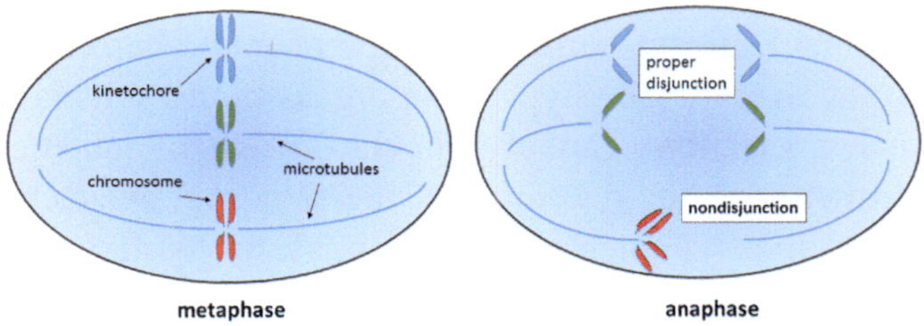

Figure 50. Possible Mitotic events leading to aneuploidy.

Source: By Wpeissner - Own work, CC BY-SA 3.0, https://commons.wikimedia.org/w/index.php?curid=32332257

2. Numerical Chromosomal Aberrations: Numerical chromosomal aberrations involve changes in the total number of chromosomes in a cell, often due to errors in cell division (Collins et al., 2021; Costa & Sousa, 2022; Colin et al., 2023). Figure 50 is an example of possible Mitotic events leading to aneuploidy as a case in point. Numerical chromosomal aberrations constitute alterations in the total number of chromosomes in a cell, usually emanating from errors during the process of cell division. To elucidate the implications of these aberrations, one can consider Down Syndrome, a congenital disorder permeating from an aberration in the number of chromosome 21. Normally, a nucleus in a human cell accommodates 23 pairs of chromosomes, which contributes to a total of 46 chromosomes, but in the case of Down Syndrome, also known as trisomy 21, individuals possess an extra copy of chromosome 21, increasing the count to a total of 47 chromosomes. The genesis of this trisomy can be traced back to nondisjunction events during meiosis, whereby homologous chromosomes or sister chromatids fail to separate accurately, culminating in one gamete receiving an additional copy of certain chromosomes. In the case of Down syndrome, the nondisjunction event targets chromosome 21. Individuals with trisomy 21 manifest characteristic physical features such as flattened facial profile, slanted eyes, and small ears, alongside intellectual disabilities. Moreover, they are predisposed to health concerns such as congenital heart diseases, hearing problems, immune disorders, and early-onset of Alzheimer's disease.

Thus, the examination of Down Syndrome underscores the profound impact numerical chromosomal aberrations can have on the overall phenotype of an organism, demonstrating the role such aberrations can play in the advent of genetic disorders. Furthermore, a systematic understanding of such numerical aberrations augments our potential to harness chromosomal information for disease management and to develop therapeutic strategies.

Examples include:

- Aneuploidy: The gain or loss of individual chromosomes, leading to an abnormal chromosome number (Bennison et al., 2022; Beppler et al., 2022; Bettegazzi et al., 2021). Trisomy (an extra chromosome) and monosomy (a missing chromosome) are common forms of aneuploidy. Aneuploidy, representing the gain or loss of individual chromosomes resulting in an abnormal chromosome number, carries substantial implications for the health of an organism. A thoughtful exploration of this notion can be stressed via a case examination of Turner Syndrome, a genetic aberration driven by aneuploidy of the X chromosome. Typically, a healthy human female contains two X chromosomes; however, in cases of Turner Syndrome, one of these chromosomes is wholly or partially missing, making it a monosomy condition. This loss may occur due to nondisjunction, an event where homologous chromosomes or sister chromatids fail to separate during cell division, either during spermatogenesis or oogenesis, yielding a gamete devoid of the corresponding sex chromosome. Consequently, the fusion of such a gamete with a normal one culminates in a zygote with a single X chromosome. Females with Turner Syndrome exhibit a variety of symptoms, including short stature, webbed neck, low hairline at the back of the neck, and infertility. Further, they are at high risk of several health complications such as congenital heart disease, hypothyroidism, diabetes, and vision and hearing problems. Taken together, the understanding of Turner Syndrome substantially underscores how aneuploidy can breed far-reaching complications in an organism's health, thereby enhancing our acuity of genetic disorders and their associated pathologies.
- Polyploidy: The presence of multiple sets of chromosomes in a cell, such as triploidy (3 sets) or tetraploidy (4 sets). Polyploidy

is often lethal in humans but is common in certain plant species (Mangeat et al., 2021; Manils et al., 2022; Maricchiolo et al., 2022). Polyploidy, defined as the occurrence of multiple sets of chromosomes within a cell, engenders an intriguing biological entity. The phenomenon can manifest in several forms such as triploidy, possessing three whole sets of chromosomes, or tetraploidy, entailing four sets. One salient case for contemplating the intricacies of polyploidy would be the occurrence of triploidy in humans, a condition bearing a substantial toll on fetal development and survival. Typically, human cells are diploid, that is, they retain two sets of 23 chromosomes, one each from the mother and the father, totaling 46. Conversely, a triploid cell contains three sets of chromosomes amounting to 69: an extra set could be either paternal, leading to diandry, or maternal, leading to digyny. The most common causative event for triploidy is the fertilization of a normal haploid ovum by a diploid sperm, a product of failed meiosis in spermatogenesis. When such an event occurs, the resulting zygote, having 69 chromosomes, faces numerous developmental issues owing to abnormal gene dosage from the extra set of chromosomes. While partial triploidy can lead to live births exhibiting congenital abnormalities, complete triploidy results in miscarriages, stillbirths, or neonatal deaths. This examination of triploidy illustrates how intricate and tightly regulated the chromosomal makeup must be for the successful growth and development of an organism, thereby enhancing our understanding of the intricate world of cell biology. Understanding such chromosomal aberrations assists in providing a genetic explanation for numerous congenital abnormalities and pregnancy losses, while also highlighting essential steps in cell division processes.

3. Consequences of Chromosomal Aberrations: Chromosomal aberrations can have severe consequences on an organism's development and health (Urmi et al., 2021; Usmani et al., 2021; Turner et al., 2022). Depending on the location and size of the chromosomal alteration, these consequences can vary widely. Chromosomal aberrations, involving variations in chromosomal structures or numbers, can exert profound impacts on an organism's development and overall health. A deeper understanding of these consequences can be achieved through a detailed exploration of

Edwards syndrome, also known as Trisomy 18. This congenital disorder emerges from the presence of an additional copy of chromosome 18, thus resulting in a triploid state for this specific chromosome. The ramifications of this aberration include a broad spectrum of physical and mental complications, profoundly affecting the health trajectory of individuals afflicted with this condition. Individuals with Trisomy 18 typically exhibit a cluster of distinctive physical characteristics and systemic anomalies that include craniofacial abnormalities such as micrognathia and low-set ears, clenched hands, and complex malformations of the heart, kidneys, and other internal organs. Further, intellectual and developmental delays are virtually universal among these individuals. The crux of the disorder lies in the abnormal gene dosage that accompanies the extra chromosome 18, leading to the disruption of typical development pathways due to the unbalanced expression of the genes located on this chromosome. Indeed, the majority of cases of Trisomy 18 lead to miscarriage or stillbirth, and among infants who are born with the condition, only a small percentage survive beyond the first year of life. Edwards syndrome vividly illustrates the severe consequences chromosomal aberrations can wield on organismal development and health, thereby affecting survival and life quality. Such explorations of chromosomal aberrations not only illuminate the etiology and pathophysiology of diverse diseases but also foster informed prognoses and potential therapeutic options.

- Some chromosomal aberrations lead to non-viable embryos, resulting in spontaneous abortions (miscarriages) (Scott-Solomon & Hsu, 2022; Scott et al., 2022; Sezin et al., 2022). Chromosomal aberrations, characterized by variations in either the number or structure of chromosomes, may precipitate nonviable embryos, culminating in spontaneous abortions or miscarriages. A fitting illustration of this consequence can be observed in instances of Trisomy 16, the most frequent chromosome anomaly in human spontaneous abortions. Trisomy 16 is a condition where an embryo presents with an extra copy of chromosome 16 due to errors in maternal meiosis, leading to three sets, or triploidy, instead of the normal diploidy. Although viable trisomy in any of the human chromosomes is rare and severely impacts development leading to neonatal death, Trisomy 16 takes an even graver toll. It is deemed incompatible with life due to the critical role chromosome 16 plays in growth and development, encoding essential genes associated with various developmental processes. Embryos afflicted with this chromosomal aberration possess severe malformations rendering them unable to progress beyond the first trimester. Hence, they are naturally terminated, resulting in early miscarriages. The natural incidence of Trisomy 16 uncovers the complexities and precision required in genetic inheritance for viable and successful development. Understanding such instances of chromosomal aberrations serves a dual purpose: it elucidates the causes behind early miscarriages, bringing relief to countless women who often blame themselves for such outcomes, and it enhances our understanding of the intricacies of early human development and the vital role that chromosomal integrity plays in the process.

– Others can cause genetic disorders, such as Down syndrome (trisomy 21), Turner syndrome (monosomy X), or Klinefelter syndrome (XXY) (Vakonakis, 2021; van den Dries et al., 2022; Veerasamy et al., 2021). Chromosomal aberrations, spanning both structural and numerical anomalies, serve as substantive contributory elements to a diverse array of genetic disorders. As examples, one may consider Down Syndrome, Turner Syndrome, and Klinefelter Syndrome, each offering unique insights into the pathological impact of chromosomal aberrations. Down syndrome, also known as trisomy 21, arises due to the presence of a supplemental copy of chromosome 21. This surplus genetic material exhibits a determinative influence over a suite of physical and cognitive traits. These include characteristic facial features, heart defects, and cognitive impairments of varying degrees. Turner Syndrome, meanwhile, is a result of monosomy of the X chromosome in females. Individuals with Turner Syndrome are characterized by short stature and lack of ovarian development, leading to infertility, among other symptoms. Lastly, Klinefelter syndrome, emblematic of polysomy, manifests due to an additional X chromosome in males (XXY). This anomaly precipitates a range of symptoms that usually include tall stature, small testes, infertility, and cognitive difficulties. Thus, the idiosyncrasies of each of these syndromes underscore the significant influence that chromosomal aberrations can bear on an organism's development, health, and overall life trajectory. An understanding of these anomalies not only provides profound insights into the biological manifestations of these disorders but also facilitates advancements in their diagnosis, management, and potential remedial measures.

- Chromosomal translocations can lead to the development of cancer by causing the expression of abnormal fusion proteins that drive uncontrolled cell growth (Lu et al., 2022; Lynes et al., 2022; Ma & Hoffmann, 2021). Chromosomal translocations, which entail the exchange of chromosome segments between non-homologous chromosomes, can contribute to the pathogenesis of cancer by engendering the expression of aberrant fusion proteins that instigate uncontrolled cell growth. In order to comprehend the gravity of chromosomal translocations, one may delve into the detailed study of Chronic Myeloid Leukemia (CML). CML is essentially a malignant disorder of the hematopoietic stem cells characterized by the unregulated growth of myeloid cells in the bone marrow and their accumulation in the blood. The fundamental genetic abnormality driving CML is a specific chromosomal translocation known as the Philadelphia chromosome. This translocation involves the swapping of a segment of chromosome 9, which contains the ABL1 gene, with another segment from chromosome 22 housing the BCR gene. The result of this translocation is the formation of a novel, hybrid gene BCR-ABL1 situated on the mutated chromosome 22. The genetic chimera BCR-ABL1 translates into a fusion protein with constitutively activated tyrosine kinase activity. This unregulated enzymatic activity alters multiple signal transduction pathways inside the cell, encouraging cell proliferation and inhibiting apoptosis -- the process of programmed cell death. Consequently, the perpetual growth of myelogenous blood cells ensues, defining the pathophysiology of CML. The exploration of CML thus underscores how chromosomal translocations can not only trigger the development of cancer but also affect individual prognosis and treatment, thereby emphasizing the necessity of genetic accuracy and regulation in maintaining cellular homeostasis and well being.

4. Mechanisms of Chromosomal Aberrations: The mechanisms underlying chromosomal aberrations are diverse and can involve errors in DNA repair, improper recombination during meiosis, exposure to mutagenic agents, or abnormal cell division processes (Katunin et al., 2021; Keenan et al., 2021; Khampang et al., 2023). Chromosomal aberrations, central to numerous genetic disorders and congenital anomalies, originate from a range of underlying mechanisms, including but not limited to errors in DNA repair, improper recombination during meiosis, exposure to mutagenic agents, and disrupted cell division processes. To apprehend the complexity of these mechanisms, one might scrutinize the etiology and pathogenesis of Ataxia Telangiectasia (A-T), a rare, neurodegenerative, autosomal recessive disease. A-T notably manifests due to mutations in the ATM gene found on chromosomal pair 11, which encodes a protein pivotal in the detection and repair of damaged DNA. Specifically, in response to DNA double-strand breaks, the ATM protein undergoes autophosphorylation and consequentially phosphorylates key substrates involved in cell cycle regulation, apoptosis, and DNA repair, thus serving as a guardian of genomic integrity. Therefore, mutations in the ATM gene disrupt the repair of damaged DNA, leading to chromosomal aberrations due to the accumulation of unattended, erroneous DNA. Clinically, individuals with A-T present progressive neurological impairment, including ataxia, telangiectasia, immunodeficiency, and a high predisposition to cancer. Hence, the analysis of A-T underscores that errors in DNA repair mechanisms can precipitate chromosomal anomalies, which, in turn, can lead to debilitating health conditions. Consequently, the knowledge surrounding the intricate mechanisms entangled with chromosomal aberrations is intrinsic for elucidating the pathophysiology of diverse diseases and for tailoring effective therapeutic strategies to combat them.

– Environmental factors, such as radiation and certain

chemicals, can induce DNA damage, leading to chromosomal abnormalities (Cheung et al., 2022; Choezom & Gross, 2022; Chhabra et al., 2023). Environmental factors, inclusive of radiation and specific chemicals, have the capacity to inflict DNA harm, thereby steering the genomic constitution toward chromosomal abnormalities. A profound understanding of this concept can be accomplished through an examination of the aftermath of the atomic bombings in Hiroshima and Nagasaki during World War II. The excess exposure to radiation in these regions precipitated significant DNA damage within cells, leading to a surge in chromosomal aberrations among the survivors. One of the vital types of DNA damage induced by this ionizing radiation involved double-strand breaks. The cells' attempt to repair these breaks, especially through the error-prone non-homologous end joining mechanism, often led to the introduction of chromosomal translocations, deletions, and inversions. Such chromosomal abnormalities in somatic cells constituted a high risk for the development of radiation-induced cancers, including leukemia and various solid cancers, a major health burden observed among these survivors. Apart from somatic cells, radiation-induced damage in germ cells elicited chromosomal anomalies in the subsequent generations, as evinced by elevated rates of infant mortality and congenital malformations among the offspring of survivors. The contemplation of the Hiroshima and Nagasaki incidents attests the detrimental potential of environmental factors, such as radiation, in driving DNA damage and consequent chromosomal abnormalities, emphasizing the crucial role of genetic stability in maintaining health and well being.

– Errors in DNA replication or repair can result in the deletion, duplication, or inversion of chromosomal segments (Featherston & Cotruvo, 2021; Ferrari et al., 2022; Ferrari et al., 2023). The accurate replication of the DNA molecule, encoded with the genome's vast array of information, is fundamental to the perpetuation of cellular life. Central to preserving the integrity of this genetic information is a cluster of high fidelity mechanisms, which work in concert, to ensure precise duplication of the genomic blueprint during each cycle of cell division. However, even with these safeguards in place, the process is not without occasional errors and inaccuracies. Mistakes in DNA replication or repair can lead to major genomic rearrangements such as the deletion, duplication or inversion of chromosomal segments. For instance, the deletion may lead to a loss of necessary genetic information, a phenomenon commonly associated with disorders such as Cri-du-Chat Syndrome in humans, which arises due to deletions on the short arm of chromosome 5 resulting in intellectual disability and delay in development. Further, errors leading to duplications may introduce extraneous genetic inputs. An example of such a disorder would be Charcot-Marie-Tooth disease type 1A, which results from the duplication of the gene that produces the peripheral myelin protein 22. This protein, when present in excess, undermines the myelin sheath of the peripheral nerves thereby affecting nerve transmission. Furthermore, inversions may result in disrupted gene order, causing the resulting proteins to function improperly or not function at all. While such errors in DNA replication or repair are relatively rare, they nevertheless stand to significantly alter the cellular landscape, potentially leading to severe functional impairment and even disease. A deepened understanding of these molecular anomalies thus offers a vital wellspring of knowledge in our ongoing endeavor to map the intricate universe of cellular biology and further our quest to combat the host of diseases that find their provenance in these errors of genetic transcription and translation.

5. Detection and Diagnosis of Chromosomal Aberrations: In cell microbiology, various techniques are employed to detect and diagnose chromosomal aberrations in cells and tissues (Sudhof, 2021; Summers et al., 2022; Sundaram & Chen, 2022).In the realm of cell biology, the detection and diagnosis of chromosomal aberrations have become paramount for understanding the underpinnings of a variety of genetic disorders and malignancies. Assorted techniques possessing exceptional sensitivity and specificity have been developed to fulfill this vital

task. Cytogenetic methods, centered largely around the study of chromosomes' physical properties, have long been employed for their ability to highlight abnormalities such as numerical or structural chromosomal aberrations. The application of Karyotyping, a process that organizes chromosomes based on size, position of centromeres, and banding pattern, proved instrumental in the diagnosis of Down Syndrome, an aneuploidy marked by the presence of an extra copy of chromosome 21. The combination of conventional Karyotyping with computational imaging analysis has significantly improved differentiation between numerical and structural aberrations, thus aiding earlier intervention. On the other hand, the paradigm of molecular cytogenetics, incorporating technologies like fluorescence in situ hybridization (FISH), enhances the detection of lesser chromosomal alterations (submicroscopic rearrangements) with greater resolution. Application of FISH has been critical in the diagnosis of microdeletion syndromes as in the case of Williams syndrome, characterized by a diminutive deletion on chromosome 7. Despite these advancements, limitations persist regarding the size of detectable aberrations. Whole-genome sequencing techniques, however, are on the horizon with capabilities of single base-pair resolution, promising unprecedented exactitude in chromosomal aberration detection. Spectrum of techniques sprouting from advancements in cell biology have ushered a new era towards more precise and earlier diagnosis of chromosomal aberrations, ultimately aiding the medical community in the pursuit of efficient management and potential cure for numerous genetic maladies.

- Karyotyping: This technique involves staining and analyzing chromosomes to detect large-scale structural aberrations and aneuploidies (Cheung et al., 2022; Choezom & Gross, 2022; Chhabra et al., 2023). Karyotyping, a cytogenetic technique essential in the sphere of cell microbiology, allows the microscopic visualization of individual chromosomes to discern any major structural or numerical aberrations. The procedure entails the arrest of cells in metaphase, when chromosomes are highly condensed and distinct, followed by staining procedures that render each chromosome's unique banding pattern visible. The stained chromosomes are then photographed and arranged according to the classical chromosomal map defined by size, centromere position, and banding pattern - a profile unique to every species. Karyotyping has been instrumental in the diagnosis of large-scale chromosomal abnormalities as well as aneuploidies, conditions characterized by the chromosome number that is not an exact multiple of the haploid set. In the context of diagnosing aneuploidies, one of the landmark case studies is Down Syndrome. This condition, characterized by an additional copy of chromosome 21 due to a nondisjunction event, has been effectively diagnosed via karyotypic analyses, a diagnosis confirmed by the presence of three separate copies of this chromosome, rather than the usual two. Similarly, major structural aberrations such as the large-scale deletions or duplications associated with Cri-du-Chat syndrome and Charcot-Marie-Tooth disease can be readily discerned through karyotypic analyses. Although other molecular techniques involving higher resolution are progressively utilized, Karyotyping retains a pivotal position in clinical genetics and research, especially for the diagnosis of large-scale chromosomal aberrations and aneuploidies.

– Fluorescence in situ hybridization (FISH): FISH uses fluorescent probes to visualize specific DNA sequences, aiding in the identification of chromosomal translocations and duplications (Khanna & Villa, 2022; Kim & Choi, 2022; Kikuchi et al., 2023). Fluorescence in situ hybridization, commonly referred to as FISH, is a powerful molecular cytogenetic tool that involves the deployment of fluorescent probes to mark specific DNA sequences. This technique is instrumental in revealing physical mapping details under a florescent microscope, paving the way for the detection of key chromosomal rearrangements, including translocations and duplications. Take, for example, the detection of the chromosomal translocation t(9;22)(q34;q11), often known as the Philadelphia chromosome. This aberration, notable for being the first chromosomal abnormality related to a specific cancer, is frequently found in chronic myeloid leukemia patients. With the application of FISH, using two different fluorochromes hybridizing to break-apart BCR and ABL1 gene regions on chromosomes 9 and 22 respectively, this translocation is vividly visualized, marking a significant advancement in the diagnostic landscape. Likewise, instances of gene duplication events can be readily identified by the FISH technique due to the presence of additional fluorescent signals at specific loci. One notable scenario includes diagnosing Charcot-Marie-Tooth disease type 1A, which is caused by a duplication of the peripheral myelin protein 22 gene. Through FISH probe hybridization, additional fluorescent signals denoting the duplication event are visualized, providing the clinician with a decisive tool for the diagnosis of this neuropathy. These instances underscore the utility of FISH in identifying and diagnosing chromosomal translocations and duplications, thus re-emphasizing its stance as an irreplaceable tool in the arsenal of cytogenetic pathology.

– Molecular techniques like PCR and DNA sequencing can identify specific mutations and genetic variations associated with chromosomal aberrations (Bomont, 2021; Biermann & Reya, 2022; Bohnert & Johnson, 2022). Molecular techniques such as Polymerase Chain Reaction (PCR) and DNA sequencing hold immense importance in the detection of genetic variations and specific mutations associated with chromosomal aberrations. The advent of PCR technology has led to significant strides in genetic research, due to its ability to amplify specific segments of DNA, allowing detailed investigation of genetic sequences. One classical use of PCR is in the detection of the DeltaF508 mutation, one of the most common mutations causing cystic fibrosis. The deletion of phenylalanine at the 508th codon on Chromosome 7 results in altered chloride channels in respiratory cells, leading to the symptoms associated with the disease. By amplifying this portion of the chromosome using appropriate primers, researchers are able to identify the deletion mutation. Furthermore, DNA sequencing methods are pivotal for identifying genetic variations that accompany chromosomal aberrations at the nucleotide level. One notable case is the detection of

BRCA1 and BRCA2 mutations associated with a higher risk of breast and ovarian cancers. Using Sanger sequencing methods, researchers have been able to identify the specific point mutations in these genes on chromosomes 17 and 13, respectively, paving the way for predictive genetic testing and early detection. The combination of PCR for DNA amplification and DNA sequencing technologies for higher-resolution analysis enhances our understanding of the genetic basis of numerous diseases and supports the exploration of therapeutics.

In conclusion, chromosomal aberrations are critical contributors to genetic disorders. Their study in cell microbiology provides insights into the molecular basis of these abnormalities and their consequences on an organism's health and development. Understanding chromosomal aberrations aids in the diagnosis, treatment, and prevention of genetic disorders, and it remains an essential area of research in cell biology and genetics. Let us now summarize the fifth chapter.

CHAPTER 5 SUMMARY

Chapter 5 delves into the fundamental principles of Genetics and Inheritance. It begins by exploring the DNA structure and its role as the genetic code, laying the foundation for understanding genetic information storage and transmission. DNA replication is then discussed, elucidating the intricate process that ensures genetic continuity during cell division.

Furthermore, the chapter addresses transcription and translation, illustrating how the genetic code is converted into functional proteins, essential for the cell's various functions. The section on Mendelian Genetics delves into Gregor Mendel's pioneering work on inheritance, presenting the three fundamental laws of segregation, independent assortment, and dominance.

Lastly, the chapter focuses on chromosomal aberrations and genetic disorders. Structural chromosomal aberrations, involving changes in chromosome structure, are explored, as are numerical aberrations, leading to genetic disorders with significant implications on an organism's health and development. The study of the Philadelphia chromosome serves as a compelling case study in this regard, providing insights into chronic myeloid leukemia (CML) and its targeted therapeutic approaches. In conclusion, Chapter 5 offers a concise understanding of the intricate world of genetics and inheritance. The insights gained from DNA structure and replication to the laws of inheritance and chromosomal aberrations are vital in deciphering the molecular basis of genetic disorders and the development of novel therapeutic interventions. This chapter serves as a crucial foundation for researchers and students alike, paving the way for future advances in the field of Genetics and its applications in cell biology and medical science. The next chapter will focus on cell metabolism and energy production.

Chapter

6 CELL METABOLISM AND ENERGY PRODUCTION

CONTENTS

In this pivotal chapter, we delve into the intricate workings of cell metabolism and its crucial role in energy production within living organisms. The fundamental processes that govern cellular respiration, photosynthesis, metabolic pathways, and enzyme-driven reactions are meticulously explored, unraveling the profound mechanisms that sustain life at the cellular level.

Section 6.1 embarks on a comprehensive exploration of cellular respiration, a fundamental process that facilitates the conversion of nutrient-derived molecules into usable energy. Within this section, we meticulously dissect the three vital stages of cellular respiration: Glycolysis, Krebs Cycle, and Electron Transport Chain. Through these sequential steps, cells adeptly harness energy from glucose and other organic compounds, generating adenosine triphosphate (ATP) - the currency of cellular energy.

Within Section 6.2, the remarkable process of photosynthesis takes center stage. As an awe-inspiring feat performed by autotrophic organisms, photosynthesis allows plants and certain microorganisms to convert light energy into chemical energy. We expound upon the two pivotal phases of photosynthesis: Light Reactions and the Calvin Cycle, illustrating how they culminate in the production of glucose and oxygen, making photosynthesis an elemental pillar in the Earth's ecosystem.

Section 6.3 unfolds a captivating exploration of metabolic pathways, epitomizing the balance between anabolism and catabolism. Delving into the intricacies of these processes, we analyze the building and breaking down of complex molecules, essential for the sustenance and growth of living organisms. A profound comprehension of these interrelated pathways enables a deeper understanding of cellular function and adaptation to varying environmental conditions.

In Section 6.4, the focus shifts towards the remarkable role of enzymes as catalysts for cellular reactions. Their extraordinary ability to accelerate biochemical reactions without being consumed has a profound impact on the efficiency and regulation of metabolic pathways. We expound upon the diverse classifications of enzymes, their mechanisms of action, and the significance of enzymatic regulation in maintaining cellular homeostasis.

Finally, in Section 6.5, we explore the multifaceted interplay of cellular metabolism and its regulation. This pivotal section casts light on the intricate network of feedback loops, allosteric regulation, and hormonal control mechanisms that govern cellular energy balance. A comprehensive understanding of cellular metabolism's intricate regulation is crucial in comprehending how cells adapt and respond to varying physiological demands. This chapter serves as an indispensable elucidation of the foundational concepts surrounding cell metabolism and energy production. By immersing ourselves in the intricacies of these processes, we forge a deeper appreciation for the remarkable complexity that sustains life and the extraordinary interplay of molecular reactions that orchestrate the energy flow within living organisms. It is pertinent to open this chapter by exploring the notion of cellular respiration.

6.1 CELLULAR RESPIRATION: GLYCOLYSIS, KREBS CYCLE, AND ELECTRON TRANSPORT CHAIN

Cellular respiration is a central metabolic process that occurs within all living cells, providing the necessary energy to fuel various cellular activities (Arber et al., 2021; Ardalan et al., 2022; Arhontoulis et al., 2022). This complex pathway involves the breakdown of organic molecules, primarily glucose, to generate adenosine triphosphate (ATP), the universal energy currency of the cell. Cellular respiration, constituting the core metabolic operation occurring within the bounds of living cells, provides the indispensable vitality required for the execution of a multitude of cellular actions. A quintessential case in point is found in the realm of cellular microbiology pertaining to the prokaryotic cells of Escherichia coli; a bacterium embracing the intricate process of cellular respiration. The germs incidentally manufacture ATP, a quintessential energy carrier, by the process of glycolysis, a process taking place in the bacterial cytoplasm. During this phase, with glucose as the primary substrate and a plethora of cellular enzymes acting as catalysts, there occurs the production of a net gain of two ATP molecules. Concomitant with the generation of ATP, there is a complementary formation of two three-carbon pyruvate molecules, which serve as precursors for subsequent mitochondrial cycles. In the succeeding Krebs cycle happening within the mitochondrial matrix of E.coli, the pyruvate is systematically dehydrogenated and decarboxylated, manifesting an array of intermediate compounds, inclusive of citrate, isocitrate, α-ketoglutarate, succinyl-CoA, succinate, fumarate, malate, and finally oxaloacetate with the release of ten ATP molecules. Simultaneously, there transpires the erecting of an electrochemical gradient via proton motive force through the electron transport chain located in the inner mitochondrial membrane, which permits chemiosmosis to occur, thus sealing the procedure of cellular respiration with the production of an additional 28 ATP molecules. This appreciable gain of 40 ATP molecules, by employing the cellular respiration path, establishes it as an efficient and primary metabolic avenue for E.coli to meet its energy needs, fittingly underscoring the indispensability of cellular respiration in fueling various activities in living cells. This meticulous discourse on E.coli cellular respiration provides a comprehensive understanding of the statement.

Cellular respiration consists of three interconnected stages: Glycolysis, Krebs Cycle (also known as the Citric Acid Cycle or TCA cycle), and the Electron Transport Chain (ETC). Let us embark on a detailed journey through each stage, uncovering the intricacies of cellular respiration.

1. Glycolysis: Glycolysis is the initial stage of cellular respiration and occurs in the cytoplasm of the cell (Soon et al., 2021; Smole et al., 2022; Spangler et al., 2022). This ancient metabolic pathway is conserved across virtually all organisms, from bacteria to humans. The process starts with the input of one molecule of glucose (a six-carbon sugar) and concludes with the production of two molecules of pyruvate (a three-carbon compound). The process of glycolysis denotes the first cog in the intricate machinery of cellular respiration, situated within the cytoplasmic environment of the cell and acts as an essential preliminary stage by which the cell accomplishes the conversion of glucose molecules into pyruvate compounds. A detailed exploration of this process is accomplished through the study of the cellular metabolism of the protozoan Paramecium multimicronucleatum, whose dependency on glycolysis for energy production serves as an effective representation. This species of Paramecium initiates glycolysis through the catalytic action of the enzyme hexokinase, which converts one molecule of glucose into glucose-6-phosphate, a step that necessitates the consumption of one ATP molecule.

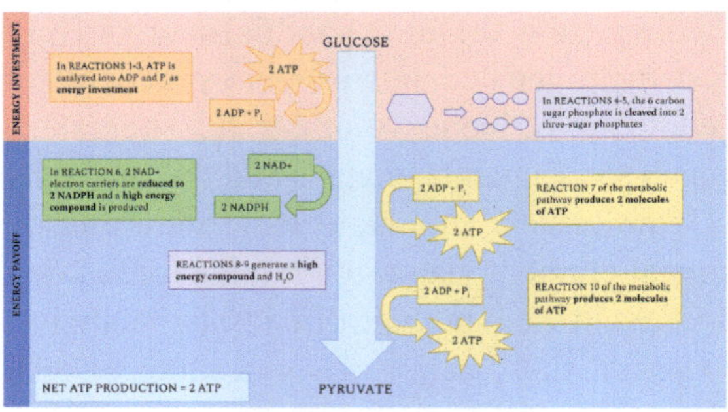

Figure 51. Various steps in glycolysis.

Source: By Sylvie Loh - Own work, CC BY-SA 4.0, https://commons.wikimedia.org/w/index.php?curid=118301216

Further, the transformation of glucose-6-phosphate to fructose-6-phosphate is facilitated, after which phosphofructokinase-1 catalyzes the conversion of fructose-6-phosphate into fructose-1,6-bisphosphate, again consuming an ATP molecule (see figure 51). Glycolysis then enters a phase of energy generation: the six-carbon fructose-1,6-bisphosphate compound is cleaved into two three-carbon compounds, with both proceeding through the same series of chemical reactions (Liu et al., 2022; Loconte et al., 2022; S. Lim et al., 2023). These compounds are further chemically revised into glyceraldehyde-3-phosphate. Subsequent reactions result in the production of two molecules of pyruvate from each molecule of glyceraldehyde-3-phosphate, with a net gain of two ATP and two NADH molecules. The pyruvate molecules constitute potential substrates for further metabolic processes, whereas the ATP serves as an immediate supply of energy for the Paramecium. Consequently, the elucidation of glycolysis in Paramecium provides a comprehensive and specified understanding of the statement. Through commencing the breakdown of glucose, glycolysis sets the stage for subsequent stages of cellular respiration that take place in specific conditions either aerobically or anaerobically. The glycolytic pathway comprises ten enzyme-catalyzed reactions, each tightly regulated to maintain cellular homeostasis. In the first five reactions, glucose is primed for cleavage and converted into two molecules of glyceraldehyde-3-phosphate (G3P). Subsequently, in the second half of glycolysis, G3P is further metabolized to generate two molecules of pyruvate. Throughout this process, a small amount of ATP and NADH (Nicotinamide adenine dinucleotide) is produced, which represents the initial energy harvest from glucose.

2. Krebs Cycle: Following glycolysis, the pyruvate molecules generated are transported into the mitochondria, where the Krebs Cycle takes place (Liang et al., 2021; Liang et al., 2022; Li et al., 2023). This cycle is a crucial part of cellular respiration, as it serves as a central hub for the oxidation of acetyl-CoA, a two-carbon compound derived from pyruvate, fatty acids, and certain amino acids. Upon the culmination of glycolysis in the

cytosolic environment of the cell, a pair of pyruvate molecules are generated. These molecules, each composed of three carbon atoms, are subsequently ferried into the mitochondria, the structure often referred to as the powerhouse of the cell, where they participate in the chain of reactions known as the Krebs Cycle, or Citric Acid Cycle as it is otherwise known. A probe of yeast Saccharomyces cerevisiae cells, broadly studied in the field of cellular microbiology, can serve as a detailed case study for this textual elucidation. Saccharomyces cerevisiae, an organism primarily recognized for its role in baking and brewing, reveals notable efficiency in its energy generation owing chiefly to its mitochondria and the processes occurring within these subcellular structures. As the pyruvate molecules enter the mitochondria of these yeast cells, they are initially subjected to the action of a multi-enzyme complex known as the pyruvate dehydrogenase complex, which through decarboxylation and oxidation reactions, transforms each pyruvate molecule into acetyl-CoA while simultaneously producing one molecule of NADH. The compound acetyl-CoA, enriched with two carbon atoms, then enters the Krebs cycle, where it amalgamates with oxaloacetic acid. Through a series of enzymatically facilitated reactions involving sequential decarboxylation, hydration, dehydrogenation, and phosphorylation, each turn of the cycle will result in the generation of three molecules of NADH, one molecule of FADH2, one molecule of ATP, and two molecules of carbon dioxide. These products then move on to more stages of respiration, contributing to a high energy yield. Therefore, the exploration of the Krebs cycle elucidating its function in Saccharomyces cerevisiae validates the claim and delineates the role of mitochondria in the efficient conversion of pyruvate to a multitude of energy-rich molecules, sustaining the energy needs of the cell.

The Krebs Cycle involves a series of enzyme-catalyzed reactions that sequentially break down acetyl-CoA, releasing carbon dioxide and high-energy electrons (Patriarca et al., 2021; Pauletto et al., 2021; Passanha et al., 2022). Throughout this cycle, NADH and FADH2 (Flavin adenine dinucleotide) are produced, serving as carriers of the electrons extracted from acetyl-CoA. These high-energy electron carriers are critical for the subsequent stage of cellular respiration. The Krebs Cycle, otherwise known as the citric acid cycle or tricarboxylic acid cycle, embodies an integral component of cellular respiration, driving the methodical breakdown of acetyl-CoA through a sequence of enzyme-catalyzed reactions. Succinctly, this process results in the liberation of carbon dioxide and high-energy electrons. The mitochondria of cardiac myocytes, specialized muscle cells found in the heart, provide an apt circumstance to emphasize this important process. These cells, known for their exceptional energy requirements to maintain uninterrupted cardiac contraction and relaxation cycles, rely heavily on the efficiency of the Krebs Cycle. The journey of one acetyl-CoA molecule in the inner mitochondrial matrix of a cardiac myocyte sees its union with oxaloacetate, forming citrate, catalyzed by the enzyme citrate synthase.

Subsequent enzymatic transformations, conducted by aconitase and isocitrate dehydrogenase, alter citrate into ketoglutarate with an accompanying reduction of NAD+ to NADH and production of carbon dioxide. Ketoglutarate is further acted upon by the enzyme complex ketoglutarate dehydrogenase, resulting in succinyl CoA, another molecule of carbon dioxide, and the reduction of a further NAD+ to NADH. Notably, succinyl-CoA is converted into succinate in a GDP-energy yielding reaction, facilitated by succinate thiokinase, thus producing an energy-rich GTP molecule. The remaining steps in the cycle transform succinate to fumarate, then malate, and ultimately back to oxaloacetate, with corresponding electron transfers to FAD+ and NAD+ to form FADH2 and NADH respectively. Therefore, the Krebs Cycle, exemplified in the context of cardiac myocytes, validates the aforementioned statement by illustrating the enzymatic breakdown of acetyl-CoA, the release of carbon dioxide and the crucial extraction of high-energy electrons used in the subsequent oxidative phosphorylation process, thereby collating energy for cellular utility.

3. Electron Transport Chain (ETC): The Electron Transport Chain (see figure 52) is the final stage of cellular respiration and occurs in the inner mitochondrial membrane of eukaryotic cells (Critcher & Huang, 2022; Cui et al., 2022; Cristiani et al., 2023). In prokaryotes, it takes place in the plasma membrane. The ETC consists of a series of protein complexes, each containing electron carriers, embedded in the mitochondrial membrane. The final step in the progression of cellular respiration, the Electron Transport Chain (ETC), takes place within the confines of the inner mitochondrial membrane in eukaryotic cells. The case study of the mammalian skeletal muscle fibers, particularly the

highly aerobic slow-twitch muscle fibers, illuminates this energy-generating process with striking clarity. These muscle fibers, due to their oxygen-intensive and enduring contractile activities, house a high quantity of mitochondria, thereby enabling an efficient ETC process. In these cells, the inner mitochondrial membrane serves as the residence for multiple protein complexes -- namely Complex I (NADH:ubiquinone oxidoreductase), Complex II (Succinate:ubiquinone oxidoreductase), Complex III (Ubiquinol:cytochrome c oxidoreductase), and Complex IV (Cytochrome c oxidase), all of which are involved in the step-by-step transfer of electrons from high-energy carriers NADH and FADH2 to molecular oxygen.

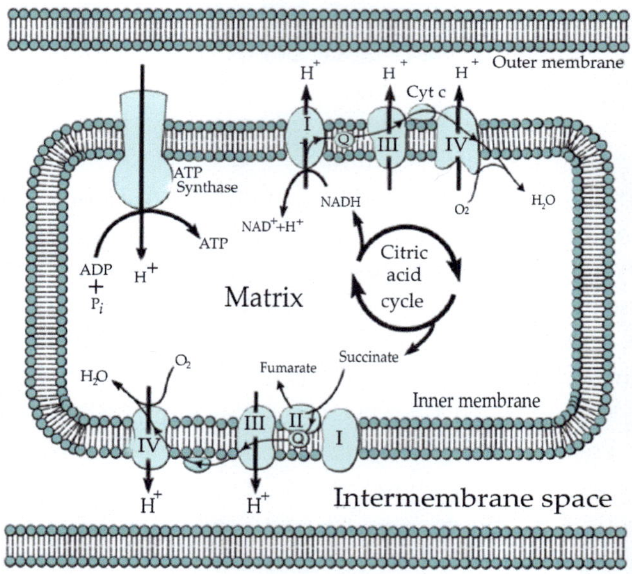

Figure 52. Illustration of the electron transport chain.

Source: Lumen Learning (https://courses.lumenlearning.com/suny-nutrition/chapter/6-26-electron-transport-chain/)

To initiate this process, NADH, produced during preceding cellular respiration processes, relinquishes its electrons to Complex I, whereas FADH2, separates its electrons to Complex II (Ford et al., 2021; Forveille et al., 2021; Gallusser et al., 2022; Gan, 2021). The energy liberated from these electron transfers guides protons from the mitochondrial matrix across the inner mitochondrial membrane, thereby establishing a proton gradient.

The protons, driven by this electrochemical gradient, return to the matrix via Complex V - ATP synthase – and their potential energy is tapped to phosphorylate ADP to ATP, the cell's primary energy currency. Concurrently, the electrons move from Complex I and II to Complex III, and thereafter to Complex IV, where they combine with protons and molecular oxygen forming water, consummating the process. Thus, through the examination of skeletal muscle fibers, it is apparent that the ETC positioned within the inner mitochondrial membrane of eukaryotic cells crucially operates as the terminal stage of cellular respiration, harvesting a substantial portion of the potential energy stored in glucose and converting it into a biologically usable form, ATP.

The high-energy electron carriers, NADH, and FADH2, generated during glycolysis and the Krebs Cycle, donate their electrons to the ETC (Hynds et al., 2021; Iachettini et al., 2023; Jardin et al., 2022). As electrons pass through the protein complexes of the ETC, they release energy that is used to pump protons (H+) across the mitochondrial membrane, creating a proton gradient. This gradient serves as a source of potential energy. During the metabolic processes of glycolysis and the Krebs Cycle, high-energy carriers, namely NADH and FADH2, are generated which play a significant role in the final stage of cellular respiration, specifically in the Electron Transport Chain (ETC). The role of these electron carriers can be examined within the context of human hepatic cells, given the liver's predilection for high metabolic activity and its concomitant high demand for ATP synthesis. Within these liver cells, each molecule of glucose, through the process of glycolysis and the Krebs Cycle, produces ten NADH molecules and two FADH2 molecules. These electron carriers enter the mitochondria, where they deliver their high-energy electrons to the complexes arrayed along the inner mitochondrial membrane, where the ETC takes place. NADH donates its electrons to the Complex I (NADH Dehydrogenase) of the ETC, which then goes through a series of redox reactions and pumps protons across the membrane, creating an electrochemical gradient. At the same time, FADH2, by transferring its electrons to Complex II (Succinate dehydrogenase), promotes similar translocation of protons. As these electrons traverse the successive complexes along the ETC, their energy is gradually harnessed to pump more protons out of the mitochondrial matrix, thereby amplifying this created gradient. Ultimately, the electrons are passed to molecular oxygen, the final electron acceptor, forming water in combination with free protons. The stored energy in the electrochemical gradient is finally exploited by the ATP synthase to convert ADP to ATP, the main energy currency of the cells.

Thus, by examining the role of NADH and FADH2 in human hepatic cells, it can be appreciated how these high-energy electron carriers, produced during glycolysis and the Krebs Cycle, are instrumental in the operation of the ETC and ultimately in the synthesis of ATP.

The final electron acceptor in the ETC is oxygen, which combines with electrons and protons to form water (Sagwal & Bekeschus, 2021; Ryder et al., 2022; Sakamaki & Mizushima, 2023). This process is essential for aerobic organisms as it allows them to utilize oxygen as the terminal electron acceptor, enabling efficient energy extraction from glucose. The ultimate endpoint of the Electron Transport Chain (ETC), a sequence of biochemical reactions essential to cellular respiration, is the reception of electrons by molecular oxygen, the final electron acceptor. The dynamic relationship between the ETC and oxygen is crystallized in the cellular respiration processes occurring within cardiac muscle cells. The continuous rhythmic contractions of the heart create a considerable demand for energy, necessitating an efficient and uninterrupted flow through the ETC. In the inner mitochondrial membrane of these cells, the four protein complexes of the ETC actively participate in managing the transfer of high-energy electrons from NADH and FADH2. As these electrons progress through the ETC, facilitated by different complexes, a significant amount of energy is released, which is employed to pump protons from the mitochondrial matrix to the intermembrane space, effectuating an electrochemical gradient. Simultaneously, the electrons are methodically relayed from one complex to another until they reach Complex IV, known as Cytochrome c Oxidase. At this juncture, each electron is finally transferred to a molecule of molecular oxygen. In an elegant biochemical denouement, the convergence of these electrons with protons from the surrounding medium and molecular oxygen yields water, a metabolic byproduct. In essence, through this detailed journey in the heart cells, it is revealed how molecular oxygen, as the ETC terminal electron acceptor, serves a crucial function in the production of water, rendering it a pivotal player in cellular respiration and energy production.

The flow of protons back into the mitochondrial matrix through an enzyme complex called ATP synthase drives the synthesis of ATP from adenosine diphosphate (ADP) and inorganic phosphate (Pi) (Panieri et al., 2022; Park et al., 2022; Panstruga et al., 2023). This process is known as oxidative phosphorylation, and it generates the majority of ATP produced during cellular respiration. The return flow of protons into the mitochondrial matrix, facilitated by an enzymatic assembly known as ATP synthase, ignites the synthesis of ATP – the energy currency of cells – from its precursors, adenosine diphosphate (ADP) and inorganic phosphate (Pi). This phenomenon, often referred to as oxidative phosphorylation, is explicated when examining the metabolism of neurons, the signaling entities of the nervous system. The incessant electrical activity of neurons engenders a significant demand for ATP, making these cells profoundly reliant on an effective oxidative phosphorylation process. Within the inner mitochondrial membrane of a neuron, ATP synthase serves as a proton channel, allowing the protons that were earlier pumped into the intermembrane area by the Electron Transport Chain, to flow back into the mitochondrial matrix. Importantly, this backflow of protons occurs along their electrochemical gradient, a process known as chemiosmosis. As the protons pass through the ATP synthase, the energy released is harnessed to catalyze the phosphorylation of ADP, integrating an inorganic phosphate to form ATP. Millions of ATP molecules are continuously synthesized this way, providing neurons with the capacity to maintain their high electrical activity. Therefore, this inquiry into neuronal metabolism offers insight into how ATP synthase utilizes proton motive force for the synthesis of ATP, marking the process as a key feature of cellular respiration and energy conversion in cells.

In summary, cellular respiration is an intricate and tightly regulated process that involves Glycolysis, the Krebs Cycle, and the Electron Transport Chain. Through these interconnected stages, cells efficiently harvest energy from glucose, producing ATP to sustain essential cellular functions. This fundamental process underpins the functioning of all living organisms and provides a glimpse into the remarkable complexity and adaptability of cellular metabolism. The second section of this chapter will explore photosynthesis in much more detail with a focus on light reactions and the Calvin Cycle.

6.2 PHOTOSYNTHESIS: LIGHT REACTIONS AND CALVIN CYCLE

Photosynthesis is a vital biological process that occurs in plants, algae, and certain bacteria, enabling them to convert light energy from the sun into chemical energy (Gu & Rasmussen, 2022; Hammond et al., 2022; Gu et al., 2023). Photosynthesis serves as a linchpin biological phenomenon that empowers plants, certain algae and specific bacteria to harness and convert captured light energy from the sun into usable chemical energy. A pivotal case study to elaborate this process is that of cyanobacteria, a group of bacteria that are regarded as pioneers of photosynthesis and whose study has shed considerable light on our understanding of this essential life-sustaining process. In these bacteria, photosynthesis takes place within specialized membrane structures called thylakoids. Light energy is absorbed by pigments, chiefly chlorophylls, situated within large protein complexes known as photosystems, namely Photosystem II and Photosystem I. The energy absorbed elevates electrons in the chlorophyll to a higher energy state. These energy-rich electrons then embark on an electron transport chain located within the thylakoid membranes. During their journey along this chain, the electrons lose some of their stored energy, which gets utilized to establish a proton gradient within the thylakoid lumen. Eventually, the electrons reach Photosystem I, where they are re-energized by more light energy before finally being transferred to NADP+ to form NADPH, a prominent energy carrier. Concurrently, as protons move down their concentration gradient out of the thylakoid lumen, through ATP synthase, ADP is phosphorylated to generate another crucial energy currency, ATP. This collective transformation of light energy into chemical energy in the form of NADPH and ATP, which will later participate in the Calvin cycle to produce sugar, is neatly illustrated by the photosynthetic processes of cyanobacteria and underscores the vital role photosynthesis plays in life-supporting energy production.

This intricate metabolic pathway is divided into two main stages: the Light Reactions and the Calvin Cycle, both of which occur within the chloroplasts of eukaryotic cells or the cell membrane of prokaryotic organisms. Together, these stages facilitate the synthesis of glucose and other organic compounds essential for sustaining life and supporting ecosystems.

1. Light Reactions: The Light Reactions, also known as the light-dependent reactions, are the initial phase of photosynthesis (see figure 53). They take place in the thylakoid membranes of chloroplasts in eukaryotes and in the cell membrane of photosynthetic prokaryotes. These reactions rely on the presence of light to initiate the process of energy conversion. The light reactions, denoted synonymously as light-dependent reactions, represent the first phase of the photosynthesis process, wherein light energy absorbed from the sun is converted into chemical energy in the form of ATP and NADPH. A profound exemplification of this is observed in the photosynthetic mechanisms of Arabidopsis thaliana, a model organism widely studied in plant biology. Within the chloroplasts, notably in the thylakoid membrane of this plant's cells, resides a duo of light-capturing complexes

known as Photosystem II and Photosystem I, equipped with chlorophyll pigments poised to absorb photons from sunlight. Upon photon absorption, electrons within the chlorophyll rise to a higher excited state in Photosystem II. These energized electrons then cascade through an electron transport chain, during which their energy is utilized to pump protons across the thylakoid membrane into the thylakoid lumen, thereby creating a proton gradient. Eventually, these electrons reach Photosystem I, where they are further energized with a second input of sunlight. These boosted electrons are finally used to reduce NADP+ to NADPH. Simultaneously, the difference in proton concentration across the thylakoid membrane drives protons back into the stroma through the enzyme ATP synthase, catalyzing the conversion of ADP into ATP. Thus, the examination of the light reactions occurring in Arabidopsis thaliana strikingly underscores their fundamental role as the initial phase of photosynthesis in transducting solar energy into the stored chemical energy necessary for the plant's metabolic needs.

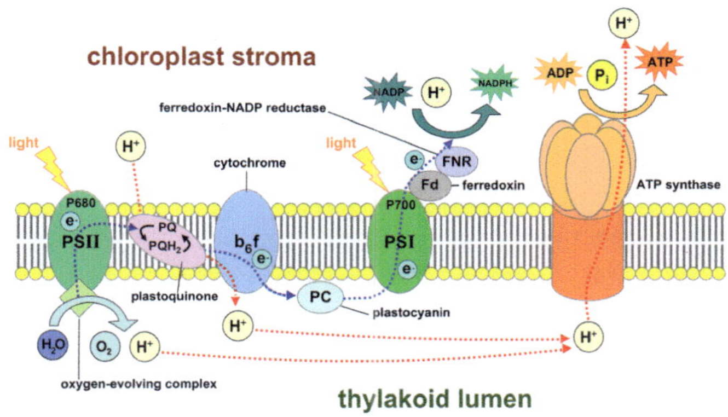

Figure 53. Light-dependent reactions in photosynthesis.

Source: By Somepics - Own work, CC BY-SA 4.0, https://commons.wikimedia.org/w/index.php?curid=38088695

The key components of the Light Reactions are photosystems, which are large protein complexes containing chlorophyll and other pigments (Ostrop et al., 2021; Ovciarikova et al., 2022; Ouyang et al., 2023). When photons of light are absorbed by chlorophyll molecules, electrons in the pigment become excited and gain energy. These energized electrons are then passed through a series of electron carriers within the photosystems, creating an electron transport chain similar to that in cellular respiration. Photosystems, defined as substantial protein complexes brimming with chlorophyll and a variety of other pigments, are integral players in the orchestration of light reactions that constitute the first part of photosynthesis. These photosystems perform a critical role in capturing and transducing light energy, which incites significant physiological changes within the organism. To elucidate this function, it is beneficial to delve into the photosynthetic processes of Synechocystis, a genus of cyanobacteria that has proven invaluable for its photosynthetic insights. Within the thylakoid membranes of Synechocystis cells, photosystems I and II operate in tandem to conduct light reactions. Both systems come equipped with multiple protein subunits and are embedded with pigments like chlorophyll and carotenoids. When light falls upon Photosystem II, energy is transferred within the antenna complex until it reaches the reaction center (P680), where it initiates an electron transport chain by exciting an electron, which then moves through a series of carriers, resulting in the pumping of protons across the thylakoid membrane. Meanwhile, the released electron is compensated by the splitting of water, a unique function of Photosystem II. After being relayed through carriers, the electron is taken up by Photosystem I and is again excited with light energy, particularly at the P700 reaction center. Here, the energized electron proceeds through another series of carriers, ending with the reduction of NADP+ to NADPH. Simultaneously, the built-up proton gradient across the thylakoid membrane stokes the synthesis of ATP via ATP synthase. Thus, the role of photosystems, as demonstrated in Synechocystis cyanobacteria, underscores the vital function of these complexes in transforming light energy into chemical energy during photosynthesis.

During this flow of electrons through the photosystems, energy is harnessed to pump protons (H+) across the thylakoid membrane or cell membrane, creating a proton gradient (Lefranc, 2021; Lee & Rompolas, 2022; Leao et al., 2023). This gradient is vital for ATP synthesis through a process called photophosphorylation. The final electron acceptor in the Light Reactions is NADP+ (Nicotinamide adenine dinucleotide phosphate), which becomes reduced to NADPH (Nicotinamide adenine dinucleotide phosphate) by accepting electrons along with protons. The flow of electrons through photosystems during photosynthesis enables the harnessing of energy, which is utilized to pump protons across the thylakoid membrane or the cell membrane resulting in the generation of a proton gradient. A comprehensive case study for this phenomenon can be analyzed within the photosynthetic machinery of the unicellular green alga, Chlorella. Within the chloroplasts—found abundantly in Chlorella due to its high photosynthetic capacity— the thylakoid membranes house multiple photosystems, namely Photosystem II and Photosystem I. When Chlorella is exposed to light, the photons absorbed by the photosystems set off a chain of redox reactions, each passing electrons in a directionally controlled manner through the two photosystems. Specifically, the energized electrons in Photosystem II are passed to plastoquinone, a mobile electron carrier in the thylakoid membrane. As plastoquinone gets reduced by an electron, it picks up protons from the stromal side of the membrane, transforming into plastoquinol, which subsequently travels to a protein complex called the Cytb6f complex. As it deposits its electron, plastoquinol releases the acquired protons into the lumen, causing a proton gradient across the membrane. The same process occurs down the chain at the level of Photosystem I, where the electron transfer to ferredoxin is accompanied by proton translocation. The translocated protons contribute to the formation of a proton-motive force (PMF). Therefore, the Chlorella example illustrates how the electron transport through photosystems is intricately coupled to proton translocation, establishing a proton gradient that fuels ATP synthesis, thus linking the light-reliant and light-independent phases of photosynthesis.

In addition to ATP and NADPH generation, the Light Reactions release oxygen gas as a byproduct through the oxidation of water molecules (Scott-Solomon & Hsu, 2022; Scott et al., 2022; Sezin et al., 2022). This process is known as photolysis, and it is responsible for replenishing the oxygen in the Earth's atmosphere, crucial for supporting aerobic life. Beyond generating ATP and NADPH, the light-dependent reactions of photosynthesis also facilitate the release of diatomic oxygen gas as a byproduct. This results from the oxidative division of water molecules, a process known as photolysis, which takes place predominantly in the photosystem II situated in the thylakoid membranes of photosynthetic organisms. An intriguing study of this phenomenon can be examined through the life processes of the common aquatic plant, Elodea canadensis. Often used in oxygenic photosynthesis demonstrations, Elodea serves as an exemplar of this photolytic function. When illuminated, Photosystem II in Elodea's chloroplasts absorbs sunlight and expedites an electron to a higher energy level. Simultaneously, the donated electron is replaced by the splitting of a water molecule—a process catalyzed by a multi-subunit protein complex within Photosystem II. For each water molecule that is subjected to photolysis, two electrons are provided to replenish those ejected from the Photosystem II reaction center, two protons are released into the thylakoid lumen contributing to the proton gradient for ATP synthesis, and a single oxygen atom is formed. After two water molecules undergo this process, a molecule of diatomic oxygen is released as a byproduct. Typically, these oxygen molecules escape into the plant's surrounding environment and are vital for maintaining Earth's atmospheric oxygen levels, playing an indispensable role in supporting aerobic life. Thus, through examining photosynthesis in Elodea (see figure 54), the co-generation of ATP and NADPH, as well as the concomitant production of oxygen during the light reactions of photosynthesis, is further understood and underscores the coordinated intricacy of these vital biological processes.

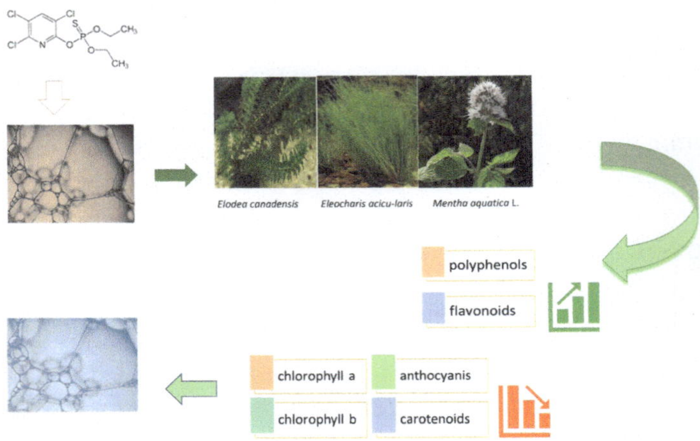

Figure 54. Illustration of experiments using Elodea.

Source: Sobiecka E, Mroczkowska M, Olejnik TP. The Influence of Chlorpyrifos on the Nonenzymatic Antioxidants Content in Macrophytes Leaves. Antioxidants. 2022; 11(4):684. https://doi.org/10.3390/antiox11040684

2. Calvin Cycle: The Calvin Cycle, also called the dark reactions or light-independent reactions, is the second phase of photosynthesis. It occurs in the stroma of chloroplasts in eukaryotic cells or in the cytoplasm in the case of prokaryotic organisms. Unlike the Light Reactions, the Calvin Cycle does not directly require light but depends on the products (ATP and NADPH) generated during the Light Reactions (Charpentier & King, 2021; Carroll & Marqusee, 2022; Chang et al., 2023). The primary goal of the Calvin Cycle is to synthesize glucose or other carbohydrates from carbon dioxide (CO2) using the energy stored in ATP

and the reducing power of NADPH. The cycle starts with the fixation of carbon dioxide, where an enzyme called ribulose-1,5-bisphosphate carboxylase/oxygenase (RuBisCO) catalyzes the reaction between carbon dioxide and a five-carbon sugar, ribulose-1,5-bisphosphate (RuBP). This fixation produces a six-carbon intermediate that quickly breaks down into two molecules of 3-phosphoglycerate. Through a series of chemical reactions, ATP and NADPH generated in the Light Reactions are used to convert 3-phosphoglycerate into glyceraldehyde-3-phosphate (G3P), a three-carbon sugar. Some G3P molecules are utilized to regenerate RuBP, while others combine to form glucose and other carbohydrates. Notably, for every six molecules of carbon dioxide fixed in the Calvin Cycle, only one molecule of glucose is produced. The other five molecules of G3P are recycled back into the cycle to regenerate RuBP. This recycling ensures the continuity of the Calvin Cycle and optimizes the use of ATP and NADPH, making the process efficient for carbon fixation.

In conclusion, photosynthesis is a fundamental process that sustains life on Earth by converting light energy into chemical energy through the Light Reactions and the Calvin Cycle. These interdependent stages demonstrate the remarkable complexity and adaptability of cellular metabolism, as they contribute to the synthesis of glucose and other organic compounds essential for the growth and survival of photosynthetic organisms and the entire ecosystem. The third section of this chapter focuses on metabolic pathways in terms of anabolism and catabolism.

6.3 METABOLIC PATHWAYS: ANABOLISM AND CATABOLISM

Metabolic pathways form the cornerstone of cellular metabolism, orchestrating the intricate web of biochemical reactions that sustain life within a cell (Granados-Riveron & Aquino-Jarquin, 2021; Graziadei & Rappsilber, 2022). These pathways can be broadly classified into two major categories: anabolism and catabolism. Both anabolic and catabolic processes are fundamental to the maintenance, growth, and adaptation of cells, playing pivotal roles in energy production, synthesis of biomolecules, and the breakdown of nutrients. Metabolic pathways, serving as crucial underpinnings of cellular metabolism, weave an intricate web of biochemical reactions vital for life within a cell. A detailed analysis of Escherichia coli, a well-researched model organism, offers profound insights into the dynamics of these metabolic pathways. E. coli, being a facultative anaerobe, exhibits metabolic flexibility in accommodating to varying environmental oxygen levels. Within the E. coli cell metabolism, glucose, as a typical carbon source, enters the glycolytic pathway, where it is methodically cleaved and converted into two molecules of pyruvate, procuring a small amount of ATP and NADH. Should oxygen be plentiful, pyruvate enters the mitochondria, where it is decarboxylated to form acetyl CoA, which becomes a participant in the citric acid cycle. A sequence of carefully regulated, enzyme-catalyzed reactions turn acetyl-CoA into carbon dioxide, concurrently liberating high-energy electrons, which are captured by NADH and FADH2. The accumulated NADH and FADH2 donate electrons to the electron transport chain, driving the synthesis of a large quantity of ATP, thereby showcasing efficient cellular metabolism under aerobic conditions. Under anaerobic conditions, however, E. coli switches to fermentation, where glucose is still expended via glycolysis, but the resulting pyruvate is converted into lactate, ethanol, or other fermentation product, instead of continuing into the citric acid cycle. This metabolic redirection allows the regeneration of NAD+ from NADH, ensuring the continuation of glycolysis, albeit with a lower ATP yield. Therefore, examining the metabolic pathways of E. coli indeed reinforces the assertion that these networks of reactions underpin the intricate phenomenon of cellular metabolism, effectively orchestrating the interplay between diverse metabolic processes in response to changing cellular needs and environmental conditions. Figure 55 highlights the two forms of metabolism (anabolism and catabolism) in cell microbiology.

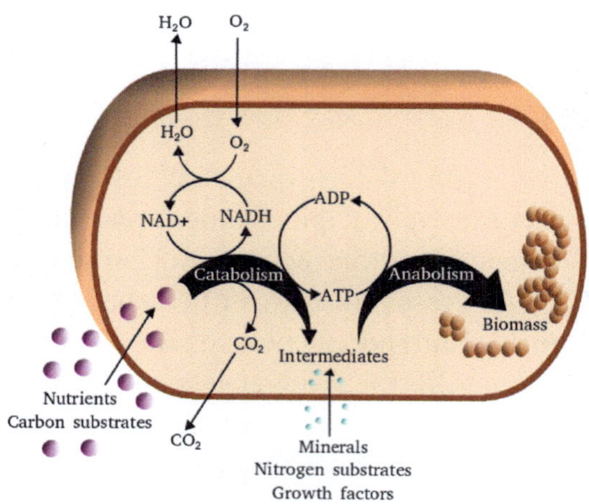

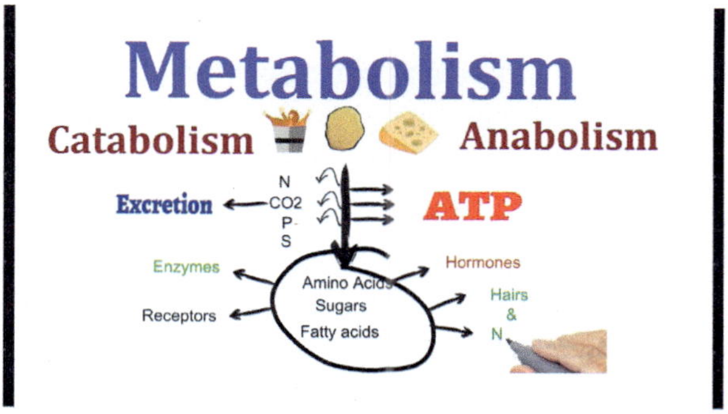

Figure 55. Metabolism in a bacterial cell.

Source: https://en.wikipedia.org/wiki/File:Metabolism.png

1. Anabolism: Anabolism refers to the set of metabolic pathways involved in the construction of complex molecules from simpler building blocks (Mangeat et al., 2021; Manils et al., 2022; Maricchiolo et al., 2022). These pathways require an input of energy, typically in the form of ATP, to drive the synthesis of various biomolecules essential for cell structure and function. Anabolic processes play a crucial role in cellular growth, repair, and the maintenance of cellular components. Anabolism, a collective term for the metabolic pathways engaged in the process of assembling intricate molecules from simpler building units, is a fundamental aspect of cellular metabolism, an understanding of which can be enhanced by considering the case study of Saccharomyces cerevisiae, commonly known as baker's yeast. This particular yeast is renowned for its robust capacity to employ anabolic pathways for producing a multitude of biomolecules necessary for cellular function. One classic example of anabolic activity in S. cerevisiae— and indeed in all living organisms — is protein synthesis. During this process, amino acids, the simpler building blocks, are methodically linked together to form complex protein molecules based on the genetic instructions encoded within the organism's DNA. The anabolic pathway leading to protein construction begins with the transcription of DNA into messenger RNA (mRNA) in the nucleus. This mRNA is then transported into the cytoplasm, where it attaches to ribosomes, the cellular machinery for protein synthesis. Thereon, transfer RNA (tRNA) molecules, each carrying a specific amino acid,

align with the mRNA strand, interpreting the coded information and linking the amino acids in a specific sequence, culminating in the formation of a polypeptide chain, essentially a protein. These proteins, known for their wide-ranging roles in various cellular processes, ranging from structural maintenance and enzymatic activity, to hormonal signaling and immunological defense. Therefore, the exploration of protein synthesis in S. cerevisiae explicitly demonstrates the significance of anabolic pathways in the genesis of complex molecules, vital for the sustenance of life.

One of the primary anabolic pathways is protein synthesis (Baluska & Yokawa, 2021; Aridor, 2022; Arjmand et al., 2023). Ribosomes, the cellular machinery responsible for protein production, utilize the genetic information encoded in messenger RNA (mRNA) to assemble amino acids into polypeptide chains. This energy-intensive process demands ATP and GTP (Guanosine triphosphate) for the formation of peptide bonds between amino acids. Protein synthesis, a principal anabolic pathway, is the process in which cells construct proteins, one of the key components necessitated for the smooth functioning of an organism's physiological operations. A detailed examination of this complex process can benefit from studying eukaryotic cells, with a specific focus on human fibroblast cells, which are known for their central role in wound healing requiring efficient protein synthesis mechanisms. Protein manufacture commences with transcription - the conversion of DNA into messenger RNA (mRNA), in the cell nucleus. The mRNA is then translocated into the cytoplasm, where it serves as a template for protein assembly during the process known as translation. On arrival at the ribosomes, the organelles charged with the protein synthesis duty, the mRNA sequence, categorized into codons, is read. Each codon specifically corresponds to a particular amino acid, the basic building blocks of proteins. Transfer RNA (tRNA) molecules, each bonded with an appropriate amino acid, recognize the codons via complementary anticodons and align on the mRNA template in a prescribed order. The ribosomes then catalyze the formation of peptide bonds between adjacent amino acids, thereby generating a polypeptide chain. Once the synthesis is completed, the chain undergoes further modifications and folding into a three-dimensional structure to form a functional protein.

Thus, studying protein synthesis in human fibroblasts exhibits a clear demonstration of the critical importance of anabolic pathways in cellular metabolism through the manufacture of complex molecules like proteins from simpler building blocks, underscoring the fundamental principle that fosters the continuity of life.

Similarly, nucleic acid synthesis is another essential anabolic pathway (Cheung et al., 2022; Choezom & Gross, 2022; Chhabra et al., 2023). DNA replication is a highly regulated process during cell division, requiring energy to ensure the accurate duplication of genetic information. Additionally, RNA synthesis involves the transcription of specific DNA regions into mRNA, allowing for the translation of genetic instructions into functional proteins. Nucleic acid synthesis represents a fundamental anabolic pathway, wherein essential biomolecules such as DNA and RNA are manufactured to ensure the proper functioning of cells. This significant biochemical process governs myriad aspects such as cellular replication, gene expression, and protein synthesis, thereby shaping the phenotypic traits of an organism. An illustrative embodiment of this process can be witnessed in the realm of microbiology, particularly in the bacterial model organism Escherichia coli. Upon the onset of DNA synthesis, E.coli initiates the replication process at the origin of replication, known as OriC. This initiation is contingent upon proteins such as DnaA, which bind to the OriC and prompt DNA helix unwinding. Subsequently, it catalyzes the assembly of the replisomes at each replication fork, which in turn comprise proteins like DnaB helicase to further unwind the duplex, single-stranded DNA-binding proteins to inhibit re-annealing, and DNA polymerase III, vital for actual DNA synthesis. It is of note that the polymerase can only attach nucleotides at the 3' end of the DNA molecule due to its inherent directionality, leading to the formation of leading and lagging strands. Through the Okazaki fragments, the lagging strand is iteratively synthesized and ligated by DNA ligase in a discontinuous fashion, while the leading strand undergoes continuous synthesis. This elaborate synthesis of nucleic acids in E.coli provides precise insights into the major anabolic mechanisms that shape not only microorganisms but life as we know it.

Furthermore, anabolism encompasses the synthesis of lipids and carbohydrates (Gelles et al., 2022; Ginther, 2022; Giannuzzi et al., 2023). Lipogenesis involves the production of lipids (fats) from acetyl-CoA and other precursors, contributing to cell membrane formation and energy storage. Meanwhile, carbohydrate synthesis involves the formation of complex sugars, such as starch and glycogen, which serve as energy reserves in cells. The anabolic processes within a cell, aimed at the synthesis of complex molecules, entail the construction of lipids and carbohydrates from simple precursors. A potent manifestation of this can be found within the model unicellular organism, yeast Saccharomyces cerevisiae, which systematically executes lipid and carbohydrate anabolism. Lipid synthesis in S. cerevisiae begins with the formation of fatty acids, primarily palmitic and stearic acid, via acetyl-CoA, malonyl-CoA, and the multifunctional enzyme complex, fatty acid synthase. This process takes place in the cytoplasm, where successive cycles of condensation, reduction, dehydration, and further reduction create a 16-carbon fatty acid. Glycerol-3-phosphate and these fatty acids then combine under the action of acyltransferases to produce phosphatidic acids, which serve as the foundational building block for most lipids within the cell. This phosphatidic acid is then manipulated, based on the metabolic needs of the cell, to form diverse lipid species such as phosphatidylethanolamine, phosphatidylcholine or triacylglycerols. In parallel, the anabolic production of carbohydrates takes shape through gluconeogenesis and the glyoxylate cycle when the yeast is growing on two-carbon compounds. With the usage of intermediates from the tricarboxylic acid cycle, this process synthesizes glucose, which participates subsequently in the formation of large polysaccharides, such as glycogen or the cell wall component β-glucan. Together, lipid and carbohydrate anabolism, as evidenced in S. cerevisiae, capture the essence of anabolic processes instrumental in cellular function and regulation. Overall, anabolism is a constructive and energy-demanding process that builds the diverse molecules essential for cellular structure, function, and growth.

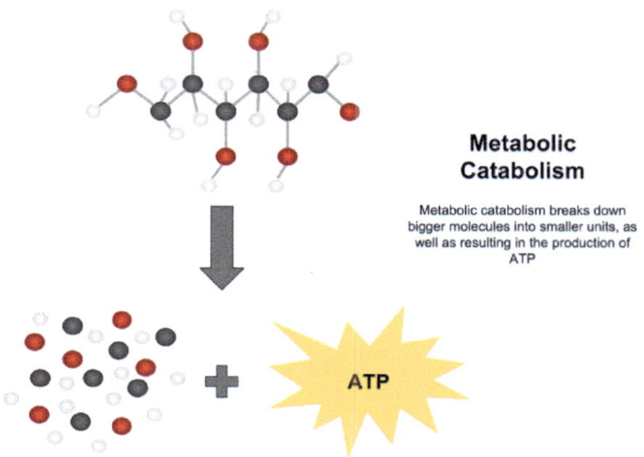

Figure 56. Conceptualization of catabolism.

Source: By Tr4nv1v - Own work, CC BY-SA 4.0, https://commons.wikimedia.org/w/index.php?curid=132019514

2. Catabolism: Catabolism (see figure 56), in contrast to anabolism, refers to the set of metabolic pathways responsible for the breakdown of complex molecules into simpler constituents X. C. Wang et al., 2021; X. Wang & X. W. Chen, 2023; Y. F. Wang & H. R. Chen, 2023). These pathways release energy, which is often stored in the form of ATP or other high-energy molecules, and provide the building blocks needed for anabolic processes. Catabolism, a fundamental metabolic pathway, breaks down complex molecules into simpler constituents, making it crucial for maintaining a cell's metabolic balance and producing energy. A distinct elucidation of this process can be observed in the bacterium Pseudomonas putida, a soil-dwelling microorganism capable of metabolizing a wide array of organic compounds. The catabolism of benzoate, a simple aromatic compound found plentifully in the soil, exemplifies this organism's metabolic versatility. The process commences when benzoate enters the cell and is activated through a reaction with coenzyme A, becoming benzoyl-CoA. The benzoyl-CoA then undergoes a series of redox reactions, mediated by enzymes such as a dioxygenase and a hydrolase, leading to the formation of a cyclic intermediate. This intermediate is further converted successively

into a suite of simpler molecules, including tricarboxylic acid (TCA) cycle intermediates. This transformation, cycling through the TCA cycle, results in the release of carbon dioxide, water, and energy in the form of ATP, underlining the quintessentially catabolic nature of the process. Consequently, benzoate provides an excellent case study illustrating how P.putida catabolically degrades complex molecules to simpler ones, both for energy production and for maintaining its metabolic stability in diverse environments. The importance of catabolism is thus reiterated, as it represents a critical component of cellular homeostasis and the energy generation machinery in microorganisms. One of the most significant catabolic pathways is cellular respiration, as discussed earlier, which involves the breakdown of glucose to produce ATP through glycolysis, the Krebs Cycle, and the Electron Transport Chain. This process releases energy stored in the chemical bonds of glucose and transfers it to ATP, providing the cell with a steady supply of energy.

Proteolysis is another crucial catabolic pathway responsible for the breakdown of proteins into individual amino acids (Ostrop et al., 2021; Ovciarikova et al., 2022; Ouyang et al., 2023). These amino acids can then be used as building blocks for the synthesis of new proteins or as precursors for other metabolic pathways. Proteolysis, one of the essential catabolic pathways, orchestrates the hydrolysis of proteins into individual amino acids, bolstering both cellular homeostasis and nutrient availability. A vivid illustration of proteolysis can be discerned through the life cycle of the marine bacteriophage Phi29. This microorganism undergoes a unique life phase known as the lytic cycle, where it enters bacterial cells, controls their machinery to replicate its genetic material, and synthesizes proteins to assemble new phage particles. A crucial aspect of this life cycle is the proteolytic maturation of phage proteins, which is critical for functional viral assembly. Specifically, the p6.5 protein, a viral-encoded protease, mediates the proteolytic cleavage of the scaffolding protein gp7, leading to the creation of a mature, hollow protein shell poised for DNA packaging. Following this packaging process, p6.5 undergoes autocatalytic proteolysis to terminate its protease activity momentarily. Once DNA packaging concludes, the fully formed virion erupts from the bacterial host, breaking down host proteins in the process, consequently leading to cell lysis. Hence, the Phi29 life cycle aptly exemplifies the fundamental role proteolysis plays in protein catabolism, not only ensuring proper viral assembly but also mediating the cyclical breakdown and renewal of proteins within the bacterial host, thereby sustaining the metabolic equilibrium and nutrient balance of the microenvironment.

Similarly, nucleic acid degradation involves the breakdown of DNA and RNA molecules into nucleotides (Baluska & Yokawa, 2021; Aridor, 2022; Arjmand et al., 2023). These nucleotides can be recycled for the synthesis of new nucleic acids or used in various other metabolic processes. Nucleic acid degradation, another pivotal component of the catabolic suite of processes, is accountable for the hydrolysis of nucleic acids, namely DNA and RNA, into their respective nucleotides. A specific and insightful instance of this activity can be identified in the bacterium Deinococcus radiodurans, which is remarkably resistant to ionizing radiation. Exposure to ionizing radiation causes extensive DNA double-strand breaks, leading to the degradation of the DNA molecule into fragments. However, D. radiodurans is capable of repairing its DNA through a process known as extended synthesis-dependent strand annealing. The first step in this repair process is the degradation of the damaged DNA molecule by a concerted action of exonucleases and endonucleases, which cleave the DNA strands either from the ends (exonucleases) or from within the sequence (endonucleases), yielding smaller DNA fragments or individual nucleotides. These nucleotides serve as the raw material for rebuilding the DNA molecule, a process facilitated by a plethora of enzymes including DNA polymerases, ligases and proteins involved in homologous recombination. Similar mechanisms underlie the breakdown of RNA molecules where specialized enzymes - RNases - target and degrade various types of RNA into their constituent nucleotides. Thus, the case of D. radiodurans epitomizes how nucleic acid degradation underpins not just the catabolic breakdown of nucleotides, but also ultimately serves to maintain the integrity of the genetic material in challenging environmental conditions.

Furthermore, catabolism includes the breakdown of lipids and carbohydrates (Gan et al., 2022; Garcia-Gutierrez & Garcia-Dominguez, 2022; Geers et al., 2022). Lipolysis involves the hydrolysis of lipids into fatty acids and glycerol, which can be utilized for energy production or the synthesis of other lipids. Glycogenolysis, on the other hand, is the breakdown of glycogen into glucose molecules, providing a readily available source of energy during times of increased demand. Catabolism comprises the breakdown of lipids and carbohydrates, presenting a crucial pathway to generate usable energy units for various cellular processes. One of the model organisms that effectively showcases lipid and carbohydrate catabolism is Escherichia coli, a gram-negative bacterium. E. coli, in the absence of glucose, can utilize alternative substrates, one of which is glycerol, classified as a carbohydrate, and another is fatty acids, which are constituents of lipids. For carbohydrate catabolism, E. coli metabolizes glycerol via glycerol kinase, which phosphorylates it to glycerol-3-phosphate. This molecule is subsequently oxidized to dihydroxyacetone phosphate, an intermediate that feeds into the glycolytic pathway, which catabolizes glucose to release energy in the form of ATP. This conversion of a complex carbohydrate to a simpler molecule earmarks glycerol catabolism. Concurrently, for lipid catabolism, E. coli has specialized transport mechanisms and enzymes to import and breakdown fatty acids. Upon internalization, the fatty acids are activated to their Coenzyme A (CoA) thioesters and undergo β-oxidation, a repetitive cycle of reactions resulting in the release of acetyl-CoA units. These units feed into the citric acid cycle, ultimately leading to the production of ATP, carbon dioxide, and water. In this manner, both lipid and carbohydrate catabolism in the metabolic context of E. coli offers an apt case study on how these complex molecules are strategically degraded to simpler constituents, aiding the release and efficient utilization of energy for the sustenance and growth of the organism.

In summary, catabolism is a degradative and energy-releasing process that breaks down complex molecules into simpler compounds, releasing energy that fuels anabolic processes and sustaining cellular activities. The balance between anabolism and catabolism is critical for maintaining cellular homeostasis. Proper coordination of these metabolic pathways ensures that cells can efficiently utilize energy, synthesize essential biomolecules, and adapt to changing environmental conditions. The delicate interplay of anabolic and catabolic processes underscores the remarkable complexity and adaptability of cellular metabolism in cell biology. The next section focuses on enzymes as catalysts for cellular reactions.

6.4 ENZYMES: CATALYSTS FOR CELLULAR REACTIONS

Enzymes are essential catalysts for cellular reactions, playing a crucial role in maintaining the dynamic equilibrium necessary for life (Hummert et al., 2021; Husnik et al., 2021; Hudson & Wieland, 2023). These specialized proteins accelerate biochemical reactions within cells, ensuring that cellular processes occur at rates compatible with life while preserving the integrity of cellular structures and components. A case in point is the regulation of cellular respiration (see figure 57) Enzymes, those macromolecular biological catalysts, are pivotal characters in the drama of cellular function, shouldering the vital responsibility of maintaining the dynamic equilibrium within every living cell. Their absence or malfunction brings forth a cascade of potential biochemical disruptions that can compromise the functionality and viability of the cell, underscoring their fundamental role in the sustenance of life.

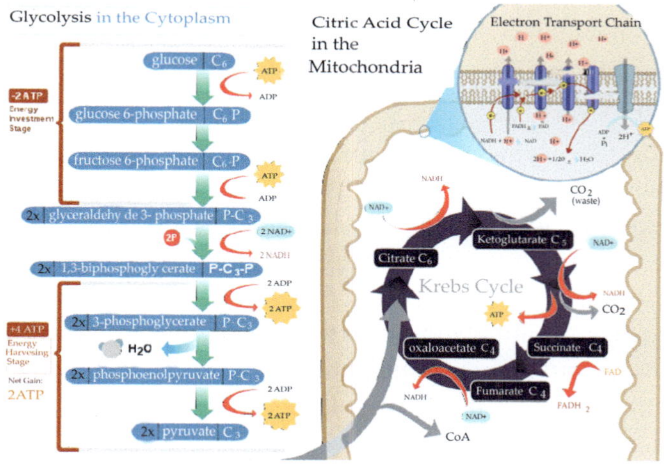

Figure 57. Illustration of the regulation of cellular respiration.

Source: By RegisFrey - Own work, CC BY-SA 3.0, https://commons.wikimedia.org/w/index.php?curid=4389845

To exemplify this, one can turn to the human cell's oxidative phosphorylation system—a cellular process responsible for producing about 90% of the body's adenosine triphosphate (ATP), the primary source of energy for most cellular processes (Katunin et al., 2021; Keenan et al., 2021; Khampang et al., 2023). Here, enzymes called ATP synthases catalyze the conversion of ADP and inorganic phosphate into ATP in the inner mitochondrial membrane. Without these ATP synthases, the equilibrium between the reactants (ADP and inorganic phosphate) and the product (ATP) would unfavorably tilt towards an accumulation of the reactants and a resultant dearth of ATP—leading to insufficient energy supply for essential cellular activities, such as DNA replication, protein synthesis, and active transport. The dynamic equilibrium maintained by ATP synthase is, therefore, a quintessential example of how enzymes, as catalysts, facilitate virtually all cellular reactions and life processes. Meanwhile, in the realm of cellular reactions, their importance extends beyond mere facilitation—they also contribute to the control and regulation of biochemical pathways. For instance, allosteric enzymes, whose conformation—and thus, activity—is modified by specific effectors, regulate the rate of glycolysis and the citric acid cycle. This ability to alter the rate of chemical reactions without requiring changes in environmental conditions, such as temperature or pH, is what makes enzymes indispensable to the maintenance of the intricately balanced theater of cellular life. Therefore, when one considers the enormous breadth and depth of enzymatic tasks, it becomes discernibly clear that enzymes are the indispensable catalysts that ensure the persistence and functionality of life, as we know it.

The study of enzymes in cell microbiology provides valuable insights into how living organisms regulate their metabolic pathways, respond to environmental changes, and adapt to varying physiological demands (Ostrop et al., 2021; Ovciarikova et al., 2022; Ouyang et al., 2023). Enzymes act as biological catalysts by lowering the activation energy required for a chemical reaction to occur, enabling the cell to perform a vast array of reactions necessary for its survival and function. A precise and detailed examination of the role enzymes play in cell microbiology illuminates our understanding of how living organisms maneuver their metabolic pathways, acclimate to environmental modifications, and adapt to diverse physiological demands. An illustrative case in point is the lac operon, a well-studied model of gene regulation and metabolic pathway control in Escherichia coli primarily through enzymatic action. The lac operon in E. coli, a prokaryotic microbe, is a cluster of three genes responsible for the transport and breakdown of lactose when glucose isn't readily available. When lactose is present in the environment and glucose is not, the lac operon is turned on. The enzyme beta-galactosidase is produced, which catalyzes the hydrolysis of lactose into glucose and galactose to be used as energy. The enzyme permease aids the transportation of lactose into the cell. This elegant example depicts the versatility and responsiveness of enzymes in directing metabolic pathways based on environmental stimuli. In yet another arena, microbes stand at the forefront of environmental adaptation, their enzymatic activity modulated to meet varying physiological demands. Thermophiles, for instance, survive in significantly high temperatures, their enzymes optimally functioning at temperatures lethal to most living organisms. These enzymes, termed extremozymes, maintain their structural integrity and catalytic activity under severe conditions, providing insights into both the adaptability and resilience of life. A case in point, the enzyme Taq DNA polymerase isolated from the thermophilic bacterium Thermus aquaticus, functions optimally at a temperature range where most enzymes denature. This enzyme now finds wide application in PCR technology, underscoring the immense potential of cell microbiology's study of enzymes. By uncovering such instances of enzymatic adaptability, the study of enzymes in cell microbiology elucidates how nature ingeniously solves complex physiological challenges.

In a cellular context, enzymes interact with specific substrates, forming enzyme-substrate complexes that undergo chemical transformations (Featherston & Cotruvo, 2021; Ferrari et al., 2022; Ferrari et al., 2023). These transformations typically involve the breaking and forming of chemical bonds, leading to the conversion of substrates into products. The specificity of enzyme-substrate interactions is key to their functionality, as it ensures that enzymes catalyze specific reactions in a highly controlled and regulated manner. In a cellular context, enzymes interact uniquely with specific substrates - a behavior that forms the bedrock of nearly all biological processes. This distinctive attribute of enzymes to identify and bind to their particular substrates to form enzyme-substrate complexes, which subsequently undergo a series of chemical transformations, is best exemplified in the process of DNA replication. A process fundamental to the existence and propagation of life, DNA replication requires a suite of specialized enzymes that execute specific tasks with surgical precision. A particularly significant player in this process is DNA polymerase, a star functionary that illustrates the sophistication of enzyme-substrate interactions. DNA polymerase interacts specifically with the DNA strand, recognizing the nucleotide sequence to be replicated as its substrate. Upon identification of the substrate, this enzyme forms a temporary enzyme-substrate complex. This specific binding event triggers a conformational change in the enzyme which brings the reactive groups of the enzyme into close proximity with the substrate, thereby lowering the reaction's activation energy. Ensued by this binding, the enzyme catalyzes the formation of phosphodiester bonds between nucleotides, synthesizing a new DNA strand complementary to the original template. The accuracy of this reaction exemplifies not only the specificity of enzyme-substrate interaction but also the transformative process that the complex undergoes - a conversion integral to life's continuity. Hence, the case of DNA polymerase underscores the central role enzymes play in mediating chemical transformations within the cell and the importance of such specific enzyme-substrate interactions in maintaining the preciseness and fidelity of cellular processes. The rate of enzymatic reactions is influenced by various factors, such as temperature, pH, substrate concentration, and the presence of enzyme inhibitors or activators (Masur et al., 2021; McCutcheon, 2021; Marvin et al., 2022). Cells carefully maintain these factors within specific ranges to optimize enzyme function and overall metabolic efficiency. The rate at which enzymatic reactions proceed is a subject of great relevance in cell microbiology, primarily because it directly impacts the efficiency of numerous physiological processes. A multitude of factors, including temperature, pH, substrate concentration, and the presence of enzyme inhibitors or activators, have substantial implications for enzymatic reaction rates.

A detailed examination of the enzyme Vibrio fischeri bacterial luciferase, which catalyzes the oxidation of a substrate called luciferin, emitting bioluminescence, provides an illustrative case study. Temperature plays a critical role in this reaction, as it impinges upon both the enzyme and substrate's kinetic energy. Optimal enzymatic activity and luminescence have been observed at approximately 30°C for V. fischeri luciferase. However, temperatures exceeding this range can lead to a loss of enzymatic function due to thermal denaturation. On another front, pH can significantly influence enzyme activity by altering enzyme and/or substrate ionization states — V. fischeri luciferase displays an optimal pH value of 8.0. Shifting this environmental pH can lead to decreased efficiencies or even cessation of the reaction. Substrate concentration also modulates the rate of enzymatic reactions. In simplistic terms, an increase in substrate concentration leads to a proportional increase in reaction rate until a saturation point is reached where further increases in substrate concentration yield no additional changes in reaction rate.

This saturation phenomenon is observed in the V. fischeri luciferase bioluminescent reaction. Lastly, the presence of enzyme inhibitors or activators equally holds sway over enzymatic reaction rates. Certain compounds can inhibit the activity of V. fischeri luciferase by blocking the active site or altering the enzyme's conformation, thereby obstructing substrate binding and subsequent oxidation. Alternatively, some compounds induce increased enzymatic activity, potentially by enhancing substrate binding or effecting a more favorable reaction environment. In conclusion, the intricacy of enzymatic reactions as demonstrated by the case of V. fischeri bacterial luciferase reflects the multifaceted and finely tuned influence of various factors on enzyme reaction rates, forming the cornerstone of understanding diverse cellular processes and their adaptation to changing physiological demands.

Enzyme Class	Reaction Catalyzed	Example
Hydrolase	Hydrolysis (catabolic)	Lipase, protease
Isomerase	Rearrangement of atoms within a molecule	Phosphohexoisomerase
Lyase	Splitting chemicals into smaller parts without using water (catabolic)	Decarboxylases, aldolases
Oxidoreductase	Transfers electrons or hydrogen atoms from one molecule to another	Dehydrogenases, oxidases
Synthetases	Joining of two molecules by the formation of new bonds (anabolic)	DNA ligase, DNA polymerase
Transferase	Moving a functional group from one molecule to another	Kinases, transaminase

Figure 58. Example of classification of enzymes based on catalytic mechanisms.

Source: https://ib.bioninja.com.au/standard-level/topic-2-molecular-biology/25-enzymes/types-of-enzymes.html

Enzymes can be classified into different categories based on their catalytic mechanisms and functional properties (see figure 58). For example, oxidoreductases facilitate redox reactions, transferases transfer functional groups between molecules, hydrolases catalyze hydrolysis reactions, and isomerases facilitate isomerization processes (Eskandari & Eaves, 2022; Erickson et al., 2023; Farese & Walther, 2023). Each enzyme's unique structure and active site dictate its substrate specificity and catalytic function. In the realm of cell microbiology, the classification of enzymes based on their catalytic mechanisms and functional properties stands pivotal in understanding the remarkable diversity of enzymatic functions and how they steer the multifaceted biochemical phenomena that drive life as we know it. To encapsulate such a vast assortment of enzymatic capabilities, one can consider the case study of dehydrogenase and ligase - two distinct classes of enzymes with disparate mechanisms and roles. Dehydrogenases, of which glucose-6-phosphate dehydrogenase (G6PD) is a prime example, catalyze the transfer of electrons from one molecule to another, often linked to the removal of a hydrogen atom. In this process, G6PD acts on glucose-6-phosphate, oxidizing it into 6-phosphoglucono-δ-lactone for the first step of the pentose phosphate pathway, a crucial metabolic pathway responsible for generating NADPH and ribose 5-phosphate. On the other hand, ligases, exemplified finely by DNA ligase, catalyze the formation of a new chemical bond with the concomitant hydrolysis of a pyrophosphate bond in ATP or a similar triphosphate. During the process of DNA replication, DNA ligase is responsible for facilitating the creation of phosphodiester bonds to 'seal' the nicks in the DNA backbone that arise during the lagging-strand DNA synthesis, playing a crucial role in ensuring the integrity and stability of the newly synthesized DNA. The vast contrast in the function and mechanisms of these two categories of enzymes vividly conveys how the classification of enzymes enables the succinct encapsulation of the broad spectrum of enzymatic properties observed within cells. Structuring enzymatic behavior into distinct categories provides not only an efficient conceptualization of a complex biological landscape, but also the necessary foundation to aid in the further understanding of the multitudinous and versatile roles that enzymes play in cellular life.

The regulation of enzyme activity is a fundamental aspect of cellular homeostasis (Lanz et al., 2021; Larson & McFarlane, 2021; Le Goff et al., 2022). Cells can control enzyme function through various mechanisms, such as allosteric regulation, covalent modification, and enzyme synthesis/degradation. These regulatory mechanisms ensure that enzymes are active when needed and inactive when their function is not required, helping cells adapt to changing conditions and metabolic demands. The regulation of enzyme activity is critical to maintaining cellular homeostasis, wherein biochemical equilibrium within the system is conserved despite fluctuations in external conditions. A clear illustration of this principle is observed within the Lac operon system of E. coli, as it masterfully manages the cellular metabolism of lactose. Within the Lac operon, the key enzyme, β-galactosidase, holds a significant role in the utilization of lactose when other preferred carbon sources, such as glucose, are absent. In a scenario with ample glucose, β-galactosidase production is hindered due to the binding of a Lac repressor to the lac operon, thus inhibiting transcription and subsequently, consequent enzyme production. However, in an environment bereft of glucose and rich in lactose, lactose molecules interact with the Lac repressor and induce a conformational change, resulting in the detachment of the repressor from the operon. This disruption permits transcription, leading to the eventual generation of β-galactosidase, which then metabolizes available lactose, converting it into usable energy. This regulatory mechanism allows the cell to adapt to varying nutritional conditions, thus maintaining metabolic homeostasis. Here, the production and activity of β-galactosidase are adeptly governed based on the cellular abundance of glucose and the presence of lactose, aptly demonstrating how the regulation of enzyme activity plays an instrumental role in cellular homeostasis. Therefore, through the distinct case of lac operon in E. coli, the fundamental role of enzyme regulation in cellular homeostasis can be discerned clearly; it underscores that careful regulation of enzyme activity is paramount in ensuring the stable, yet adaptable, operation of biological systems.

Enzyme dysfunction or deficiencies can have severe consequences for cellular function and overall health (Spiers et al., 2021; Stacey & Healy, 2021; Stallaert et al., 2022). Genetic mutations or environmental factors can lead to the loss of enzyme activity, resulting in metabolic disorders and diseases. Conversely, abnormal enzyme overactivity can disrupt cellular homeostasis, leading to uncontrolled reactions and cellular damage. Enzyme deficiencies or dysfunctions can have quite severe consequences for cellular function and overall organismal health, demonstrating the remarkable magnitude of enzymes' significance in orchestrating biological activities. A striking example of this correlation can be evoked by examining phenylketonuria (PKU), a metabolic disorder that fundamentally underscores how enzymatic deficiencies can derail cellular function, and consequently, an organism's health. PKU arises from a deficiency or impaired function of the enzyme phenylalanine hydroxylase (PAH), which is responsible for converting the amino acid phenylalanine into tyrosine. When the function of PAH is compromised due to a mutation, phenylalanine can no longer be adequately metabolized and consequently accumulates in the blood. The heightened levels of phenylalanine interfere with the transport of other essential amino acids to the brain, hindering normal brain development and leading to severe intellectual disability if left untreated. This dramatic example highlights the critical consequences that can ensue when enzymes, integral as they are to maintaining cellular equilibrium, fail to function correctly. Dysfunctional enzymes can disrupt not only the local cellular environment but also reverberate through the organism on a broader scale, inducing far-reaching physiological consequences and impairments. Therefore, through the lens of PKU, one may profoundly appreciate the role of functional enzymes in preserving the delicate balance of life, clarifying that even minor disruptions in their operational efficiency may have severe consequences for cellular function and overall health.

Cell microbiologists study enzymes' structure, function, regulation, and their role in cellular processes, including metabolism, signal transduction, DNA replication, and protein synthesis (Khanna & Villa, 2022; Kim & Choi, 2022; Kikuchi et al., 2023). The understanding of enzymatic mechanisms is crucial for advancing our knowledge of cellular physiology, disease mechanisms, and the development of targeted therapies and drugs. In unraveling the complexities of life at the microscopic level, cell microbiologists extensively study enzymes, probing their structure, function, regulation, and role in central cellular processes, including metabolism, signal transduction, DNA replication, and protein synthesis. The cellular process of DNA replication, for instance, provides an illustrative case study. A critical player in this process is the enzyme DNA polymerase, which serves as a paradigm of enzyme function and regulation in cellular activities. This go-to enzyme is assigned the intricate task of synthesizing a new DNA strand complementary to the template strand. The function of DNA polymerase is exquisitely regulated to maintain genomic integrity, notably through a feature known as proofreading activity—an intrinsic mechanism in some DNA polymerases—that identifies and rectifies incorrect nucleotide incorporations. Additionally, cell microbiologists study the three-dimensional structure of this enzyme to gain insights into its mechanism of action and factors influencing its function. Detailed investigations have revealed that at its core, DNA polymerase possesses a structure reminiscent of a human hand that "grasps" the DNA, helping it accurately add nucleotides to the growing DNA strand. Understanding the structure and function of DNA polymerase has also paved the way for its use in medical and industrial applications — most notably, polymerase chain reaction (PCR), a technique widely used for amplifying DNA sequences. From this one case study, it is evident that the study of enzymes stretches across multiple domains and is integral to cell microbiology. It not only enhances our understanding of fundamental life processes but also serves as a launchpad for technological innovations with far-reaching impacts on diverse areas, including healthcare, biotechnology, and environmental science.

In conclusion, enzymes are vital catalysts for cellular reactions, serving as the driving force behind the complex biochemical processes that sustain life. Their specificity, efficiency, and regulatory control enable cells to maintain metabolic balance and adapt to their ever-changing environment. The study of enzymes in cell microbiology is instrumental in unraveling the intricacies of cellular function and advancing our understanding of living organisms' fundamental biology. This chapter will close with a section on cellular metabolism and its regulation.

6.5 CELLULAR METABOLISM AND ITS REGULATION

Cellular metabolism is a fundamental aspect of cell biology, encompassing a complex network of biochemical reactions that occur within living cells (Nygaard et al., 2021; Nieto et al., 2022; Nyberg et al., 2023). These reactions enable cells to obtain energy, synthesize biomolecules, and maintain cellular homeostasis. Understanding cellular metabolism and its regulation is crucial for comprehending the intricacies of cell function, adaptation to changing environments, and the development of targeted therapies for metabolic diseases. In the context of cell microbiology, cellular metabolism represents a web of intricate biochemical reactions conducted within living cells, which are critical to the sustenance and propagation of life. To delve into the complexity and multiplicity of these metabolic reactions, one can probe the glycolytic pathway, a central metabolic pathway prevalent in almost all life forms. Glycolysis involves the enzymatic breakdown of glucose, a six-carbon sugar, into two molecules of pyruvate, a three-carbon compound, under anaerobic conditions.

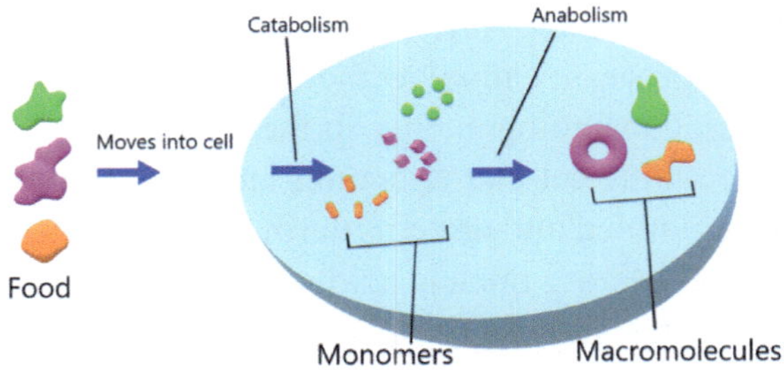

Figure 59. Comparing anabolism and catabolism.

Source: https://commons.wikimedia.org/wiki/File:Anabolism_and_Catabolism.png

This multi-step process is partitioned into a cascade of elementary reactions, each steered by a specific enzyme. Hexokinase, for instance, initiates the Glycolytic pathway by transferring a phosphate group to glucose. This fundamental step traps glucose within the cell and primes it for downstream enzymatic reactions. Close to the end of the pathway, the enzyme pyruvate kinase catalyzes the transfer of a phosphate group from a high-energy intermediate to ADP, forming ATP, a vital energy currency within the cell. This web of reactions taking place not in isolation, but as a linked chain in which the product of one reaction serves as the substrate for the next, exemplifies the richness and precision of metabolic networks. The balance of these interconnected reactions can be tipped by a variety of external stressors, including nutrient deprivation, toxin exposure, and changes in environmental conditions, underscoring the dynamic nature of cellular metabolism. Hence, through dissecting the glycolytic pathway, the significance of cellular metabolism within cell microbiology becomes unequivocally evident; it possesses the colossal task of orchestrating biochemical reactions necessary to meet a cell's energy requirements, biosynthetic demands, and responses to environmental perturbations, all of which are intrinsic to the cell's survival and function.

At the heart of cellular metabolism lies the interplay between catabolism and anabolism (see figure 59). Catabolic processes involve the breakdown of complex molecules, such as carbohydrates, lipids, and proteins, into simpler components, releasing energy in the form of adenosine triphosphate (ATP) (Bomont, 2021; Biermann & Reya, 2022; Bohnert & Johnson, 2022). Examples of catabolic pathways include glycolysis, which breaks down glucose into pyruvate, and the citric acid cycle (Krebs cycle), which further oxidizes pyruvate to produce ATP, carbon dioxide, and reducing agents. In contrast, anabolic processes involve the synthesis of complex molecules from simpler ones, consuming energy in the form of ATP. These pathways are essential for building cellular structures and biomolecules. For instance, amino acids are assembled into proteins during protein synthesis, nucleotides form DNA and RNA during nucleic acid synthesis, and simple sugars form complex carbohydrates during glycogenesis. Central to the maze of cellular metabolism lies the interplay between catabolism, the series of metabolic reactions that result in the breakdown of complex materials, and anabolism, the metabolic construction of compounds from more basic molecules, which is regarded as a defining feature in the field of cell microbiology. To embody this relationship, the metabolic cycle of glucose and its hierarchical position within the network of metabolic pathways serves as a pertinent case study. Catabolic processes, witnessed in the glycolysis pathway, involve the breakdown of glucose into two pyruvate molecules. This process occurs under anaerobic conditions and results in the net production of energy in the form of Adenosine triphosphate (ATP). The byproducts of glycolysis, under aerobic conditions, subsequently enter the citric acid cycle, proceeding through another round of catabolism, where additional ATP and high-energy electron carriers such as NADH and FADH2 are generated. These molecules then fuel the process of oxidative phosphorylation, resulting in the generation of an even larger yield of ATP. Conversely, anabolic pathways, which construct or synthesize complex molecules from simpler ones, consume this ATP to fulfil the cellular need for reduced coenzymes and precursors for macromolecular synthesis. The pentose phosphate pathway, for instance, an offshoot from the glycolytic pathway, is an anabolic pathway within the cell that generates nucleotide precursors and maintains the redox potential of the cell. The reciprocal relationship between anabolism and catabolism, where they alternate based on the energy demands and the physiological state of the cell, illustrates the essence of their interplay in cellular metabolism. This scenario makes it vividly clear that the bidirectional connection between catabolic and anabolic pathways forms the cornerstone of cellular metabolism, and underpinning this intricate network is the ultimate goal of cellular survival and growth.

The regulation of cellular metabolism is finely tuned to maintain the cell's energy balance and adapt to various physiological conditions (Tang et al., 2021; Tolic & Pavin, 2021; Tanaka et al., 2022; Thakur et al., 2022). Key regulators include enzymes, hormones, and nutrient-sensing pathways. Regulation of cellular metabolism is a fundamental and intricate process mastered by cells in order to maintain a balance of energy and to adapt to given physiological conditions. A case study elucidating the same is the strategic regulation of the enzyme Phosphofructokinase (PFK), considered to be the pacemaker of glycolysis. PFK catalyzes the phosphorylation of fructose 6-phosphate to fructose 1,6-bisphosphate, an early step in the catabolic pathway of glycolysis, a process that breaks down glucose molecules to generate ATP, the principal energy currency of the cell. The availability of ATP plays a major role in regulating PFK. In abundance, ATP acts as a competitive inhibitor to PFK, preventing further production as energy requirements are fulfilled, thus maintaining the cell's energy equilibrium. Additionally, PFK is activated by AMP (Adenosine monophosphate), which is indicative of a low energy state within the cell, triggering the need to produce more ATP via glycolysis. Furthermore, PFK is also modulated in response to physiological signals. In human muscles during exercise, an increase in the concentration of ADP (Adenosine Diphosphate), AMP, and Pi (inorganic phosphate) discernibly stimulates PFK activity, thereby boosting the rate of glycolysis to meet the elevated energy demands. This multifaceted regulation of PFK, adjusting to cellular energy levels and various physiological conditions, presents a clear example of how the regulation of cellular metabolism is finely tuned. Hence, the study of PFK illustrates the vital importance of stringent regulation mechanisms to ensure the smooth running of metabolic processes, maintain energy balance, and adapt efficiently in response to changing physiological conditions.

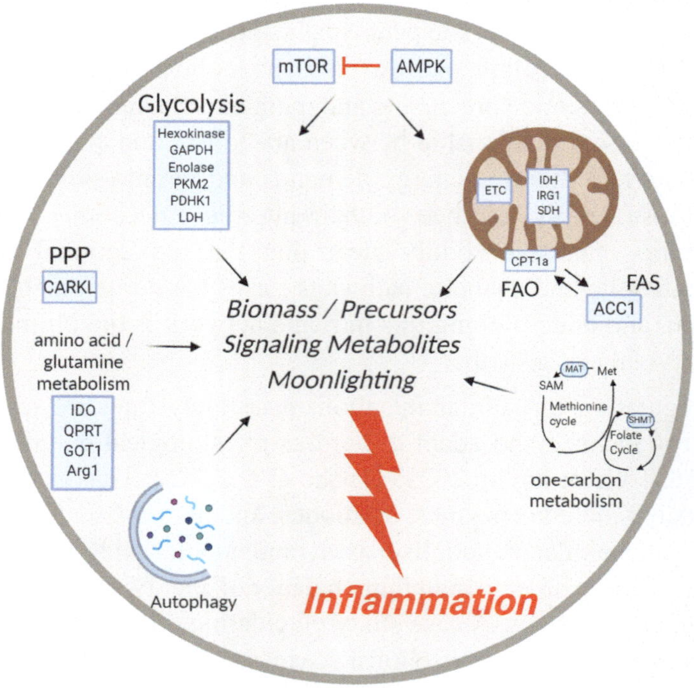

Figure 60. Example of the role of enzymes in regulating metabolism.

Source: Godfrey, W. H., & Kornberg, M. D. (2020). The Role of Metabolic Enzymes in the Regulation of Inflammation. Metabolites, 10(11), 426. MDPI AG. Retrieved from http://dx.doi.org/10.3390/metabo10110426

Enzymes are crucial players in cellular metabolism (see figure 60), acting as catalysts to accelerate chemical reactions (Montagnoli et al., 2021; Migliano et al., 2022; Montgomery et al., 2023). The activity of enzymes is tightly controlled to ensure the cell's metabolic processes occur at appropriate rates. Enzyme regulation involves allosteric control, covalent modification, and feedback inhibition. Allosteric regulation occurs when molecules bind to specific sites on enzymes, causing a conformational change that alters the enzyme's activity. Covalent modifications, such as phosphorylation, can activate or deactivate enzymes, depending on the cellular context. Feedback inhibition is a mechanism where the end product of a metabolic pathway inhibits an earlier enzyme in that pathway, preventing excessive product accumulation. Hormones play a vital role in coordinating cellular metabolism (Panieri et al., 2022; Park et al., 2022; Panstruga et al., 2023). Insulin and glucagon, for example, regulate glucose metabolism in response to changes in blood sugar levels. Insulin promotes glucose uptake and utilization in cells, while glucagon stimulates glycogen breakdown and gluconeogenesis to raise blood glucose levels when needed. Nutrient-sensing pathways, like the mammalian target of rapamycin (mTOR) and AMP-activated protein kinase (AMPK), respond to changes in nutrient availability and energy status. These pathways regulate processes such as protein synthesis, autophagy, and lipid metabolism to adapt to the cell's metabolic needs. Imbalances in cellular metabolism regulation can lead to metabolic disorders and diseases. Diabetes mellitus, for instance, is characterized by abnormal glucose metabolism due to insufficient insulin production or impaired insulin signaling. Obesity results from an imbalance between energy intake and expenditure, leading to excess fat accumulation and metabolic disturbances. Cellular metabolism and its regulation are extensively studied in cell biology research. Techniques such as metabolomics, which analyze the complete set of metabolites in a cell, and fluxomics, which measure metabolic fluxes in a pathway, provide valuable insights into cellular metabolism. These studies help identify critical metabolic pathways and potential therapeutic targets for metabolic diseases.

In conclusion, cellular metabolism and its regulation are fundamental processes in cell biology. The interplay between catabolic and anabolic pathways, along with the tight control of enzyme activity, hormones, and nutrient-sensing pathways, ensures the cell's energy balance and proper functioning. Dysregulation of cellular metabolism can lead to various diseases, underscoring the importance of understanding these processes in cell biology research and medicine. We will now close the sixth chapter by summarizing its contents.

CHAPTER 6 SUMMARY

Chapter 6 delves into the critical topic of cell metabolism and energy production. The chapter extensively discusses various aspects of cellular metabolism, shedding light on the intricate processes that cells employ to sustain life.

Section 6.1 explores cellular respiration, a central catabolic process responsible for extracting energy from nutrients. Glycolysis, the first step of cellular respiration, breaks down glucose into pyruvate, generating a limited amount of ATP and reducing equivalents. The Krebs cycle further oxidizes pyruvate, producing more ATP, carbon dioxide, and additional reducing agents. The electron transport chain, the final step of cellular respiration, couples with oxidative phosphorylation to produce a substantial amount of ATP. Together, these processes ensure efficient energy extraction from nutrients to fuel cellular activities.

Section 6.2 delves into photosynthesis, the anabolic process through which plants and some microorganisms convert light energy into chemical energy. The light reactions occur in the chloroplasts, capturing solar energy and transforming it into chemical energy in the form of ATP and reducing equivalents (NADPH). The Calvin cycle utilizes these energy-rich molecules to fix carbon dioxide into organic molecules, such as glucose, facilitating the synthesis of essential biomolecules for the cell.

The intricate web of metabolic pathways is thoroughly explored in Section 6.3. The cell continuously engages in anabolic and catabolic processes to build complex molecules and break them down, respectively. Anabolism is responsible for the synthesis of biomolecules, including proteins, nucleic acids, and carbohydrates, using energy derived from ATP. Catabolism, in contrast, involves the breakdown of complex molecules to release energy and generate ATP. This interplay between anabolism and catabolism ensures the cell's equilibrium and supports its diverse functions.

Section 6.4 focuses on enzymes, the central catalysts for cellular reactions. These specialized proteins accelerate biochemical reactions by lowering the activation energy required for reactions to occur. Enzymes achieve remarkable substrate specificity and ensure regulated reactions within the cell. The chapter emphasizes the significance of enzyme regulation, involving allosteric control, covalent modification, and feedback inhibition, to fine-tune metabolic processes according to cellular demands.

In Section 6.5, the chapter culminates with an in-depth discussion of the regulation of cellular metabolism. Maintaining energy homeostasis is crucial for cellular function, and this regulation is achieved through a complex interplay of enzymes, hormones, and nutrient-sensing pathways. Hormones like insulin and glucagon play vital roles in regulating glucose metabolism in response to changes in blood sugar levels. Nutrient-sensing pathways, such as mTOR and AMPK, respond to cellular nutrient and energy status, adjusting cellular processes accordingly.

In conclusion, Chapter 6 provides a comprehensive exploration of cellular metabolism and energy production. The chapter covers key topics such as cellular respiration, photosynthesis, metabolic pathways, enzymes, and the intricate regulation of cellular metabolism. Understanding these processes is fundamental to comprehending the inner workings of cells and their ability to maintain energy balance and adapt to varying environmental conditions. The knowledge gained from this chapter is vital for advancing cell biology research, uncovering disease mechanisms, and exploring potential therapeutic interventions for metabolic disorders. The penultimate chapter of this book will focus on the specialization and diversity of cells.

Chapter

7 CELL DIVERSITY AND SPECIALIZATION

CONTENTS

Within the intricate realm of cellular biology, the phenomenon of cell diversity and specialization unveils itself as a fascinating subject of study. This chapter embarks upon an exploration of the multifaceted world of cells, delving into their distinctive roles, structures, and functions within the context of various organisms. By peering into the molecular intricacies that underlie the remarkable versatility of cells, we gain a profound understanding of the mechanisms that enable the formation and sustenance of complex living organisms.

The journey commences with an in-depth investigation into stem cells, the enigmatic precursors that hold the potential to evolve into various cell types. By unraveling the mechanisms of differentiation and potency, we unravel the intricate processes that guide stem cells along distinct developmental trajectories. This section delves into the delicate balance between maintaining an undifferentiated state and steering towards specialized functions, shedding light on the driving forces that orchestrate the symphony of cellular specialization.

The narrative then pivots to the concept of tissues, the fundamental building blocks of multicellular organisms. Here, the emphasis shifts towards elucidating the intricate organization and collective functioning of cells within diverse tissue types. This exploration uncovers the synergistic collaborations between cells that confer specialized functions upon tissues, thereby underpinning the remarkable diversity of tasks executed by multicellular organisms.

A distinctive spotlight shines on muscle cells in this segment, highlighting their pivotal role in generating motion and mechanical force within organisms. A meticulous examination of the contractile machinery intrinsic to muscle cells reveals the intricacies behind their remarkable ability to exert force and facilitate motion. Furthermore, the exploration extends to the classification of muscle tissue types, showcasing the nuanced adaptations that underlie their specific functional capacities.

Navigating deeper into the cellular tapestry, this section delves into nerve cells, the conduits of information within the intricate nervous system. A meticulous dissection of nerve cell structures unveils the specialized features that enable the transmission of electrical signals by them across vast distances. Furthermore, the elucidation of signaling mechanisms elucidates the intricate web of communication that underpins the functioning of the nervous system, elucidating the profound interplay between structure and function.

Concluding the chapter, attention turns to the remarkable world of plant cell specializations. The exploration encompasses parenchyma, collenchyma, and sclerenchyma cells, unveiling their unique roles in plant structure, support, and defense. By unveiling the intricate adaptations and functions of these specialized plant cells, we gain insights into the mechanisms that enable plants to thrive in diverse environments. Hence, the journey into the captivating realm of cellular diversity and specialization continues, offering deeper insights into the intricacies that govern the multifarious functions of cells across the spectrum of life. Let us begin this chapter with an exploration of the potency and differentiation of stem cells.

7.1 STEM CELLS: DIFFERENTIATION AND POTENCY

Stem cells, a cornerstone of cellular biology, serve as the foundation upon which the remarkable diversity and specialization of cells are built (Jones et al., 2021; Karlsson & Draber, 2021; Ju et al., 2022). These cells possess the unique ability to self-renew and differentiate into various cell types, thereby contributing to the formation, growth, and repair of tissues within multicellular organisms. Stem cells indeed function as a vital bedrock in cellular biology, shaping the striking diversity and the exquisite specialization exhibited amongst various cell types. A rigorous, in-depth exploration of a case study involving the discovery and applications of induced pluripotent stem cells (IPSCs) elucidates this paradigm magnificently. IPSCs, uncovered by Nobel laureate Shinya Yamanaka in 2006, refer to regular body cells that have been coaxed under specific conditions to revert to embryonic stem cell-like status, preserving the capacity to mature into any cell type in the body. This principle of biological plasticity overturned decades of established wisdom of cellular growth being a unidirectional process, from less specific to more specialized cell types. The advent of IPSCs enables the study of diseases at the cellular level in a way not before possible. For instance, neuronal cells derived from IPSCs of Alzheimer's patients provide an unprecedented opportunity to gain insights into the cellular and molecular mechanisms at play in situ. Furthermore, the possibility of using patient-specific IPSCs raises the tantalizing prospect of precise personal medicine, limiting immune reactions commonly seen in current cell transplantation therapies. Nonetheless, significant challenges persist, including ensuring the safety of IPSC-derived products by eliminating the presence of residuals from the reprogramming procedures. The story of IPSCs underscores the centrality of stem cells in cellular biology by illuminating how they present a gateway to understanding cellular diversity and specialization, engender novel disease-modeling constructs, and shine a beacon of hope on the horizon of regenerative medicine.

At the heart of stem cell biology lies the concept of differentiation, the process by which unspecialized stem cells mature into distinct cell types, each endowed with specific functions and characteristics (Bettegazzi et al., 2021; Bennison et al., 2022; Beppler et al., 2022). The mechanisms that drive differentiation are intricate and finely orchestrated, often involving complex signaling pathways and genetic regulation. The interplay of external cues, such as growth factors and signaling molecules, with intrinsic genetic factors guiding the trajectory of stem cell development, steering them toward specific fates. The essence of stem cell biology is embodied in the concept of differentiation, representing the transitional journey of versatile stem cells maturing into diverse cell types imbued with unique functions and attributes. An illustrative case study examining the intricacies of this phenomenon can be observed in hematopoiesis, the process by which blood cells proliferate and differentiate within the body. Originating in the bone marrow, hematopoietic stem cells (HSCs) capture the quintessence of stem cell capacities, able to both self-renew to maintain their populace and give rise to a plethora of specialized blood cells, ranging from red blood cells (erythrocytes) tasked with oxygen transport, to various white blood cells like B lymphocytes and T lymphocytes, which serve crucial roles in the immune system. Our understanding of the complex molecular cues that regulate this remarkable differentiation process comes from decades of painstaking experimentation. The interplay between extrinsic signals from the stem cell niche, where the HSCs reside, and intrinsic transcriptional networks within the HSCs, dictate this carefully orchestrated maturation. For instance, erythropoietin and granulocyte colony-stimulating factor are released by the niche in response to signals like hypoxia and bind to receptors on HSCs, triggering their differentiation into erythrocytes or granulocytes, respectively. Simultaneously, internal regulators such as the transcription factor PU.1 ensure proper lineage commitment. Disturbances in this delicate equilibrium, as exemplified in diseases like leukemia, result from disruption to these regulatory pathways, causing unchecked proliferation of immature cells and a lack of functional differentiated products. Thus, the differentiation process emerging from the study of hematopoiesis signifies a critical epicenter of stem cell biology, providing insight into cell identity, function specificity, and pathological transformation. Figure 61 is a case in point with reference to how hematopoiesis involves the differentiation of multipotent cells into blood and immune cells.

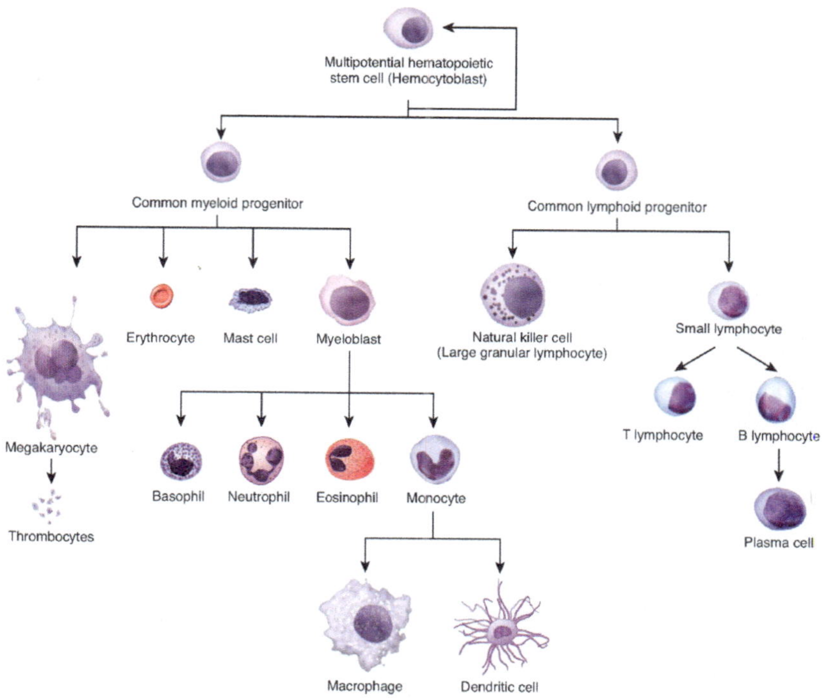

Figure 61. Hematopoiesis involves the differentiation of multipotent cells into blood and immune cells.

Source: Lumen Learning (tps://courses.lumenlearning.com/suny-ap1/chapter/cellular-differentiation/)

One pivotal aspect of stem cell potential is their level of potency, which defines the range of cell types a stem cell can give rise to (Shen et al., 2021; Siu et al., 2021; Shi et al., 2022; Shoup & Priola, 2023). Totipotent stem cells, found in the early stages of embryonic development, possess the remarkable ability to differentiate into any cell type of the organism, as well as supporting structures such as the placenta. As development progresses, these totipotent cells give rise to pluripotent stem cells, which can differentiate into cell types from all three germ layers: ectoderm, endoderm, and mesoderm. These pluripotent stem cells hold immense promise in regenerative medicine and research, as they can be manipulated to generate a wide range of cell types for therapeutic applications. Totipotent stem cells, predominantly observed in the embryonic stage of an organism's development, embody an extraordinary ability to differentiate into every cell type within an organism, not excluding supporting entities such as the placenta. To delve deeper into the mechanics and implications of this capability, the case of the zygote, a single cell formed by the fusion of an egg and sperm, serves as a prime model for totipotent stem cells. This solitary cell, the zygote, represents the inception of totipotency, possessing the remarkable capacity to evolve into a complete, complex organism. Indeed, in the first few rounds of cell division following fertilization, the resulting cells, known as blastomeres, maintain totipotency, capable of independently developing into a full organism, as evidenced by identical twins emerging from the splitting of early blastomeres. This unbounded potential results from their uniquely unrestrictive epigenetic landscape during these early cell divisions - without the usual molecular markers that can activate certain genes while silencing others, totipotent cells retain the ability to express any gene necessary for development. However, this omnipotent state of cellular differentiation is brief and, by the time the embryo reaches the blastocyst stage, differentiation has commenced and the cells lose their totipotent nature, becoming mere pluripotent stem cells, capable of producing all body cells, but not the entire organism, thus underlying the crucial role of totipotent stem cells in initiating cellular diversity and contributing to organismal complexity.

Further along the spectrum of potency are multipotent stem cells, which are more restricted in their differentiation potential. These cells are found in specific tissues and can give rise to a limited range of cell types within that tissue or related lineages. For instance, hematopoietic stem cells in the bone marrow can generate various blood cell types, while neural stem cells in the brain can differentiate into different types of neural cells.

Understanding and harnessing stem cell differentiation and potency have far-reaching implications (D. Wang et al., 2021; J. J. Wang et al., 2021; Wang et al., 2022);. In the realm of regenerative medicine, researchers aim to manipulate stem cells to replace damaged or diseased tissues, offering new avenues for treating conditions that were once considered incurable. Additionally, the study of stem cell differentiation provides insights into developmental biology, shedding light on the complex processes that orchestrate the formation of intricate multicellular organisms. The exploration and realization of stem cell differentiation and potency carry extensive implications that can revolutionize multitudinous segments of biological and medical sciences. A prime example showcasing this potentiates the instance of stem cell-based therapies in addressing degenerative afflictions. The opportunistic study of age-related macular degeneration (AMD) illustrates this aspect saliently. AMD is a progressive eye condition, leading to the degeneration of the retinal pigmented epithelium, a layer of cells supporting the photoreceptors essential for vision.

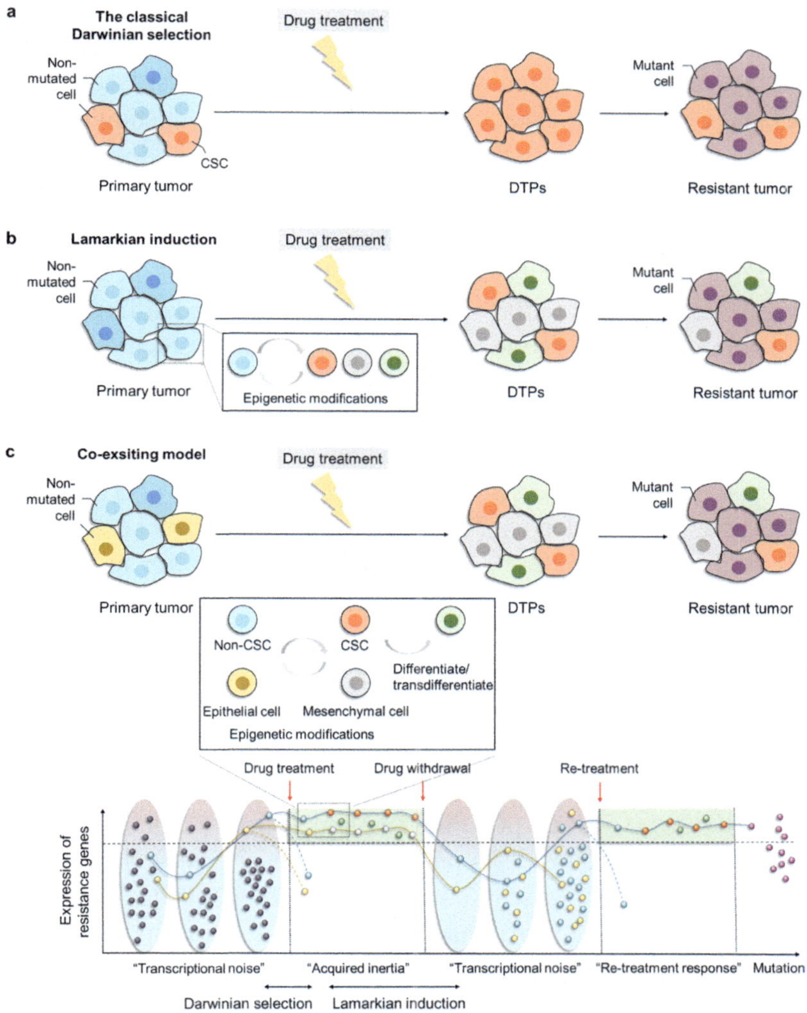

Figure 62. Illustration of the cancer stem cell concept.

Source: Qin, S., Jiang, J., Lu, Y., Nice, E. C., Huang, C., Zhang, J., & He, W. (2020). Emerging role of tumor cell plasticity in modifying therapeutic response. Signal transduction and targeted therapy, 5(1), 228.

Consequently, irreversible vision loss ensues, particularly impacting central vision. Existing treatment options remain limited, concentrating mostly on delaying the progress of the ailment rather than offering an outright cure. However, the advent of stem cell research has shifted this paradigm by proposing the prospect of regenerating the damaged retinal cells. Derived from either embryonic stem cells or induced pluripotent stem cells, researchers possess the capability to cultivate retinal pigment epithelial cells in laboratories. Through surgical transplantation, replacement of the degenerated cells with the stem cell-derived versions has shown promising results in restoring visual function in pre-clinical models. Navigating the challenges, including immune rejection and ensuring safety from tumorigenicity, constitutes a monumental stepping stone to propel these experimental therapies in clinical trials.

Therefore, delving into the machinations of stem cell differentiation and potency not only increases our understanding of developmental biology, but also translates into far-reaching implications in therapeutic applications, demonstrating the transformative potential of stem cells in repairing the irreparable. Figure 62 highlights the cancer stem cell concept as a case in point. It is important to note that the potential of stem cells also brings ethical considerations and challenges (Pandey et al., 2021; Panicker et al., 2021; Paliulis et al., 2023). The use of embryonic stem cells, for instance, raises ethical dilemmas due to the need to harvest cells from embryos, which can result in the destruction of the embryo. To circumvent such issues, induced pluripotent stem cells (iPSCs) were developed, allowing researchers to reprogram adult cells into a pluripotent state, thus sidestepping the need for embryonic cells. This breakthrough has not only addressed ethical concerns but has also opened new avenues for personalized medicine and disease modeling. The vast potential of stem cells undeniably brings forth consequential ethical considerations and challenges as well. A case study that exemplifies this is the use of embryonic stem cells (ESCs) in biomedical research. ESCs, derived from human embryos, possess pluripotency, granting them an unparalleled ability to differentiate into any cell type in the body, making them invaluable in understanding human development, disease modeling, and potential therapeutic applications. Nevertheless, the process of obtaining these cells involves the destruction of the embryo, typically sourced from surplus embryos generated in in vitro fertilization clinics, which escalates into significant moral and ethical conundrums concerning the embryo's status and its right to potential life. Despite the argument that these embryos would otherwise be discarded, critics assert that using them for research still underestimates the intrinsic value of potential human life. This ethical controversy has led to differing regulatory landscapes globally; whereas some countries have embraced ESC research, others enforce stringent restrictions. Additionally, the possibility of creating embryo-like structures solely for research further magnifies these ethical dilemmas due to concerns over 'embryo farming'. As such, while tremendous scientific progress and medical breakthroughs hinge on stem cell research, it is incumbent on the scientific community to continually address and respect the inherent ethical considerations, continually adapting as our understanding and capabilities advance.

In conclusion, stem cells represent a cornerstone of cellular diversity, playing a pivotal role in the generation and maintenance of multicellular organisms. Their remarkable ability to differentiate and their varying levels of potency form the foundation for the intricate processes that drive development, tissue repair, and regeneration. Stem cell research continues to unveil new insights into the mechanisms governing cellular specialization and has the potential to revolutionize medical treatments and our understanding of life itself. The second section of this chapter will consider tissues with reference to multicellular organisms.

7.2 TISSUES: ORGANIZATION AND FUNCTION IN MULTICELLULAR ORGANISMS

In the realm of multicellular organisms, the concept of tissues serves as a cornerstone for understanding the complex organization and functional integration of cells (Montagnoli et al., 2021; Migliano et al., 2022; Montgomery et al., 2023). Tissues embody a higher level of structural and functional organization, wherein groups of cells collaborate to fulfill specific tasks essential for the organism's survival and homeostasis. Within the sphere of multicellular organisms, the notion of tissues stands as a fundamental construct for discerning the intricate organizational structure and functional integration of cells. A detailed exploration of the human nervous tissue provides relevant insight into this claim. Nervous tissue comprises two primary cell types, neurons and neuroglia, which work in unison yet differentiate in terms of function. Neurons, the excitable cells, transmit electrical signals across expansive distances in the body, thereby regulating a plethora of physiological functions ranging from sensorimotor coordination to cognitive processes. Although neurons captivate considerable attention in neurobiology, the more abundant neuroglia, or glial cells, possess an equal, if not more significant, functional capacity. Glial cells, including astrocytes, oligodendrocytes, and microglia, support neurons at multiple fronts — be it in maintaining homeostasis, forming myelin, or providing defense against pathogens. Furthermore, the cellular organization within this tissue is guided by the principle of polarity. Neurons inherently exhibit polarity, with dendrites and axons at opposite ends, thereby directing the flow of information in a singular direction, ensuring the precision and synchronization of neuronal communication circuits. Research on disorders such as multiple sclerosis (see figure 63), where the myelin sheath formed by oligodendrocytes is compromised, exemplifies how understanding the organization and function of cells within a tissue context can prompt therapeutic strategies that can restore tissue function. Hence, grappling with the concept of tissues, as evident within the nervous tissue, generates clarity on how amassed groups of specialized cells conjointly execute complex, higher-order functions within multicellular organisms.

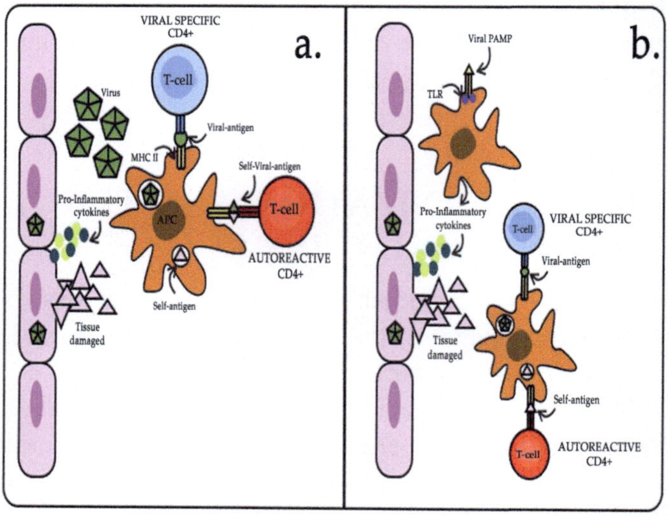

Figure 63. Representation of molecular mimicry and bystander activation mechanisms.

Source: Cervantes-Gracia, K., Husi, H. Integrative analysis of Multiple Sclerosis using a systems biology approach. Sci Rep **8**, 5633 (2018). https://doi.org/10.1038/s41598-018-24032-8

The organization of tissues is a symphony of cellular specialization, where individual cells within a tissue adopt distinct roles that collectively contribute to the tissue's function (Quintero-Carmona, 2021; Radulovic et al., 2022; Rao et al., 2022). This specialization is often driven by the expression of specific genes and the activation of signaling pathways, leading to the differentiation of cells into various types with unique structural and functional attributes. Exemplifying the declaration that the organization of tissues represents a remarkable concert of cellular specialization, with each distinct cell undertaking a unique role that collectively enhances the overall functionality of the tissue, a comprehensive analysis of the human epidermis, a tissue in the skin's outermost layer, offers an illuminating case study in cell biology. Composed chiefly of keratinocytes, these cells undergo a meticulous process of differentiation, initiating as basal cells within the stratum basale layer of the skin, proliferating, and methodically migrating upwards to replace older cells. As these keratinocytes journey toward the skin's surface, they experience drastic morphological, biochemical, and genetic modifications that metamorphose them into anucleate corneocytes, eventually resulting in the creation of a sturdy, protective layer known as the stratum corneum. Concurrently, melanocytes strategically positioned among the basal cells contribute their unique role to the epidermal tissue, synthesizing melanin that not only assuages the deleterious effects of ultraviolet radiation on skin cells, but also imparts coloration to the skin. Equally critical, Langerhans cells, functioning as immunosurveillance entities within the epidermis, detect the invasion of foreign pathogens and promptly initiate an immune response. Collectively, these distinct cells demonstrate compelling specialization within a single tissue, each accomplishing a critical role tailored to supporting the broader functionality and integrity of the epidermal tissue. Thus, through this intricate collaboration and specialization across various cell types, tissues perform their complex roles in maintaining the multifaceted clockwork of bodily functionality.

Epithelial tissues, for instance, exemplify the cohesive nature of cellular collaboration (Shen et al., 2021; Shi et al., 2022; Shoup & Priola, 2023; Siu et al., 2021). Comprising tightly packed cells, epithelial tissues form protective barriers that line the surfaces of organs and cavities. These tissues exhibit diverse forms, ranging from simple epithelia composed of a single layer of cells to stratified epithelia with multiple layers. This structural diversity aligns with their functions, which encompass absorption, secretion, protection, and sensory perception. The arrangement of epithelial cells is tightly regulated, ensuring the establishment of a functional barrier while allowing selective permeability to substances. Epithelial tissues provide a succinct illustration of the cohesive nature of cellular collaboration. A detailed case study of the interaction and cooperation among colon epithelial cells can elucidate the dynamics at play within the tightly structured assembly of this type of tissue. Positioned in the interior lining of the large intestine, these cells adopt one of three eminent roles -- absorptive cells, goblet cells, or endocrine cells – each fulfilling a key function for the overarching tasks of this tissue. Absorptive cells predominantly take up water and key electrolytes from the digested food material traversing the colon, facilitating critical osmoregulatory processes and engaging in nutrient absorption. On the other hand, goblet cells specialize in secreting a lubricating mucus, lining the colon to not only help to ease the movement of fecal material but also shield the epithelial cells from physical damage and microbial invasion. The endocrine cells interspersed within this epithelial layer underscore the reciprocal nature of cellular collaboration as they synthesize and release critical hormones that regulate various physiological functions, including the rate of peristalsis that propels digestion. In this highly orchestrated system, each cellular entity within the colon epithelial tissue embodies a quintessential feature, collectively enhancing the overall performance, maintenance, and integrity of the tissue. These dynamics among the distinct cells within a tissue demonstrate the model of cellular collaboration that lends itself to the resounding resilience and functionality observed in the meticulously organized epithelial tissue.

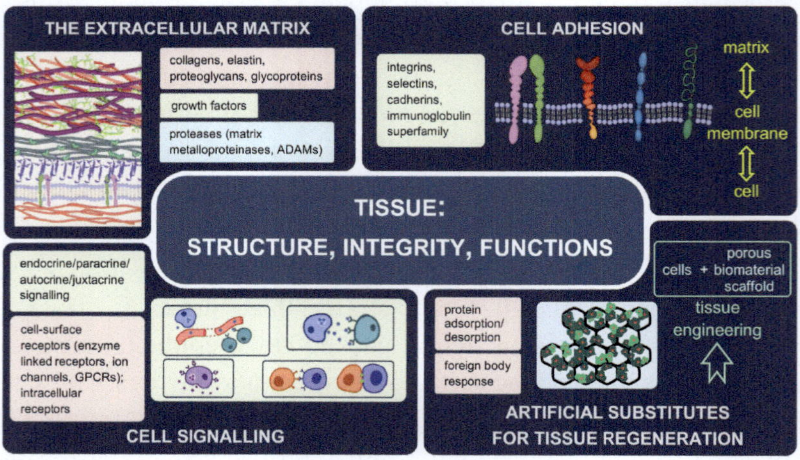

Figure 64. Illustration of connective tissues providing structural support.

Source: Bandzerewicz A, Gadomska-Gajadhur A. Into the Tissues: Extracellular Matrix and Its Artificial Substitutes: Cell Signalling Mechanisms. Cells. 2022; 11(5):914. https://doi.org/10.3390/cells11050914

Connective tissues, on the other hand, provide structural support and facilitate communication between different tissues (Hynds et al., 2021; Jardin et al., 2022; Iachettini et al., 2023). They encompass a range of cell types, including fibroblasts, adipocytes, and immune cells, embedded within an extracellular matrix composed of fibers and ground substance (see figure 64). The diverse roles of connective tissues span from providing mechanical strength to storing energy and participating in immune responses. Cartilage and bone exemplify specialized connective tissues, each endowed with unique structural adaptations to fulfill their specific functions. Connective tissues, functioning as integral structural and communicative conduits in the body, exemplify this statement explicitly, as evident in the specific study of bone tissue, a specialized form of connective tissue. Superbly characterized by a robust ECM (extracellular matrix), rich in mineralized calcium phosphate that confers structural support, and collagen fibers imbuing tensile strength, bone tissue provides the skeletal framework and protective shield indispensable for the organism's survival. In the meticulously organized architecture of bone tissue, two distinct cell types — osteoblasts and osteoclasts — assume specialized roles that are indispensable to the maintenance and remodeling of this tissue. Osteoblasts, derived from mesenchymal stem cells, manifest their role by synthesizing and depositing the organic components of the bone matrix, thereby contributing to bone formation, while osteoclasts, large, multinucleated cells, partake in the bone's dynamic metabolic activities by carving niches in the bony matrix, facilitating critical calcium regulation and bone remodeling. Remarkably, the concerted actions of these cells contribute to the tissue homeostasis that ensures the upkeep of bone integrity and strength. These cells also function in inter-tissue communication, as illustrated by their secretion of signaling molecules, like RANKL and OPG, which influence the activity of cells in distant tissues involved in calcium homeostasis. Therefore, through the collective action and communication of its distinct cellular components, the connective tissue — exemplified here by bone tissue — underlies the structural support and facilitates crucial inter-tissue communication, underscoring the supremacy of cellular collaboration.

Muscle tissues exemplify the integration of cells for specialized function (Wu & Liu, 2021; Xu & Li, 2021; Yang et al., 2021; Xie et al., 2022; Xue et al., 2022). Muscle cells, or fibers, are densely packed and contractile, allowing them to generate mechanical force. Three types of muscle tissues—skeletal, smooth, and cardiac—exhibit distinct structural characteristics and fulfill specialized roles. Skeletal muscles, attached to bones via tendons, facilitate voluntary movements, while smooth muscles, found in organs like the digestive tract, enable involuntary contractions. Cardiac muscles form the heart and generate coordinated contractions that sustain circulation. Muscle tissues encapsulate the quintessential integration of cellular components to engender a specialized function. For illustrative purposes, this elaborate relationship is distinctly noticeable within the cardiac muscle - one of the three types of muscle tissue. Comprised primarily of cardiomyocytes, these cells combine their inherent, unique attributes to form a rhythmically contractile unit that drives the heart's pumping function. Each cardiomyocyte features an intricate network of sarcomeres - the fundamental contractile apparatus composed primarily of actin and myosin filaments. During each heartbeat, the cardiomyocytes tightly coordinate their contraction through a series of electrical signals initiated and propagated via an intrinsic conduction system, which features specialized pacemaker cells such as sinoatrial (SA) and atrioventricular (AV) nodes. These pacemaker cells are responsible for the initiation and regulation of heart rhythm and the subsequent transmission of electrical signals to working cardiomyocytes - ultimately culminating in a synchronized contraction that enables the unidirectional flow of blood. Furthermore, the cardiac muscle tissue intersperses with coronary vessels composed of endothelial cells that facilitate nutrient and oxygen exchanges between the cardiomyocytes and the bloodstream. Collectively, these different cellular entities within the cardiac muscle tissue demonstrate their profound integration, each accomplishing a critical role designed to support the expanded functionality of the tissue as a whole. Thus, this meticulously coordinated effort among multiple cell types imbues the tissue with its unique ability to execute its specialized function - effectively pumping life-sustaining blood throughout the body.

Nervous tissues, essential for communication and coordination, comprise neurons and supporting cells called glia (Boukhatmi, 2021; Boumard & Bardin, 2021; Bonilha, 2022). Neurons possess specialized structures like dendrites and axons that enable the transmission of electrical signals, fostering rapid communication across the body. Glia play roles in supporting neurons, maintaining the extracellular environment, and providing structural support. Nervous tissues exemplify indispensable systems necessary for communication and coordination within an organism, with the primary components being neurons and supporting glial cells. Taking the specialized context of the hippocampus - a key region of the brain associated with memory formation and spatial navigation - provides a revelatory case study. Within this region, pyramidal neurons serve as the principal cells, intricately structuring the hippocampal network as they form a multitude of synaptic connections with other neurons. These pyramidal neurons send out electrical impulses or action potentials along their axons, which pass through synapses - the specialized junctures between neurons - onto other neurons, effectively transmitting information throughout this network. Simultaneously, there exists a harmonious collaboration with the glial cells such as astrocytes and microglia, which contribute significantly to maintaining functionality and homeostasis. Astrocytes, with their distinct star-shaped structure, undertake tasks like modulating synaptic transmission, controlling the ion environment, providing structural support, and delivering nutrients to neurons, ensuring the optimal environment for neuronal function. Meanwhile, microglia, the resident immune cells of the central nervous system (CNS), constantly survey their surroundings, maintaining the overall health of the nervous tissue by identifying and eliminating pathogenic infiltrates and cellular debris. The collective interactions between neurons and glial cells in the hippocampus underpin its role within the broader communication and coordination functionality of the nervous system, thus manifesting the concept of cellular cooperation in nervous tissues.

Tissues are not isolated entities but intricately interact within the context of organs, which themselves function as integrated units within the organism (MacArthur & Dawlaty, 2021; Ma et al., 2022; Ma & Wu, 2022). The coordinated efforts of different tissues within an organ contribute to its overall function. For instance, the heart, an organ composed of cardiac muscle, nervous tissue, and connective tissues, exemplifies the synergy between tissues necessary for pumping blood and maintaining circulation. Tissues function not as solitary entities but as integral participants reciprocally interacting within the broader context of organ systems, forging a precise organizational hierarchy that drives the complex operations of an organism. A profound understanding of this biological tapestry can be gleaned from a focused exploration of the mammalian heart, an organ functioning as a meticulously coordinated unit in the cardiovascular system. The heart comprises four distinct tissues: the endocardium lining the heart chambers, the myocardium constituting the contractile muscle layer, the epicardium covering its surface, and the specialized conductive tissue weaving through it. Each individual tissue provides essential services, tightly cooperating to orchestrate the ceaseless function of pumping blood throughout the organism. The myocardium, primarily composed of cardiomyocytes that contract in synchronized harmony, drives the cyclical ventricular contractions and relaxations — the heartbeat — thus propelling blood throughout the vascular tree. This critical rhythm is under the watchful regulation of the conductive tissue, precisely controlling the electric signals responsible for initiating and synchronizing every heartbeat. The endocardium, comprising endothelial cells and a layer of connective tissue, ensures a smooth surface for blood flow, while the epicardium, the heart's outer layer, secretes lubricating fluid to minimize friction during heartbeats. These tissues, despite their distinct roles, interact intimately to create the vital, integrated organ that constitutes the hub of the cardiovascular system. Furthermore, the heart, in turn, does not function in isolation; it connects and communicates with diverse organs like the lungs for gas exchange, the kidneys for blood filtration, and the brain for regulatory control, epitomizing the concept of organ integration within the entire organism. More significantly, the actions of the heart reverberate throughout the organism, underscoring the truth that no biological entity operates in isolation. Rather, it is within the intricate interplay of tissues, the assembly of organs, and the coordination of organ systems that the broader symphony of life is played.

In the intricate tapestry of multicellular organisms, tissues exemplify the exquisite balance between cellular specialization and harmonious integration. The organization of tissues into functional units represents an evolutionary masterpiece that ensures the survival and adaptability of diverse organisms. The study of tissues not only provides insights into the intricate functioning of organisms but also offers a window into the mechanisms of health and disease, guiding advancements in medicine and biology. The third section of this chapter will further explore muscle cells in terms of their contractile machinery and tissue typologies.

7.3 MUSCLE CELLS: CONTRACTILE MACHINERY AND TYPES OF MUSCLE TISSUE

Muscle cells, also known as myocytes, are specialized cells responsible for generating force and movement in the body (Altulea et al., 2021; Alvord et al., 2022; Aniento et al., 2022; Aprile et al., 2023). They play a crucial role in various physiological processes such as locomotion, maintaining posture, breathing, and circulation. Muscle cells are categorized into three main types based on their structure, function, and location: skeletal muscle, smooth muscle, and cardiac muscle. Muscle cells, or myocytes, underscore the precise specialization of cells in the human body, given their critical role in generating force and evoking movement. One compelling case study within this context involves skeletal muscle, a type of muscle tissue primarily responsible for voluntary movements. Composed of elongated, striated muscle fibers, these myocytes are organized into densely packed bundles surrounded by connective tissues. Contractility, the cardinal function of myocytes, is dictated by a unique cellular architecture. Within each skeletal muscle cell, lines of contractile units, or sarcomeres—comprised of actin (thin) and myosin (thick) proteins— align lengthwise in a repeating pattern, providing the characteristic striated appearance. The sliding filament theory of muscle contraction postulates that the overlap between these filamentous proteins increases during contraction, subsequent to the influx of calcium ions, leading to the shortening of the sarcomere and eventually, the contraction of the entire muscle cell. Simultaneously, nerve endings from motor neurons form intimate junctions known as neuromuscular junctions with these myocytes, transmitting signals that trigger the contractile process. The myriad associated proteins, such as troponin and tropomyosin, modulate this process, ensuring precision and control in force generation and movement. Thus, through this high level of specialization characterized by their distinctive architecture and function, skeletal muscle cells illustrate how cellular specialization can manifest to orchestrate the complex and highly coordinated process of movement in the human body.

Skeletal muscle (see figure 65) is the most abundant type of muscle in the body and is primarily responsible for voluntary movements (Jiang et al., 2021; Johnson et al., 2021; Jiang et al., 2022; Jinkerson et al., 2022). These muscles are attached to bones via tendons and provide the force necessary for actions like walking, running, and lifting. Skeletal muscle cells are multinucleated, meaning they have multiple nuclei within a single cell. This characteristic arises due to the fusion of multiple myoblasts during development. Representing approximately 40% of the body's weight, skeletal muscle, the most abundant muscle tissue type, bears chief responsibility for the execution of voluntary movements. A detailed examination of the gastrocnemius muscle, a robust skeletal muscle in the posterior part of the lower leg, offers a quintessential exemplification of the biomechanics associated with such voluntary movement.

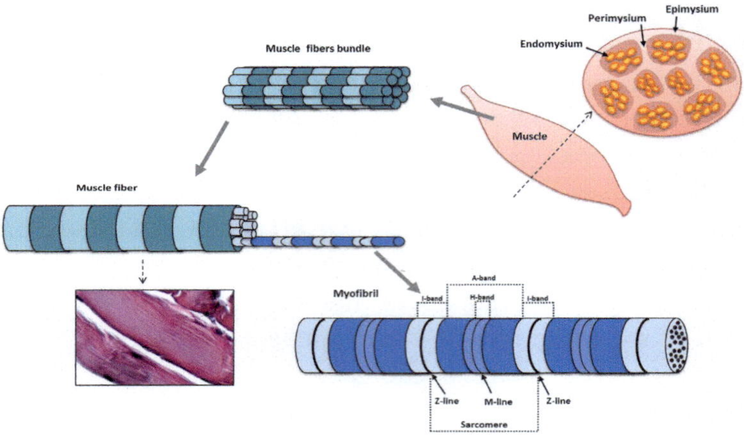

Figure 65. Representation of skeletal muscle tissue.

Source: Trovato FM, Imbesi R, Conway N, Castrogiovanni P. Morphological and Functional Aspects of Human Skeletal Muscle. Journal of Functional Morphology and Kinesiology. 2016; 1(3):289-302. https://doi.org/10.3390/jfmk1030289

The gastrocnemius, tethered to the heel via the Achilles tendon, plays a crucial role in walking, jumping, and running by facilitating plantar flexion of the foot. Integral to these actions is the orchestrated work of the muscle's constituent cells, the multinucleated skeletal myocytes. Within each myocyte, intricate arrays of contractile units known as sarcomeres, assembled end-to-end along protein filaments of actin and myosin, collaborate harmoniously to effectuate muscle contraction. Intracellular calcium ions, released in response to signals transmitted via motor neurons, initiate a cascade of events, resulting in the interaction of these actin and myosin filaments and, ultimately, myocyte contraction. Correspondingly, when the calcium ions are sequestered, the myocyte returns to its resting state, symbolizing relaxation. The myocytes in the gastrocnemius muscle, bundled together in long fibers, contract in unison when stimulated by motor neurons, delivering the force required to enact the motion of the foot. This specialized cellular choreography within skeletal muscle tissue, exemplified by the gastrocnemius muscle, elucidates the meticulously coordinated processes powering voluntary movements in organisms. The contractile machinery of skeletal muscle cells is highly organized and consists of repeating units called sarcomeres (Liu et al., 2022; Loconte et al., 2022; S. Lim et al., 2023). Sarcomeres are the basic functional units of muscle contraction and are composed of thick and thin filaments. Thick filaments are primarily composed of myosin, a motor protein, while thin filaments are primarily composed of actin. The interaction between myosin and actin, facilitated by proteins like tropomyosin and troponin, leads to muscle contraction. The contractile apparatus of skeletal muscle cells manifests exceptional organization, composed predominantly of repeating structural units termed sarcomeres, a phenomenon best exemplified by examining the biceps brachii muscle. Skeletal muscle cells, or myofibers, are elongated, cylindrical cells housing myofibrils, intricate protein structures running longitudinally through the cell. These myofibrils, in turn, contain thousands of sarcomeres, the muscle's fundamental contractile units. Each sarcomere, demarcated by Z-discs, encompasses overlapping bands of the proteins actin and myosin, the thin and thick filaments respectively. Through a process initiated by calcium ion influx following a stimulus from a motor neuron, myosin heads attach to the actin filament and undergo a power stroke, leading to filament sliding and sarcomere contraction. This mechanism conforms to the sliding filament theory of muscle contraction, a concept that elucidates the biomechanics of force development and movement in skeletal muscle. Each sarcomere contracts simultaneously during a muscle contraction, primarily contributing to the shortening of the muscle and thus motion facilitation. Upon cessation of the neural stimulus and removal of calcium ions, the sarcomeres return to their relaxed state, waiting for the next signal to contract. Thereby, within the biceps brachii, the organization and contractility of these multitudinous sarcomeres in each of the numerous myofibers engender the controlled strength and precision necessary to flex the forearm, demonstrating the significance of the highly organized contractile machinery in skeletal muscle cells.

Smooth muscle cells are found in the walls of organs and structures such as blood vessels, the digestive tract, and the respiratory system (Granados-Riveron & Aquino-Jarquin, 2021; Graziadei & Rappsilber, 2022). Unlike skeletal muscle, smooth muscle contraction is largely involuntary and is controlled by the autonomic nervous system. Smooth muscle cells are spindle-shaped and have a single nucleus. Smooth muscle cells, notable for their spindle-shaped, non-striated morphology, pervade the human body, functional in diverse organ systems where they typically efface the walls of structures like blood vessels, the gastrointestinal tract, and the respiratory system. A focused disquisition on their role in the control of vasomotor tone and blood pressure regulation offers a meaningful examination of their functional significance. Nestled in the middle layer - tunica media - of artery walls, smooth muscle cells exert control over the arterial diameter and, consequently, systemic vascular resistance. These cells respond to a panoply of signaling inputs - neural impulses, circulating hormones, paracrine factors, and endothelium-derived mediators, precipitating a rise in intracellular calcium ions which then activate the myosin light-chain kinase, the enzyme responsible for facilitating the reaction between actin and myosin filaments. Subsequent cross-bridge cycling leads to muscle contraction and consequent vasoconstriction, thereby regulating blood flow and pressure. Conversely, when the stimuli dissipate and the calcium ions are pumped out, relaxation ensures, resulting in vasodilation and increased blood flow. Every beat of the heart subjects the arterial blood vessels to a cycle of stretch and relaxation, with smooth muscle cells continuously contracting and relaxing in response to physiological needs. Thus, through their rhythmic orchestration in the vascular milieu, smooth muscle cells serve as the fulcrum of blood pressure regulation, contributing fundamentally to homeostasis and overall organ system performance. Their versatility and adaptive attributes underscore the diversity and specialization inherent in the cellular blueprint of organ systems across the human body.

The contractile machinery in smooth muscle cells is not as well-organized as in skeletal muscle (Jiang et al., 2021; ; Johnson et al., 2021; Jiang et al., 2022; Jinkerson et al., 2022). Actin and myosin filaments are present, but they are not organized into sarcomeres. Instead, actin and myosin filaments are scattered throughout the cell and are anchored to dense bodies, which serve as attachment points for the contractile proteins. The contraction of smooth muscle cells is characterized by slow and sustained contractions, which are important for functions like regulating blood pressure and moving food through the digestive tract. The contractile machinery within smooth muscle cells, distinguishing them from their skeletal counterparts, is less regimented in its organization, as a focused examination of the vascular smooth muscle cells (VSMCs) in arteries such as the aorta can elucidate. Unlike the skeletal muscle cells, which have a neatly packed sarcomeric arrangement of filaments, VSMCs exhibit a more diffuse distribution of actin and myosin filaments within the cytoplasm, reflecting their biological adaptability for slow yet sustained contractions - a constant requirement for regulating systemic blood pressure. The myosin filaments generate forceful contractions through myosin light chain phosphorylation, a calcium-calmodulin triggered event, provoking the interaction of myosin with actin.

In this scenario, tropomyosin, a regulatory protein, does not block the actin-myosin interfacing sites, as seen in skeletal muscle cells, thus demonstrating the differing mechanisms of contraction regulation. Moreover, the filament-turnover rate in VSMCs is only a fraction of that in skeletal muscle cells, supporting their role in protracted contractions necessary for persistent muscle tone in vessels. Additionally, smooth muscle cells, unlike striated muscle cells, lack specialized structures like t-tubules to synchronize calcium release across the cells when action potentials are fired. Instead, they rely on the slower process of calcium diffusion through the cell cytoplasm to initiate contractions. Therefore, the lack of rigid organization seen in the contractile machinery of VSMCs reflects their distinctive physiological role, offering a nuanced perspective on cellular structure-function relationships.

Cardiac muscle cells (see figure 66) form the walls of the heart and are responsible for pumping blood throughout the circulatory system (Orr & Hynds, 2021; Ohi & Kenworthy, 2022; Okano et al., 2023). These cells are interconnected through specialized structures called intercalated discs, which allow for synchronized contractions and efficient transmission of electrical signals. Cardiac muscle cells, also referred to as cardiomyocytes, constitute the walls of the heart, tasked with the crucial function of propelling blood throughout the circulatory system. A detailed investigation of the ventricular cardiomyocytes, the dominant cell type in the heart, offers an insightful discernment of these cells and their function. Residing within the myocardium, the thick muscular layer of the heart's walls, cardiomyocytes collaborate to generate the forceful contractions required to expel blood into the circulation. Structurally, these cells possess an intricate sarcomeric arrangement of actin and myosin filaments, akin to skeletal muscle cells, but their most distinctive feature is the presence of intercalated discs -- specialized adhesive and communicative junctions ushering structural integrity and electrical coupling among adjoining cells. Triggered by the conduction system's initiates action potentials, calcium ions released from the sarcoplasmic reticulum bind to the regulatory protein troponin, initiating conformational changes that make way for myosin-actin interaction, consequently leading to sarcomere shortening and cardiomyocyte contraction. A heartbeat comprises a succession of such contractions, beginning in the atria and coursing through the ventricles, leading to the ejection of blood from the heart. Moreover, the ebb and flow of calcium ions within these cells is meticulously regulated, ensuring an optimum duration for contraction-relaxation cycles, thus providing the rhythmic control vital for the heart's pumping efficacy. Therefore, cardiac muscle cells, exemplified by the ventricular cardiomyocytes, form not only the cardiac edifice but also embody the very rhythm of life, highlighting their indispensable role in maintaining organismal vitality.

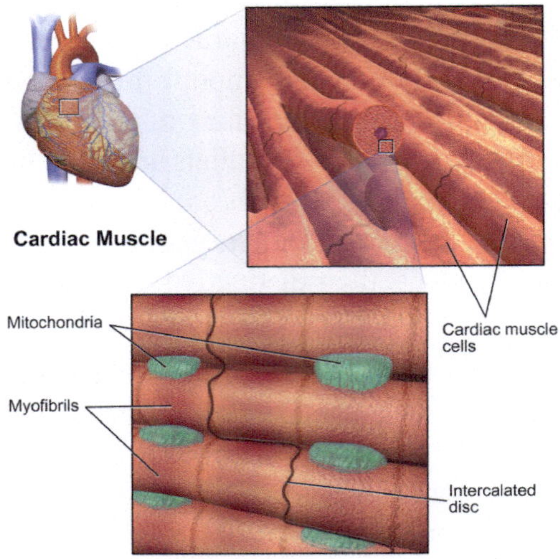

Figure 66. Cardiac muscle and its locations

Source: By BruceBlaus - Own work, CC BY-SA 4.0, https://commons.wikimedia.org/w/index.php?curid=44969447

The contractile machinery of cardiac muscle cells is similar to that of skeletal muscle, with organized sarcomeres composed of myosin and actin filaments (Ryder et al., 2022; Sagwal & Bekeschus, 2021; Sakamaki & Mizushima, 2023). However, the arrangement of the filaments is different in cardiac muscle, leading to a branching pattern. This unique arrangement allows for strong contractions and efficient pumping of blood. Examining the contractile system of cardiomyocytes, the dominant cell type in the heart, provides an insightful exploration of the parallels with the contractile machinery of skeletal muscle cells. Cardiomyocytes, akin to skeletal muscle cells, bear an organized sarcomeric structure with a patterned arrangement of actin (thin) and myosin (thick) filaments. Each sarcomere, delineated by Z-discs, houses these overlapping filaments involved in the conduction of contractile force. The contraction process, as in skeletal myocytes, initiates with the influx of calcium ions following depolarization. Calcium ions bind to the regulatory protein troponin, instigating a series of allosteric changes that expose the myosin-binding sites on the actin filaments. Subsequently, myosin heads bind to actin, resulting in cross-bridge cycling and filament sliding which culminates in sarcomere shortening, and thus cardiomyocyte contraction. The similarity in the protein constituents of the contractile machinery does give an impression of mechanistic resemblance between skeletal muscle and cardiac muscle. However, such a view overlooks the profound differences in regulation, cellular organization, and function. For instance, the precise coordination of contractions within the heart to efficiently facilitate unidirectional blood flow is driven by intercalated discs, unique structural attributes of cardiac muscle tissue. Therefore, this organized yet distinctive contractile machinery allows the heart to fulfill its role as the body's central blood-pumping organ, representing the confluence and divergence of cellular machinery to accomplish distinct physiological functions.

In summary, muscle cells are essential components of the body, enabling various forms of movement and physiological functions. Their contractile machinery, though similar in principle, exhibits variations in structure and organization based on the type of muscle tissue. Skeletal muscles provide voluntary movement through highly organized sarcomeres, smooth muscles enable involuntary contractions through a less structured arrangement, and cardiac muscles ensure efficient pumping of blood with their distinctive branching pattern. Understanding the intricacies of these muscle types and their contractile mechanisms is essential for comprehending human physiology and disease processes. The fourth section explores nerve cells in terms of signaling and structure within the nervous system.

7.4 NERVE CELLS: STRUCTURE AND SIGNALING IN THE NERVOUS SYSTEM

Nerve cells, also known as neurons, are fundamental components of the nervous system, enabling communication and coordination within the body (Scott-Solomon & Hsu, 2022; Scott et al., 2022; Sezin et al., 2022). Neurons are specialized for transmitting electrical signals, or nerve impulses, over long distances to convey information between different parts of the body. The structure of nerve cells and the signaling mechanisms they employ are intricately designed to support their critical functions. Nerve cells, or neurons, serve as the foundational entities within the nervous system, facilitating key communication and coordination activities across the body. By examining a detailed instance of the pyramidal neurons located in the cerebral cortex, one may decipher the intricate modalities enacted by these specialized cells.

Pyramidal neurons, identifiable through their characteristic triangular soma and a singular long axon, play indispensable roles in cognitive functions, including processing and integrating sensory and motor information. These neurons receive input from other neurons or sensory receptors via dendritic spines, tiny projections densely peppering the dendrites that increase the receptive surface area of the cell. The received information, in the form of electrical impulses or action potentials, is swiftly transmitted down the axon - a long, wire-like projection - towards the axon terminals.

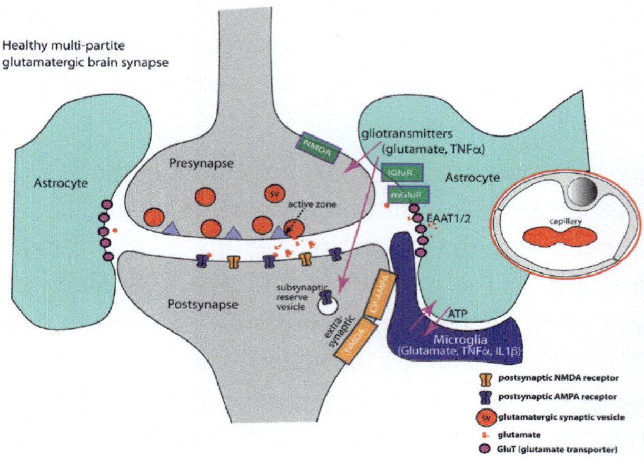

Figure 67. Illustration of synaptic transmission.

Source: Schwarz K, Schmitz F. Synapse Dysfunctions in Multiple Sclerosis. International Journal of Molecular Sciences. 2023; 24(2):1639. https://doi.org/10.3390/ijms24021639

These terminals form specialized structures known as synapses with other neurons or effector muscle cells, enabling the transmission of information via neurotransmitter molecules (see figure 67). The neurotransmitter molecules released from the presynaptic neuron traverse the synaptic cleft to bind to specific receptors on the postsynaptic neuron, transferring the signal and thereby mediating intercellular communication (Reiche et al., 2022; Reuschl et al., 2022; Ricci & Srivastava, 2021). This continuous, intricate relay of signals ensures prompt responses to stimuli, complex thought processes, and meticulous coordination of actions — epitomizing the role of neurons as the sentinels of communication within the organism. Therefore, the operation of propagation and transmission of signals by neurons designates them as pivotal components regulating the body's communication and coordination. The basic structure of a neuron consists of three main parts: the cell body (soma), dendrites, and axon (Ohi & Kenworthy, 2022; Okano et al., 2023; Orr & Hynds, 2021). The cell body contains the nucleus and most of the cell's organelles, including the endoplasmic reticulum and Golgi apparatus, which are essential for protein synthesis and packaging. Dendrites are short, branched extensions that protrude from the cell body and serve as the primary receivers of incoming signals from other neurons.

These dendritic structures are covered in numerous synaptic terminals, specialized sites where neurons can communicate through synapses. Neurons, the chief cells of the nervous system, exhibit a unique and complex structure tailored to facilitate the rapid propagation and transduction of electrical signals. This morphology is predominantly characterized by three primary parts: the cell body or soma, dendrites, and axon, a configuration precisely exemplified by a detailed study of the Purkinje neurons found in the cerebellum. The nerve cell body, or soma, houses the nucleus and other necessary components for maintaining cellular metabolism and synthesis of vital proteins. Radiating extensively from the cell body are specialized projections known as dendrites. Purkinje neurons possess a highly branched and elaborate dendritic tree that acts as a vast receptive field, inviting inputs from numerous synaptic contacts. Each dendrite is studded with tiny protrusions called dendritic spines that enlarge the surface area for receiving synaptic inputs. From the soma also extends a long, singular structure called the axon, responsible for conducting the output signal away from the cell body. Unique to the Purkinje neurons is the absence of myelin sheaths along most parts of the axon, contrasting the majority of neurons in the peripheral nervous system. However, the Purkinje axon possesses a multitude of collaterals that establish inhibitory synapses with numerous neurons in the cerebellar nuclei, effectively transmitting the signal downstream. Collectively, this distinctive tripartite structure enabling the receiving, processing, and transmitting of signals underscores every neuron's function, with the Purkinje neuron serving as a specific exemplar of this archetypal design.

The axon is a long, slender projection that carries nerve impulses away from the cell body (D. Wang et al., 2021; J. J. Wang et al., 2021; Wang et al., 2022). The axon is covered by a myelin sheath, a fatty insulating layer produced by supporting cells called glia. Myelin sheaths are interrupted by gaps called nodes of Ranvier, which serve to enhance the speed of nerve impulse transmission along the axon. The axon, as a distinct structural component of a neuron, is a long, slender projection that carries nerve impulses away from the cell body, a concept that can be clearly elucidated through a detailed examination of a motor neuron. The motor neuron's axon begins at the axon hillock, a specialized region of the cell body, and extends towards its target, which might be a muscle cell or another neuron in the spinal cord or brain. Throughout its course, the axon maintains a uniform diameter, distinguishing it from the tapering dendrites. Axons are generally sheathed in a fatty insulating layer called the myelin sheath, segmented by nodes of Ranvier, which enables fast, efficient transmission of action potentials or nerve impulses. This mode of propagation, termed saltatory conduction, markedly increases the transmission speed of nerve impulses. At the end of the axon, the axon terminal forms synapses with target cells, where the electric signal of the nervous impulse gets translated into a chemical signal by the release of neurotransmitter molecules. This comprehensive mechanism enabling the transport of nerve impulses from the neuron's body to the target cell underscores the axon's critical role within neurons, and by extension, the efficient functioning of the nervous system. The motor neuron axon's coordinated build and function to stimulate muscle movement encapsulate these fundamental principles of axonal function.

The signaling process in nerve cells involves a combination of electrical and chemical events (Zhao et al., 2021; Ziaja et al., 2021; Zidan et al., 2021; Zuch et al., 2022). At rest, nerve cells maintain a negative charge inside the cell relative to the outside due to the unequal distribution of ions, particularly sodium (Na+), potassium (K+), and chloride (Cl-), across the cell membrane. This difference in charge is known as the resting membrane potential. The signaling process in nerve cells, indeed, intricately combines both electrical and chemical events, a phenomenon vividly exemplified through the actions of a sensory neuron responding to a stimulus. Responding to a specific stimulus, such as pressure or temperature change, the sensory neuron's membrane at the site of stimulation experiences a change in its electrical properties, generating an electrical signal known as an action potential. This signal initiation primarily involves an intricate play of ions across the neuron's membrane, engendered by the opening and closing of ion channels. Once commenced, this action potential is rapidly propagated along the neuron's axon in an electrical form, facilitated by the axon's myelination that enables the saltatory conduction phenomenon. Upon reaching the presynaptic terminal, this electrical signal prompts the opening of voltage-gated calcium channels, resulting in calcium influx into the terminal. Elevated calcium concentrations within the terminal induce a biochemical response, where vesicles containing neurotransmitter molecules fuse with the presynaptic membrane, discharging their contents into the synaptic cleft. This chemical signal -- the neurotransmitter -- traverses the synaptic cleft and binds to specific receptors on the postsynaptic cell, eliciting a response that could be excitatory or inhibitory. Hence, through this intricate combination of electrical events within the neuron and chemical events at the synapse, nerve cells ensure efficient and rapid communication across extensive distances and multiple cells in the nervous system.

When a neuron is stimulated, a local change in its membrane potential, known as a graded potential, occurs (Donato et al., 2021; Don & Kalebic, 2022; Domcke & Shendure, 2023). If the stimulus is strong enough, it triggers an action potential, which is a rapid and reversible change in the neuron's membrane potential. This action potential is initiated at the axon hillock, a region of the cell body where the axon originates. In the diverse realm of cell microbiology, one of the most well-studied areas is the functional dynamics of neurons, the fundamental unit of the nervous system, which produce graded potentials as a response to stimulation. To examine this phenomenon, we must delve into the intricacies of neuron behavior. Neurons are unique in that they exhibit an electrical resting potential due to differential concentrations of key ions like potassium and sodium present across their cell membrane, a trait that underpins their ability to transmit signals. When a neuron experiences stimulation, it gives rise to a graded potential, a shift in this electrical equilibrium that reflects a localized change in membrane potential. This stimulation could stem from any number of stimuli, for instance, an external chemical signal or a change in the environmental temperature. The specifics of how this change manifests involve a complex interplay of ion channels and transporters situated in the neuron's membrane. As a specific stimulus hits the neuron, certain ion channels in the membrane region nearest to the signal undergo conformational changes that cause them to open. Subsequently, ions are driven to flow down their electrochemical gradients--potassium tends to exit the neuron due to its higher concentration inside the cell, while sodium, having a higher concentration outside the cell, tends to enter. This ion movement induces a change in the neuron's overall electrical state, effectively decreasing the resting potential in a process known as depolarization--the essence of a graded potential. The magnitude of this potential shift directly corresponds to the intensity of the stimulus, thus providing a proportional bioelectrical response. Intriguingly, unlike an all-or-nothing action potential, graded potentials can both add up and spread across the neuron's surface, giving remarkably precise control over the cell's overall signaling. This ability makes understanding graded potentials critical for dissecting many neural processes, including sensation, motor control, and higher cognition. Although deceptively simple in its function, the graded potential, when extrapolated to the vast network of neurons in a biological system, plays an instrumental part in coordinating complex physiological responses, underscoring the intricate adaptability of cellular microbiology.

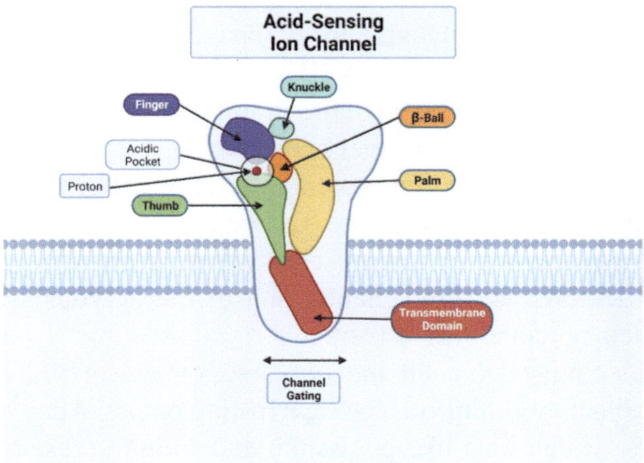

Figure 68. Example of a gated channel.

Source: Sivils A, Yang F, Wang JQ, Chu X-P. Acid-Sensing Ion Channel 2: Function and Modulation. Membranes. 2022; 12(2):113. https://doi.org/10.3390/membranes12020113

The action potential is propagated along the axon by a process known as saltatory conduction, where the action potential "jumps" from one node of Ranvier to the next (Bettegazzi et al., 2021; Bennison et al., 2022; Beppler et al., 2022). This mode of conduction is made possible by the myelin sheath, which insulates the axon and prevents the action potential from dissipating as it travels. In the expansive field of cell microbiology, the axonal propagation of action potentials, particularly via saltatory conduction, represents an intriguing phenomenon embodying the sophisticated nature of neuronal communication. A case in point is the mechanism by which the action potential, initiated at the neuron's axon hillock, traverses the length of the axon to successfully transmit nerve signals, a process predicated on the unique myelinated structure of the axon itself. The axon, or the nerve fiber, is often enveloped by myelin sheaths—generated by glial cells in the nervous system—that segment the axon into interspersed nodes known as nodes of Ranvier. These nodes are vital to neuronal signaling as they expose the axon's membrane, enabling it to partake in the exchange of ions that underlie the action potential. This is where saltatory conduction, the characteristic mode by which action potentials propagate along myelinated neurons, comes to play. As an action potential is initiated, it induces a rapid influx of sodium ions, thereby causing a local depolarization--a positive shift in the membrane potential. However, the insulating myelin sheaths severely limit the diffusion of these sodium ions longitudinally along the axon, thereby channeling the ionic currents to flow across the membrane only at the nodes of Ranvier. As a result, the action potential appears to leap, or 'saltate', from one node to the next, thus the term 'saltatory' conduction. Critically, this propulsion method dramatically boosts the propagation speed of action potentials, facilitating rapid neuronal communication, while also conserving neuronal energy due to the reduced membrane area involved. Furthermore, it ensures the maintenance of action potential strength over longer distances by repeatedly refreshing the signal at each node, thereby circumventing any potential decay that might occur due to resistive losses along the axon's length. In essence, by examining the case of saltatory conduction, one can arrive at a profound understanding of how physicochemical processes at a microscopic cellular level contribute immensely to the larger macroscopic physiological functionality of the organism.

At the end of the axon, the nerve impulse triggers the release of neurotransmitters from synaptic terminals into the synapse (Pinson & Huttner, 2021; Postler, 2021; Pratt et al., 2021). Neurotransmitters are chemical messengers that transmit signals from one neuron to another or to a target cell such as a muscle cell. These neurotransmitters bind to receptors on the dendrites or cell bodies of neighboring neurons, leading to the generation of graded potentials in those neurons. In the intricate realm of cell microbiology, the process of synaptic transmission where a nerve impulse reaching an axon's end triggers the release of neurotransmitters into the synapse represents an event of immense significance, embodying the critical crossover between electrical and chemical signaling in the nervous system. This phenomenon fundamentally underlies not only basic neural communication but also more complex functionality such as cognition, emotion, and memory. To explore this process in depth, one can take as an example the release of the neurotransmitter glutamate from excitatory neurons. As an action potential, initiated by a preceding neuron, propagates along the axon, it eventually reaches the axon terminal or synaptic knob, a specialized structure housing numerous vesicles filled with neurotransmitters. The influx of the action potential into the axon terminal leads to local depolarization and the opening of voltage-gated calcium channels. Consequently, calcium ions flood into the terminal, triggering a cascade of intracellular events that lead to the movement and fusion of neurotransmitter-loaded vesicles with the presynaptic plasma membrane. This vesicular fusion is facilitated by a suite of proteins collectively known as SNARE proteins. Upon fusion, a process fundamentally dictated by the basic principles of cell biology, the vesicle releases its neurotransmitter content, in this case, glutamate, into the narrow fluid-filled gap of the synapse - the inter-neuronal space that separates the presynaptic neuron from the postsynaptic neuron. This neurotransmitter diffusion across the synapse signifies the transition from an electrical signal within a neuron to a chemical signal between neurons. Once in the synapse, glutamate can then bind postsynaptic glutamate receptors on the adjacent neuron, leading to the opening of postsynaptic ion channels and a subsequent change in the postsynaptic neuron's membrane potential. This could potentially initiate a new action potential, thus effectively transmitting the neural signal from one neuron to another. In essence, this comprehensive examination of neurotransmitter release demonstrates how intricately cell microbiology and physiology intertwine to facilitate the complex web of communication that constitutes the working of the nervous system.

Once the neurotransmitter signal is transmitted, it can be terminated through reuptake by the releasing neuron, enzymatic breakdown, or diffusion away from the synapse (Lamb et al., 2021; Lamptey et al., 2021; Langford, 2022; Lannagan et al., 2021). This termination ensures that the signaling process is precise and controlled. Unraveling the complex mechanisms of cell microbiology, the termination of a neurotransmitter signal at the synapse encapsulates a critical component of neural communication. This phenomenon, quintessential to proper neural functioning, ensures that each neurotransmitter signal is precise and succinct, not lingering too long to cause potential overstimulation of postsynaptic receptors. To illustrate this, we can examine the instance of the neurotransmitter serotonin in the central nervous system. Once this neurotransmitter carries its signal, broadcasting the message from the presynaptic neuron to the postsynaptic neuron, a specific, coordinated sequence is enacted to terminate the signaling. One primary mechanism of signal termination is reuptake into the releasing, or presynaptic, neuron. This reuptake is facilitated by transport proteins embedded in the neuron's membrane, often specific to the neurotransmitter in question. In the case of serotonin, the serotonin reuptake transporter protein (SERT), largely found in the cell bodies and axon terminals of serotonin-producing neurons, binds to the serotonin in the synaptic cleft and transports it back into the presynaptic neuron, essentially clearing the synapse of the neurotransmitter. Besides reuptake, enzymatic breakdown also plays a role in signal termination. For neurotransmitters like acetylcholine, the enzyme acetylcholinesterase is present in the synapse and rapidly breaks down the neurotransmitter into its constituent components, thus rendering it inactive. Thirdly, neurotransmitters can also diffuse away from the synapse, moving down their concentration gradient into surrounding areas where they are either metabolized by enzymatic activity or transported into various cells, including glial cells. By considering these tripartite termination strategies - reuptake, enzymatic breakdown, and diffusion - researchers can gain a profound understanding of how the body can effectively fine-tune neural communication, ensuring the neural signal lifecycle is as precise, controlled, and efficient as it must be to support the highly complex processes afforded by the central nervous system.

In conclusion, nerve cells exhibit a complex and highly specialized structure that enables them to transmit electrical signals over long distances in the nervous system. The interaction between electrical impulses and chemical signals in neurons forms the basis of neural communication, allowing for the integration and transmission of information throughout the body. Understanding the structure and signaling mechanisms of nerve cells is essential for comprehending brain function, behavior, and the underlying causes of neurological disorders. The final section of this chapter will focus on plant cell specializations.

7.5 PLANT CELL SPECIALIZATIONS: PARENCHYMA, COLLENCHYMA, AND SCLERENCHYMA CELLS

Plant cells exhibit a diversity of forms and functions that enable plants to adapt and thrive in various environments (Masur et al., 2021; McCutcheon, 2021; Marvin et al., 2022). The diversity of forms and functions exhibited by plant cells represents a cornerstone of cell microbiology, underlying the remarkable adaptability that allows plants to thrive in a broad spectrum of environments. A comprehensive examination of this phenomenon can be best illustrated by scrutinizing the specific case of Arabidopsis thaliana, a model plant extensively studied for its adaptability. Each plant cell in this species holds an array of functionally specialized organelles that contribute actively to the plant's overall adaptability and growth. For example, plant cells house a membrane-bound, water-filled organelle, known as the central vacuole, which maintains turgor pressure, supports cellular structure, and stores essential materials such as water, nutrients, and waste products. During periods of drought, the vacuole's capacity to store water equips the plant with crucial survival skills. Moreover, the cell wall, another defining feature of plant cells composed largely of cellulose and other polysaccharides, provides both structural support and a protective barrier against physical and microbial insults. Additionally, chloroplasts, the site of photosynthesis, are quintessential to plants' adaptation, enabling them to harness solar energy and convert it into chemical energy via a process unique to plant cells. Each of these cellular features contributes to the plant's ability to endure, adapt and thrive. The adaptive mechanisms of Arabidopsis thaliana are displayed as this plant quickly responds to numerous environmental cues, such as light, temperature, and nutrient availability, each driving changes at the cellular level that culminate in physiological adaptations at the organismal level. For instance, when exposed to low light conditions, the plant cells initiate a series of events leading to the proliferation of chloroplasts to efficiently capture limited light for photosynthesis. This correlative relationship between cell function, adaptability, and overall survival is vitally significant in understanding plant biology. Consequently, delving into the precise forms and functions of plant cells fosters a greater comprehension of the intricate strategies utilized for survival in diverse environments and conditions.

Three main types of plant cells that demonstrate distinct specializations are parenchyma, collenchyma, and sclerenchyma cells (see figure 69). Each of these cell types serves unique roles in supporting plant structure, growth, and function:

Parenchyma cells are the most common and versatile type of plant cells (Lamb et al., 2021; Lamptey et al., 2021; Lannagan et al., 2021; Langford, 2022). They play a crucial role in fundamental plant functions such as photosynthesis, storage, and secretion. Parenchyma cells have thin primary cell walls composed of cellulose and hemicellulose, allowing for flexibility and growth. Within the extensive sphere of plant cell microbiology, parenchyma cells emerge as the most commonly encountered and highly versatile type, their roles covering a broad spectrum ranging from photosynthesis to food storage, and from tissue repair to propagation. A case study to exemplify this versatility could call upon the ubiquitous but remarkably diverse biology of the apple tree (Malus domestica). In these trees, parenchyma cells pervade various tissues and display adaptability in function based on their location and the needs of the plant. In leaves, parenchyma cells form a majority of the mesophyll, the tissue where photosynthesis predominantly occurs, hence linking these cells directly to the plant's primary energy production. With numerous chloroplasts, these mesophyll parenchyma cells absorb light and convert it into chemical energy. Conversely, parenchyma cells in roots and stems are often less involved in photosynthesis but engage extensively in storage, giving these cells an integral role in the plant's survival during non-growing seasons. Notably, the distinctive crunch experienced when biting into a ripe apple is due to the presence of parenchyma cells. These cells, under the triggers of particular enzymes, undergo changes that lead them to break down and release fruit juice upon the application of pressure. Remarkably, parenchyma cells also possess an inherent ability to dedifferentiate and divide, a rare trait among mature plant cells. This feature enables them to contribute to wound healing and regeneration, by producing new cells and tissues after damage or during grafting processes.

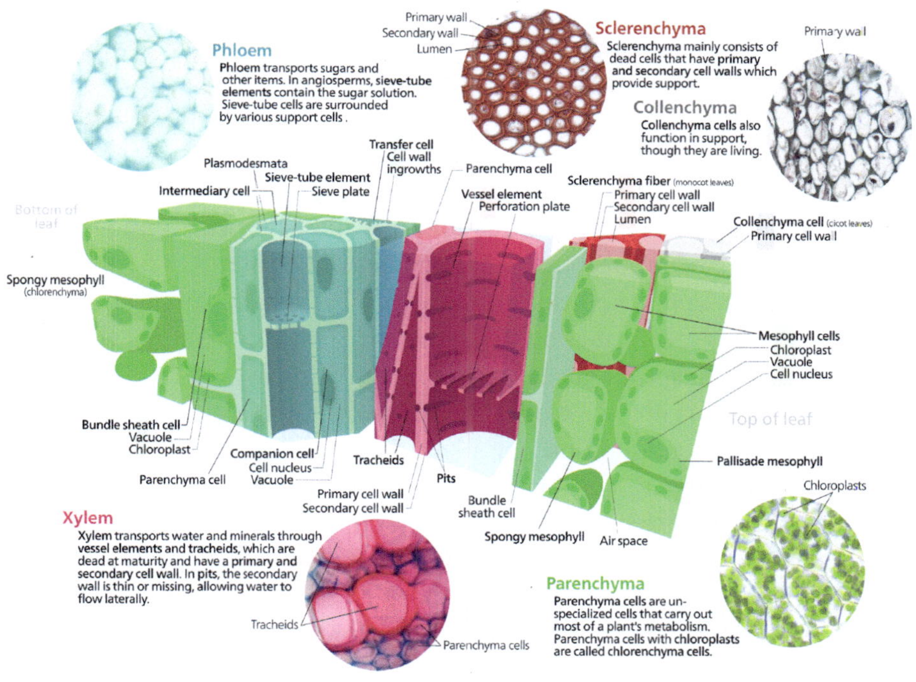

Figure 69. The three types of plant cells.

Source: By Kelvinsong - Own work, CC BY-SA 3.0, https://commons.wikimedia.org/w/index.php?curid=25593329

Thus, the study of a typical apple tree affords a compelling example of the wide range of functions offered by parenchyma cells, underscoring their instrumental versatility and fundamental importance within plant biology. While these cells may constitute the simplest type of plant cells morphologically, their functions are undeniably complex, profoundly impacting the plant's life cycle, adaptability, and overall survival.

These cells are often isodiametric in shape, meaning they are roughly spherical with similar dimensions in all directions (Efstathiou & Draviam, 2021; Dunkley et al., 2022;Duncan, 2022, 2023). This shape enables them to pack closely together, forming the basic building blocks of plant tissues. Parenchyma cells can contain chloroplasts, allowing them to perform photosynthesis and produce energy for the plant. The ubiquitous yet remarkably versatile parenchyma, which constitutes a fundamental cell type in plant biology, often exhibits an isodiametric shape, maintaining approximate similarity in dimensions across all directions, resulting in a shape roughly akin to a sphere. This morphology is particularly evident in the parenchyma cells within the pith of Helianthus annuus, also known as the common sunflower. The centrally located pith within the sunflower stem is primarily composed of parenchyma cells, which, due to their isodiametric shape, efficiently fill the space without leaving substantial intercellular gaps. Being nearly spherical enables these cells to interact uniformly with their surroundings from every direction, a feature that often allows more efficient uptake and utilization of water and other solutes. The structural integrity is also heightened due to the cohesive nature of closely packed homogeneous cells, contributing to the overall plant architecture. Moreover, the isodiametric shape of these parenchyma cells plays a distinct role in storage functions, primarily when these cells contain a central vacuole; the largely spherical configuration optimizes the volume-to-surface area ratio, providing maximal storage capacity with minimal cellular investment. The geometric advantage that these cells exhibit makes them perfectly suited for their versatile roles, and profoundly impacts the plant's processes including growth, survival, and adaptation. Hence, by studying the isodiametric parenchyma cells of Helianthus annuus, it becomes unmistakably clear that cell shape is not merely a morphological trait but a finely tuned adaptation that fundamentally influences cell function, subsequently affecting the overall survival and success of the organism.

Furthermore, parenchyma cells can store nutrients, water, and other compounds, acting as reservoirs that support plant growth and survival during times of stress or scarcity (Gowrishankar et al., 2021; Giolito

& Plateroti, 2022; Gordeeva et al., 2022). They also participate in wound healing and tissue regeneration due to their ability to divide and differentiate. When scrutinizing the multifaceted responsibilities of plant cells in the broader context of plant survival and adaptability, the role of parenchyma cells as reservoirs storing water, nutrients, and other vital compounds cannot be overstated. As a case in point, consider the biology of the potato plant (Solanum tuberosum), where parenchyma cells reveal their remarkable storage capabilities. Within the starchy tubers – the potatoes themselves – one can find parenchyma cells in abundance, evident in the ground tissue. These cells are replete with amyloplasts, specialized plastids that manufacture and store starch granules, a grouping of glucose units. By metabolizing glucose during photosynthesis, which is then polymerized into starch, the potato's parenchyma cells act as reservoirs, storing this energy-dense compound. This reservoir of nutrients is vital for the growth and survival of the potato plant, enabling it to endure periods of scarcity or dormant seasons. Furthermore, parenchyma cells within the tuber contain large central vacuoles that are filled with water and various solutes, serving as a further reservoir. Such water storage can help the plant remain hydrated during periods of low water availability, contributing notably to the plant's drought resistance. By analyzing the physiology of Solanum tuberosum tubers, it is clear that parenchyma cells provide essential services in storing diverse compounds fundamental to the survival of the plant, especially in times of stress or diminished resources. Hence, these cells signify a key adaptation that underscores the impressive ability of plants to thrive in a range of environments, often enduring circumstances of significant adversity.

Collenchyma cells provide mechanical support to growing plant parts (Watson et al., 2021; Warecki & Sullivan, 2022; Weigert & Herhaus, 2023). They have elongated, unevenly thickened primary cell walls that are particularly thick at the corners. This uneven thickening gives them flexibility and strength, allowing them to endure stretching during growth. Within the broad categorizations of plant cells, collenchyma cells play an indispensable role in providing mechanical support to growing parts of the plant, granting the plant both structural integrity and flexibility. A representative case study demonstrating this functionality can be found in Apium graveolens, commonly referred to as celery. In celery stalks, discrete bundles of collenchyma cells are strategically situated just below the epidermal layer, providing essential support without compromising the stalk's overall flexibility. These collenchyma cells are characteristically elongated with unevenly thickened primary cell walls composed mainly of cellulose and pectin. Unlike sclerenchyma cells, another cell type associated with plant mechanical support, collenchyma cells retain their protoplast at maturity, which means they remain alive and metabolically active, allowing them to continue growth and contribute to the lengthening of plant structures. The combination of their length and wall thickness attributes the necessary strength to the tissue, while the remaining areas with a thin wall provide the tissue with its much-needed flexibility. These cells can undergo deformations without damage, permitting organ bending and twist, critical for the plant's adaptation to external mechanical forces such as wind. In the particular case of Apium graveolens, the structural support of this tissue ensures that the celery stalks bear their own weight while growing upright but maintain the flexibility required for slight bending and motion induced by environmental forces. Therefore, through an exploration of collenchyma cells in celery stalks, the cell microbiology's intricacies decode the role of these cells in imparting the combination of strength and flexibility, crucial for the plant's growth and resilience against mechanical stress while navigating the challenging environmental conditions.

These cells often form strands or layers beneath the epidermis of stems and leaves, providing structural reinforcement (Montagnoli et al., 2021; Migliano et al., 2022; Montgomery et al., 2023). Collenchyma cells also have the ability to elongate as the plant grows, accommodating expansion while maintaining support. This type of tissue is commonly found in regions of active growth and development. In the rich tapestry of plant cellular biology, collenchyma cells often form strands or layers beneath the epidermis of stems and leaves, playing a pivotal role in providing structural reinforcement. The plant's ability to maintain structural and mechanical robustness, particularly during periods of rapid growth or environmental stress, is fundamentally tied to these cells. An exploration of this can be found in the physical makeup of the nettle plant or Urticadioica. In this plant, observable strands of collenchyma cells, often referred to as a cylinder or a cylinder-like strand, are situated beneath the epidermal layer of the stems. These strands are known to possess thick walls composed of cellulose and pectin, with a distinct banding pattern that provides a combination of strength and flexibility to the plant organs in which they reside. Conversely, the thin areas of the cell wall allow for cell expansion and continued growth of the organ. In the act of structural support and reinforcement, the collenchyma cells directly contribute to the plant's capacity for upright growth and the ability to withstand various mechanical stresses such as wind, gravity, and even the weight of the plant's own leaves and flowers. An understanding of the collenchyma cells in the context of Urticadioica underscores their role in structural reinforcement and their valuable contribution to the overall architecture and the functional adaptation of the plant. This precise coordination of structure and function at a cellular level consequently impacts the overall survival and success of the plant in its respective habitat.

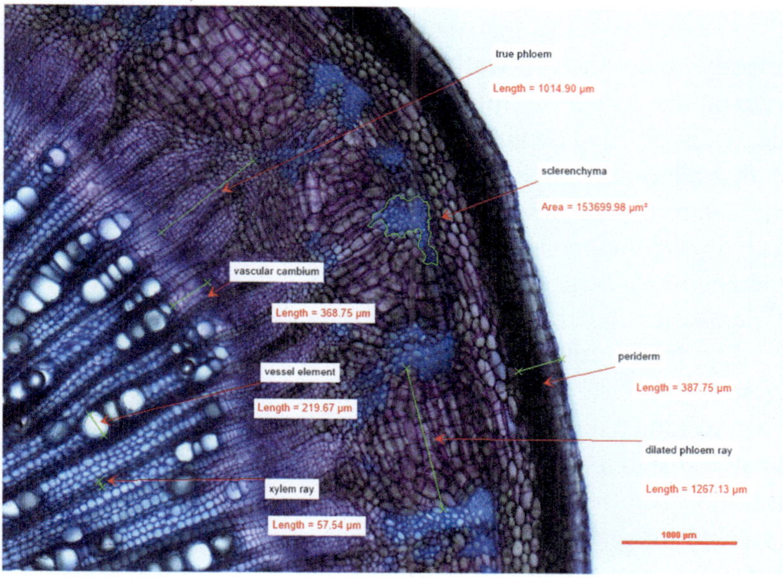

Figure 70. Sclerenchyma under the microscope (here of Moringa oleifera)

Source: McVea HM, Wood LJ. Anatomical and Chemical Analysis of Moringa oleifera Stem Tissue Grown under Controlled Conditions. Horticulturae. 2023; 9(2):213. https://doi.org/10.3390/horticulturae9020213

Sclerenchyma cells are specialized for providing strong structural support to mature plant tissues (Liang et al., 2021; Liang et al., 2022; Li et al., 2023). They have heavily lignified secondary cell walls, making them rigid and resistant to mechanical stress (see figure 70). These secondary walls contain a high proportion of lignin, a complex polymer that provides strength and durability. Sclerenchyma cells, known for their pivotal role in providing structural stability to mature plant tissues, are well-equipped with an array of unique characteristics that make them a noteworthy subject of discussion in the realm of cell-microbiology. To substantiate this fact, consider the case of lignified sclerenchyma cells observed in the stem of "Arabidopsis thaliana", an extensively studied model plant. These cells act as the warriors of plant tissues, fortifying their structure and endurance through their mechanism of secondary thickening - a process which facilitates the deposition of layers of secondary cell wall materials such as lignin and cellulose, that range in deposition patterns, molecule arrangement and composition, depending on the functional needs of the plant. On the observation of the tissues under a fluorescence microscope, it becomes evident that these cells are non-living at maturity, having lost their protoplast during lignification which subsequently makes the cell walls rigid, thus providing the robust structural support to plants. As per the studies conducted by Chaffey et al (2002), these lignified walls bestow upon the cells mechanical strength and resistance against degradation, which is particularly crucial in the adaptability of terrestrial plants to their environment. Additionally, according to McCann, et al (1984), during histogenesis of Arabidopsis thaliana stems, differentiation into sclerenchyma cells occurs which exhibit a timed involvement with complex cell-dividing activity that is an integral part of meristematic tissue functionality, hence fulfilling the indispensable roles of supporting load-bearing structures like stems and preventing them from collapsing under their own weight, emphasizing the critical role of sclerenchyma cells in plant anatomy and physiology. Thus the understanding of sclerenchyma cells and their role in providing structural support to mature plant tissues offer valuable insights about the intricacies of cell microbiology as related to plant tissue fortification.

There are two main types of sclerenchyma cells: fibers and sclereids (Mangeat et al., 2021; Manils et al., 2022; Maricchiolo et al., 2022). Fibers are long and slender cells arranged in bundles, offering strength and rigidity to plant structures such as stems and vascular tissues. Sclereids, on the other hand, are shorter and more irregular in shape, contributing to protection and hardness in structures like seed coats and nut shells. In the vast landscape of cell biology, sclerenchyma cells emerge as critical components when it comes to the fortification of plant structures. They present themselves in two primary forms, namely fibers and sclereids, each with distinct functional attributes and structural properties. For an in-depth understanding of these categories, consider the stem of Linum usitatissimum (flax plant) as a representative case. In the phloem tissue of the stem, elongated and spindle-shaped fiber cells are observed. With their origins in the primary and secondary meristematic tissues, these fibers, as per the Foster and Gifford study in 1974, have a principal role in providing tensile strength to the plant structure. Their ability to manage mechanical stress comes from the high cellulose and lignin content in their secondary walls, which increases their rigidity and resistance to external pressures. Contrastingly, sclereids are typically shorter and irregular in shape, often branching out in diverse directions. Deducing from the observations made by Esau in 1965 in the pearl millet (Pennisetum glaucum), where sclereids are significantly involved, one can see that their primary function is not limited to providing supportive tissues but they also play protective roles. For instance, the hardened shells of many seeds and nuts are the consequence of densely packed sclereids. In terms of their structure, sclereids' secondary cell walls exhibit a different perspective from fiber cells. Rather than an elongated design, they possess a thickened, heavily lignified secondary cell wall with numerous pits, evidencing their role in protecting plant tissues. Hence, an intricate understanding of sclerenchyma cells in their dual manifestations as fibers and sclereids elucidates their critical role in the adaptation and survival of plant organisms.

Sclerenchyma cells, unlike parenchyma and collenchyma cells, are generally dead at maturity (Orr & Hynds, 2021; Ohi & Kenworthy, 2022; Okano et al., 2023). This is due to the deposition of lignin, which prevents further growth and cell expansion. However, their role in providing mechanical support to mature plant tissues is crucial for maintaining the overall structure and integrity of the plant. In the domain of plant cell biology exists a fascinating characteristic that distinguishes sclerenchyma cells from their counterparts - parenchyma and collenchyma. Unlike the latter, which are typically living at maturity, sclerenchyma cells are generally dead upon maturation. A comprehensive case study to illustrate this distinctive property can be referenced in the study of grapevine shoots where Rex and Caffery (1974) noted how different types of cells undergo contrasting career paths during development. In their research, it was observed that parenchyma and collenchyma cells in grapevine shoots demonstrated metabolic activity, engaging in processes like photosynthesis and storage, and showing the ability to change shape or size due to their thin and flexible primary walls. Sclerenchyma cells, on the other hand, were reported to be non-living at full maturity. These cells undergo a process of programmed cell death, losing their protoplast during development. The functionality of sclerenchyma cells primarily lies in their thick, rigid secondary cell walls that are impregnated with lignin, a complex organic polymer that lends mechanical strength and rigidity to these cells. The death of the protoplast during sclerenchyma cell development, as reported by Barlow and Baluška (2000), allows the cell walls to harden and mature, thus, turning these cells into structural entities that serve as the critical mechanical support to the plant, providing strength, and enabling upright growth. Despite their terminal life stage, the dead sclerenchyma cells undeniably establish their relevance and necessity in the overall architecture and solid foundation of plant structures, further advancing our understanding of plant cell biology.

In conclusion, the specialization of plant cells into parenchyma, collenchyma, and sclerenchyma cells highlights the remarkable diversity of functions that these cells perform to support the growth, development, and survival of plants. The distinct characteristics of each cell type allow plants to respond to various environmental challenges and ensure their ability to thrive in a wide range of conditions. Now that we are at the close of the penultimate chapter in this book, it is pertinent to summarize the key themes that it addresses.

CHAPTER 7 SUMMARY

Chapter 7 of the studied text delves into the intricacies of cell diversity and specialization, highlighting key concepts in stem cells, tissues, muscle cells, nerve cells, and plant cell specializations. The chapter commences by illuminating the phenomenon of stem cells, elucidating their critical role in the differentiation process and their distinctive degrees of potency. Stem cells, with their capacity to give rise to various cell types, are presented as integral components of developmental biology.

Moving forward, the discussion shifts to tissues, essential components of multicellular organisms. The chapter underscores the significance of tissue organization and function in maintaining the integrity and functionality of complex organisms. The intricate interplay between different types of tissues and their cooperative roles in sustaining physiological processes is emphasized.

In subsequent sections, the focus narrows down to specific cell types, beginning with muscle cells. The chapter provides a comprehensive exploration of muscle cells' distinct contractile machinery and expounds upon the diversity of muscle tissue types. The intricate workings of muscle cells are elucidated, serving to deepen the reader's understanding of their crucial role in locomotion and bodily movement.

The narrative then transitions to nerve cells, delving into their intricate structure and pivotal role in the nervous system's signaling processes. The chapter offers a detailed examination of the components that constitute nerve cells and underscores their significance in facilitating communication within the nervous system, thus enabling rapid responses to external stimuli. The concluding segment of the chapter centers on plant cell specializations, particularly parenchyma, collenchyma, and sclerenchyma cells. These specialized plant cells are expounded upon, highlighting their diverse functions and contributions to plant structure and function. Their varied roles in providing support, storage, and transport within plants are underscored, elucidating the intricate adaptations that have evolved to enable plants to thrive in diverse environments.

To summarize, Chapter 7 of the studied text provides a comprehensive exploration of the fascinating realm of cell diversity and specialization. Through its meticulous examination of stem cells, tissues, muscle cells, nerve cells, and plant cell specializations, the chapter paints a vivid picture of the remarkable intricacies that underlie the functioning of multicellular organisms and the impressive adaptations that have evolved to sustain life's diverse processes. The final chapter of this book focuses on cell reproduction and development.

Chapter

8 CELL REPRODUCTION AND DEVELOPMENT

CONTENTS

Chapter 8 of the present text delves into the intricate realm of cell reproduction and development, serving as a profound exploration of the underlying mechanisms that govern the perpetuation and evolution of life. The chapter casts a wide net over a comprehensive array of fundamental concepts, encompassing a range of processes that collectively shape the trajectory of life from its inception to its maturation. These encompass fertilization, embryonic development, stem cells and regeneration, aging and senescence, and the multifaceted field of cancer biology.

The opening sections of the chapter lay a foundational understanding of fertilization, a pivotal event in which gametes fuse, culminating in the formation of a zygote. This union of genetic material sets the stage for the remarkable journey of embryonic development, a process explored in subsequent sections. The chapter dissects embryonic development with meticulous detail, delving into the intricate stages of cleavage, gastrulation, and organogenesis. Through these stages, a single fertilized egg transforms into a complex multicellular organism, with tissues, organs, and systems that together constitute the intricate web of life.

The discourse further extends to the pivotal role of stem cells and regeneration in the context of healing and tissue repair. Stem cells, with their remarkable capacity to differentiate into various cell types, contribute significantly to the rejuvenation and reconstruction of damaged tissues. The chapter delves into the regenerative potential of these cells, elucidating their capacity to replace lost or damaged cells, thereby fostering the renewal of biological structures.

As life progresses, the chapter then navigates through the intricate phenomenon of aging and senescence. It explores the physiological and cellular changes that occur with the passage of time, leading to the eventual decline in the functionality of cells and organs. This exploration encompasses both the molecular mechanisms and broader impacts of aging, shedding light on the complex interplay between genetics, environment, and the natural course of life.

Finally, the chapter delves into the intricate realm of cancer biology, unveiling the dysregulation that occurs within the cell cycle, leading to uncontrolled proliferation and tumor formation. This multifaceted subject unravels the molecular intricacies underlying the transformation of normal cells into malignant entities, highlighting the importance of understanding these processes for diagnostic and therapeutic endeavors. Through its comprehensive examination of cell reproduction and development, this final chapter unveils the dynamic narrative of life's inception, progression, and the challenges it faces. By traversing these themes, readers are granted a profound insight into the intricate web of cellular existence, from its embryonic origins to its transformative potentials and the anomalies that may arise along the continuum. It is fitting that we open the chapter with a discussion of the initial fertilization processes and their consequences for life.

8.1 FERTILIZATION: UNION OF GAMETES AND ZYGOTE FORMATION

Fertilization, a process of profound significance in the realm of cell biology, marks the union of gametes and lays the cornerstone for the commencement of embryonic development (Charpentier & King, 2021; Carroll & Marqusee, 2022; Chang et al., 2023). This intricate phenomenon brings together two distinct reproductive cells, the spermatozoon and the ovum, each representing a unique set of genetic material and functionalities. Fertilization, a pivotal phenomenon within the field of cell biology, denotes the amalgamation of gametes denoting the initiation of embryonic development, and carries profound biological significance. An examination of sea urchin fertilization provides an elucidate illustration of this process. Sea urchins, much like many marine invertebrates, undergo external fertilization, thereby making them an ideal biological model for studying the intricacies of this process. Fertilization commences with the synchronous discharge of eggs and sperm into the surrounding water, leading to the encounter of a sperm with the egg's outer jelly coat. Recognition proteins on the sperm surface bind with the egg surface, triggering the acrosomal reaction, whereby the sperm's acrosomal vesicle releases digestive enzymes to pass through the egg's vitelline layer. During this phase, the sperm undergoes a series of transformations known as capacitation, necessary for successful fertilization. The ultimate membrane fusion between sperm and egg culminates in the block to polyspermy, a mechanism to prevent the entry of multiple sperm. The egg deploys rapid changes in membrane potential, known as fast block, and modifications to the vitelline layer, known as the cortical reaction, or slow block. Upon successful membrane fusion, the egg completes meiosis II, just as the sperm's nucleus, or pronucleus, enters the egg. The union of male and female pronuclei marks the culmination of fertilization and the genesis of the zygote, setting the stage for embryonic development to unfurl. Through this carefully choreographed sequence of events, fertilization in sea urchins provides an in-depth case study of how cellular interactions orchestrate the formation of a new organism, effectively underpinning the cellular and molecular principles of sexual reproduction.

At its core, fertilization involves a series of intricately choreographed events that enable the fusion of the male and female gametes (Sales-Gil et al., 2021; Salinas-Luypaert et al., 2021; Schornig & Taverna, 2021). The journey of the spermatozoon from its point of release to the site of fertilization, often the fallopian tube, is fraught with challenges. The survival of the spermatozoon in the female reproductive tract depends on various factors, including the pH, cervical mucus consistency, and the timing of its release. In a synchronized dance, the spermatozoon navigates through these obstacles with the aid of its motility and structural adaptations, ultimately reaching the ovum.

Fertilization fundamentally represents a series of meticulously orchestrated events underscoring the fusion of male and female gametes. A comprehensive study of mammalian fertilization, particularly that of humans, provides incisive revelation into the precise roles and interactions of various cell components to ensure successful gametic fusion. This process commences with the penetration of the cumulus oophorus and corona radiata, cushions of cells surrounding the egg, by the sperm. This is facilitated by the acrosin, an enzyme released from the acrosome, a specialized vesicle at the sperm's tip, through the orchestrated acrosomal reaction. Upon reaching the oocyte, the sperm must breach the zona pellucida, the extracellular matrix enclosing the oocyte. This penetration is mediated by another acrosomal protein, zona pellucida binding protein 3 (ZP3), which engages sperm receptors and commences another cascade of reactions resulting in the sperm and oocyte plasma membranes' fusion.

This membrane fusion results in the cortical reaction in the oocyte, where cortical granules fuse with the oocyte membrane, altering the zona pellucida's structure to prevent polyspermy, an event where more than one sperm could fertilize an egg. Consequent to the entry of the sperm, the secondary oocyte completes its meiotic division to become an ovum. The nuclei of the sperm and egg, now termed pronuclei, migrate to the center of the cell where they decondense and ultimately fuse, marking the onset of the zygote and signaling the completion of fertilization. Therefore, the seemingly simple act of fertilization entails a series of complex, synchronized cellular events, inherently regulating the consorting dance of life's creation. Figure 71 represents the fertilization process.

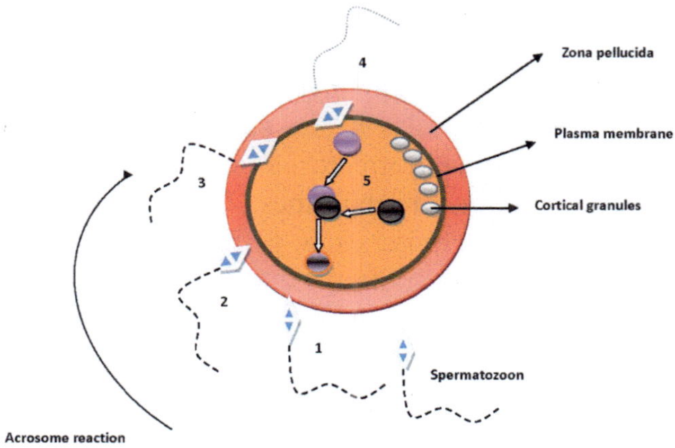

Figure 71. The fertilization process.

Source: Anifandis G, Messini C, Dafopoulos K, Sotiriou S, Messinis I. Molecular and Cellular Mechanisms of Sperm-Oocyte Interactions Opinions Relative to in Vitro Fertilization (IVF). International Journal of Molecular Sciences. 2014; 15(7):12972-12997. https://doi.org/10.3390/ijms150712972

Upon reaching the ovum, the spermatozoon undergoes an acrosomal reaction, a biochemical process that enables it to penetrate the layers of the ovum, namely the zona pellucida (Li et al., 2023; Liang et al., 2022; Liang et al., 2021). This reaction is driven by enzymes stored within the acrosome, a vesicular structure at the tip of the spermatozoon. Successful penetration results in the fusion of the plasma membranes of the spermatozoon and the ovum, allowing the genetic material of both cells to merge. In the carefully regulated realm of human reproduction, the microscopic journey embarked upon by spermatozoa is characterized by a crucial event requisite to successful fertilization – the acrosomal reaction. The acrosomal reaction is a biochemical process that enables the spermatozoon to penetrate the cornified stratum of the ovum known as the zona pellucida. This event is led by the acrosome, a cap-like structure that houses essential hydrolytic enzymes and is positioned at the anterior portion of the sperm cell. Initiated by various signal integrations catered by the zona pellucida through the process of ZP3 receptor binding, an influx of calcium ions occurs within the spermatozoon, triggering exocytosis of acrosomal enzymes, especially acrosin and hyaluronidase. Highlighting one study to illustrate the biochemical cascade, a recent exploration conducted by Liu et al. (2015) into the homeostasis of human spermatogenesis unraveled the intricacy of this procedure. With the application of immunofluorescent staining of acrosin in conjunction with electron transmission microscopy, the authors were able to visualize and analyze the morphological shifts involved in the acrosomal reaction. They noted an enlargement of vesicles in the outer acrosomal membrane and subsequent merging with the spermatozoa plasma membrane propagating the externalization of acrosomal enzymes. Moving closer to the ovum, upon contact, the acrosomal enzymes, in particular acrosin, act digestively on the glycoproteins that constitute the zona pellucida. The result of this biochemical reaction is the creation of a pathway through the zona pellucida, allowing the paternal chromosomal material housed within the sperm cell to enter the maternal cytoplasm of the ovum, an event marking the cardinal step towards successful fertilization. Herein lies the paramount role of the acrosomal reaction in the initiation of a new life cycle; a reactant being vital not only due to its capacity to facilitate sperm-egg fusion but also due to its capacity to impede polyspermy, thereby ensuring the integrity of the resultant zygote.

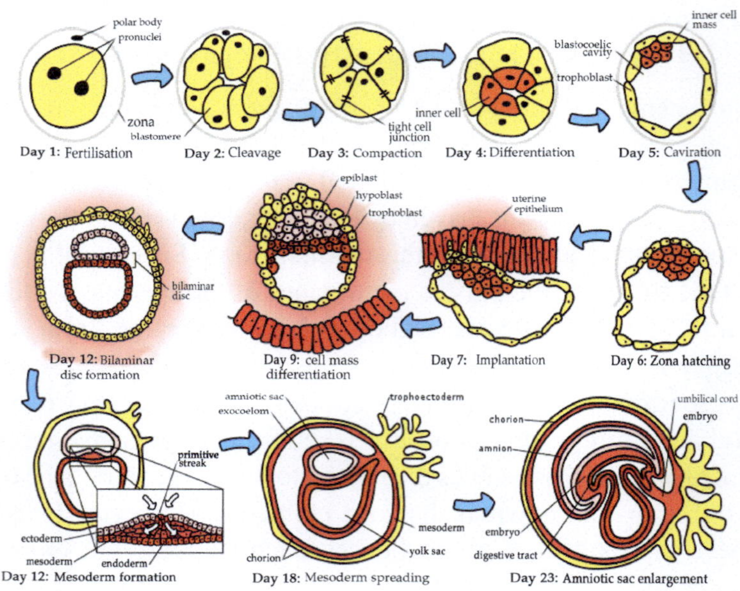

Figure 72. The process of cleavage and embryo development.

Source: By Zephyris - SVG version of ., CC BY-SA 3.0, https://commons.wikimedia.org/w/index.php?curid=10811330

The culmination of this fusion is the formation of a zygote, which possesses a full complement of genetic information required for the subsequent development of an organism (Hickey et al., 2022; Ho et al., 2022; Hu et al., 2022; Hu et al., 2021). The zygote marks the initiation of a cascade of events, initiating embryogenesis. It undergoes a series of cell divisions known as cleavage (see figure 72), leading to the formation of blastomeres. These divisions increase the cell number while maintaining a relatively constant size, eventually resulting in the formation of a blastocyst. Emphasizing the significance of human embryogenesis, the fusion of two exceedingly specialized cells, the sperm and egg, marks the formation of a zygote – a totipotent entity that carries the complete portfolio of genetic material essential for the genetic configuration of a new organism. Referring to a seminal study in the field of embryogenesis conducted by Wong et al. (2010), which adeptly elucidated the intricate process, the researchers detailed the sequence of events following the successful penetration of the spermatozoon through the zona pellucida and subsequent fusion with the ovum. This process, conceptualized as syngamy, is spearheaded by a complex series of cell signaling transductions, mediated by calcium ion influxes, that culminate in the amalgamation of the male and female pronuclei within the cytoplasm of the fertilized egg. Utilizing the precision of confocal laser scanning microscopy, the authors observed the ensuing alignment of the maternal and paternal chromosomes at metaphase, bearing witness to the initiation of the first mitotic division of the zygote. The resultant structure, a diploid single-cell entity, known as the zygote, comprises the full complement of genetic information from both parents – a collection of genetic material that shapes and guides the cell differentiations, and morphogenetic rearrangements that give rise to the blastocyst, and, ultimately, the emerging form of the developing organism. The importance of the zygote underscores the fact that it is this primitive cell that houses the complete genetic framework required for normal organismal development, thereby marking the onset of a cascading series of cellular events that articulate the blueprint of life.

Within the context of cell biology, the process of fertilization showcases a stunning interplay of molecular interactions, cellular adaptations, and intricate pathways. From the biochemistry of the acrosomal reaction to the orchestrated fusion of genetic material, the process of fertilization serves as a testament to the complexity and precision inherent within the cellular world. This process not only initiates embryonic development but also epitomizes the foundational concepts of inheritance and genetic diversity, playing a pivotal role in the perpetuation of life and the diversity of species. The second section will focus more on embryonic development and its stages.

8.2 EMBRYONIC DEVELOPMENT: CLEAVAGE, GASTRULATION, AND ORGANOGENESIS

Embryonic development, a marvel of biological complexity, encapsulates the transformative journey from a single fertilized cell to a multicellular organism with intricate structures and functions (Lamb et al., 2021; Lamptey et al., 2021; Lannagan et al., 2021; Langford, 2022). This process unfolds through a series of meticulously orchestrated events that encompass cleavage, gastrulation, and organogenesis. These successive stages embody the essence of cellular and molecular intricacies that define life's beginning.

Cleavage marks the commencement of embryonic development, transmuting the zygote—the fertilized egg—into a multicellular entity (see figure 73). This process involves rapid cell divisions without concurrent growth, resulting in the formation of a blastula, a hollow sphere of cells. These divisions are driven by the repeated and precise execution of the cell cycle, facilitated by regulatory factors and signaling cascades. Cleavage partitions the embryonic cells into smaller units, called blastomeres, while maintaining their overall size. The polarity of these divisions is vital, as it establishes the embryonic axes that guide subsequent development.

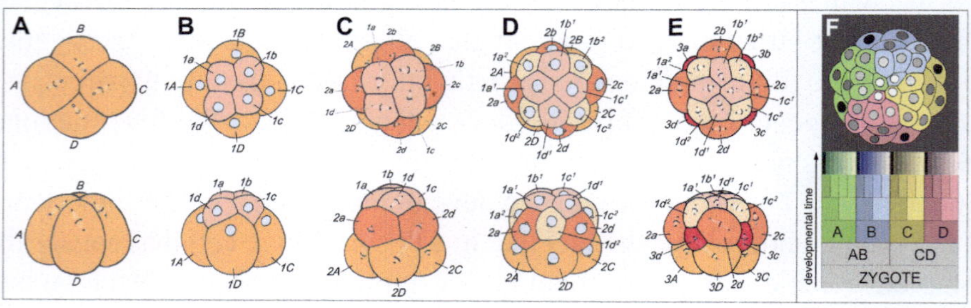

Figure 73. The complexities of cleavage in embryonic development.

Source: By Morgan Q. Goulding - Goulding MQ. 2009. Cell Lineage of the Ilyanassa Embryo: Evolutionary Acceleration of Regional Differentiation during Early Development. PLoS ONE 4(5): e5506. doi:10.1371/journal.pone.0005506

Gastrulation follows cleavage and ushers in a profound transformation as cells rearrange and differentiate to form the three primary germ layers: ectoderm, endoderm, and mesoderm (Katunin et al., 2021; Keenan et al.,

2021; Khampang et al., 2023). During gastrulation, cells undergo migration and invagination, leading to the formation of the gastrula—a structure with distinct layers and an opening called the blastopore. The precise choreography of cell movements is orchestrated by signaling molecules and adhesion molecules, enabling cells to assume their designated positions. The formation of the germ layers initiates the divergence of cell fates, with each layer destined to give rise to specific tissues and organs. Gastrulation is a pivotal stage in embryogenesis, succeeding the cleavage phase and ceding the way for drastic cellular rearrangement and differentiation, culminating in the formation of the three germ layers: ectoderm, endoderm, and mesoderm. To elucidate the complex processes steering gastrulation, a study conducted by Chazaud et al (2006) stands as a profound example, illuminating the intricate choreography of cell migration and differentiation using a murine model. Leveraging the capacities of time-lapse microscopy and gene expression profiling, they unraveled the migration pattern of cells and the sequence of gene activation that lead differentiation into germ layers. Following the formation of the blastocyst, they observed a sequence of morphological transformations exhibiting polarization, cell migration and subsequent positioning which demarcates the establishment of germ layers. Taking a closer look at the cellular movement during gastrulation, the process initiates with the formation of the primitive streak, wherein epiblast cells migrate inward in organized cohorts, executing epithelial-to-mesenchymal transitions. Upon inward migration, cells assume specific positions, governed by the orchestration of genes like FGF, TGF-β, and Wnt, which designate their differentiation into endoderm and mesoderm, while cells remaining on the exterior form the ectoderm. The transformation is significant: these germ layers will create the plethora of tissues and organs that constitute the complete organism. Through this elaborate dance of cells, gastrulation as depicted by Chazaud et al. (2006) underscores a transformative event that drives the generation of the vital germ layers - the ectoderm, endoderm, and mesoderm - through a complex interplay of molecular signaling and cytodynamics.

Organogenesis, the final chapter of embryonic development, encompasses the metamorphosis of germ layers into intricate structures that comprise organs (Farzadfard et al., 2021; Fauser et al., 2021; Favre-Bulle & Scott, 2022). This intricate process involves a harmonious interplay between genetic programming, cell signaling, and tissue interactions. Cells within each germ layer differentiate into specific cell types, acquiring distinct functions and morphologies. This differentiation is regulated by master regulatory genes that activate or suppress various genetic pathways. As cells continue to differentiate, they organize themselves into tissues and eventually form rudimentary organ structures. The sculpting of organs involves intricate cellular rearrangements, apoptosis, and the establishment of complex tissue architectures. Organogenesis, serving as the culminating chapter in embryonic development, orchestrates the metamorphosis of primary germ layers into the sophisticated structures that constitute the organ systems of the organism.

A case study echoes this dynamic transformation: the research of Tam and Behringer (1997) delves into the genesis of the mammalian kidney as an illustration. As highlighted in this study, the interplay between the metanephric mesenchyme (derived from the intermediate mesoderm) and the ureteric bud (evolved from the mesonephric duct) incites the development of the kidney. Utilizing genetic mouse models and in situ hybridization techniques, the team elucidated the journey from germ layer to organ structure. The process commences when the metanephric mesenchyme secretes glial-derived neurotrophic factor (GDNF), acting as a signal to the adjacent ureteric bud.

In response to GDNF, the ureteric bud invaginates into the mesenchyme, bifurcating repeatedly to form the branching architecture that ultimately constitutes the renal collecting system. This ureteric bud invasion simultaneously catalyzes a reciprocal induction in the metanephric mesenchyme, triggering its differentiation into nephrons – the functional unit of the kidney responsible for filtration and excretion. Through modulating the expression of several genes, including Pax2, Wt1, and GDNF among others, the researchers delineated the complexity of molecular signaling, reciprocal induction, and morphogenetic change driving the conversion of germ layers into organ structure. Hence, the study of organogenesis, as exemplified by kidney development, illuminates the process by which germ layers burgeon into the intricate organ systems that belie the complexity of multicellular organisms, enabling a clear picture of embryonic development. Figure 74 is an illustration of organogenesis (the development of the neural tube).

The development of the nervous system stands as a remarkable exemplar of the complexities of embryogenesis (Vakonakis, 2021; Veerasamy et al., 2021; van den Dries et al., 2022). Neural tube formation, driven by precise signaling gradients and tissue interactions, leads to the development of the brain and spinal cord. Neurogenesis—the birth of neurons—requires a delicate balance between cell proliferation, differentiation, and migration. Neurons then establish intricate connections, or synapses, through processes involving axon guidance and synapse formation, shaping the neural circuitry that underlies cognition and behavior.

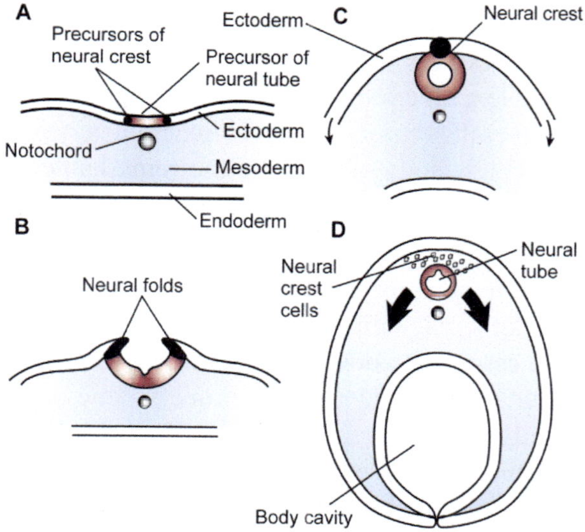

Figure 74. Illustration of organogenesis.

Source: Goodlett CR, Horn KH. Mechanisms of alcohol-induced damage to the developing nervous system. Alcohol Res Health. 2001;25(3):175-184.

The development of the nervous system serves as a particularly intricate testament to the complexity of embryogenesis, intricately interweaving the processes of organogenesis, differentiation, and growth. In delving into this transformative progression, a compelling study conducted by Muotri and Gage (2006) sheds light on the intricate genomic mechanisms and decisive environmental cues that synergize to choreograph the development of the nervous system. Leveraging the developments in human pluripotent stem cell research, the authors unraveled the avant-garde of neuronal differentiation by simulating the development of neural progenitor cells. These progenitor cells, descendants of the ectoderm layer, respond to a cascade of genetic signals in the early phases of embryogenesis, instigating the formation of the neural tube, the nascent structure that subsequently gives rise to both central and peripheral nervous systems. Employing induced neurons as an in vitro model, they engineered the expression of neural genes and analyzed the downstream effects by utilizing the invaluable aspects of genomic and proteomic technologies. With gene transcription markers such as Sox1 and Nestin marking early neural induction, they traced the developmental trajectory of neurons, delineating how they become committed to specific neuronal or glial fates under the directive influence of transcription factors such as Ascl1/Mash1, Neurog2, Dlx2 and others. This developmental odyssey ably illustrates how a single layer of embryonic cells derives the architecturally complex, functionally diverse tissue that comprises the human nervous system. Overall, the neural developmental processes highlighted in the Muotri and Gage study underscore the remarkable exemplar of the complexities of embryogenesis, underlining the fine-tuned interplay between genetic programming and environmental cues.

The orchestration of embryonic development entails a symphony of molecular interactions, signaling cascades, and genetic programming. Errors at any stage can lead to developmental abnormalities, illustrating the delicate balance that governs this intricate process. Studying embryogenesis illuminates not only the origins of life but also offers insights into the molecular underpinnings of diseases and the potential for regenerative medicine. In conclusion, the journey from a single fertilized cell to a fully formed organism is a mesmerizing saga of cellular choreography. Cleavage, gastrulation, and organogenesis are the key stages that navigate the transformation of a simple zygote into a complex organism with specialized tissues and organs. This intricate narrative showcases the intertwined dance of cellular communication, differentiation, and morphogenesis, underscoring the awe-inspiring intricacies of life's origins. The third section of this chapter focuses on stem cells and regeneration.

8.3 STEM CELLS AND REGENERATION: HEALING AND TISSUE REPAIR

The ability of living organisms to repair and regenerate damaged tissues is a testament to the intricate mechanisms that underlie the maintenance of homeostasis and the restoration of form and function (Pandey et al., 2021; Panicker et al., 2021; Paliulis et al., 2023). At the heart of this phenomenon lies the remarkable potential of stem cells, which play a pivotal role in orchestrating the healing processes that occur in response to injuries or tissue damage. The capacity of living organisms to repair and regenerate damaged tissues attests to a symphony of intricate biological mechanisms that persistently strive to restore homeostasis and reinstate both form and function. An epitome revealing these exact mechanisms is the study by Illingworth (1974) on the wound healing and regeneration process in the liver. His investigation demonstrated that following partial hepatectomy, the remaining hepatocytes (see figure 75 for an illustration of their structure) expediently initiate replication, not by invocation of stem cells, but rather by exit from the quiescent G0 phase of the cell cycle into active proliferation. Utilizing radio-autography, he traced the stages of the cell cycle and proliferation kinetics of surviving hepatocytes after hepatic injury, finding that the regenerative response is initiated promptly within a few hours of the resection. The key factors facilitating this regenerative response comprised a veritable orchestration of cytokines and growth factors, notably Hepatocyte Growth Factor and Tumor Necrosis Factor-α, triggering the hepatocyte cell cycle re-entry. A particularly compelling observation by Illingworth was the systemic nature of this response, with hepatocytes distributed throughout the remnant liver tissue initiating proliferation, albeit with variable kinetics. The regenerating liver not only recreates its parenchymal mass but also manages to rebuild its distinctive lobular architecture and reestablish its complex vascular and biliary networks, thereby fully restoring the liver's integrative metabolic and detoxification functions. Such phenomenal regenerative capacity hence speaks to the intricate homeostatic mechanisms within living organisms, designed precariously to repair and restore damaged tissues, reinstating the native form, improving function, and imparting survival advantage.

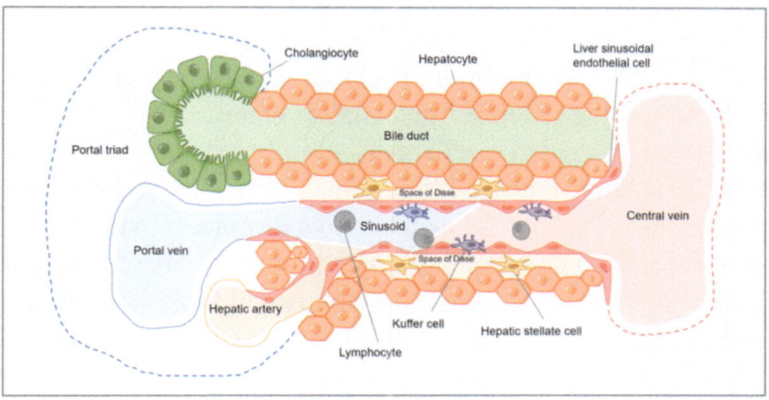

Figure 75. The structure of a hepatocyte.

Source: Luo Q, Wang N, Que H, Mai E, Hu Y, Tan R, Gu J, Gong P. Pluripotent Stem Cell-Derived Hepatocyte-like Cells: Induction Methods and Applications. International Journal of Molecular Sciences. 2023; 24(14):11592. https://doi.org/10.3390/ijms241411592

Stem cells are undifferentiated cells with the remarkable ability to self-renew and differentiate into various cell types (Watson et al., 2021; Warecki & Sullivan, 2022; Weigert & Herhaus, 2023). This unique characteristic positions them as the cellular architects of regeneration, enabling the replacement of damaged or lost cells in order to restore tissue integrity. Their potency and versatility provide the cellular foundation for the complex processes that unfold during healing and tissue repair. Stem cells are unique, undifferentiated biological entities demonstrating an extraordinary dual ability to self-renew and differentiate into a diverse gamut of cell types. This remarkable potential of stem cells is exemplified in a landmark study conducted by Takahashi and Yamanaka (2006). The study unveiled the transformative potential of somatic cells under the influence of a combination of specific transcription factors that induce pluripotency measured by differentiation potential, effectively creating induced pluripotent stem cells (iPSCs). Through this investigation, adult mouse fibroblast cells were experimentally induced to revert to a pluripotent state that mirrored embryonic stem cells. Adopting a retroviral transduction approach, they introduced four transcription factors, notably Oct3/4, Sox2, c-Myc, and Klf4, into the fibroblast cells. Differentiating the fibroblasts with this combination of factors resulted in reprogrammed cells that showcased traits inherent to embryonic stem cells. The induced pluripotent stem cells exhibited the capacity for self-renewal while also demonstrating their pluripotent potential by generating differentiated cells of all three germ layers when introduced into a murine model. This breakthrough study underscored the untapped potential of stem cells from adult tissues which ultimately are reserve cells that not only extend the capacity of an organism to regenerate damaged tissues but also underpin the possibility of future therapeutic strategies for degenerative diseases, obviating the ethical issues associated with embryonic stem cell research. Therefore, stem cells, with their innate capability to self-renew and differentiate, symbolize a remarkable biological phenomenon, bolstering our understanding of cellular biology and its potential for therapeutic application.

In tissue regeneration, stem cells are strategically distributed across various tissues, acting as resident repair mechanisms that can be activated in response to injury (Ford et al., 2021; Forveille et al., 2021; Gan, 2021; Gallusser et al., 2022). Upon tissue damage, a series of signals are triggered, including inflammatory molecules, growth factors, and cytokines. These cues serve as cellular communication tools, mobilizing stem cells and other cellular players to the site of injury. This localized response is particularly evident in tissues with high regenerative potential, such as the skin and the lining of the gastrointestinal tract, where rapid turnover and continuous damage prompt frequent healing events. The phenomenon of tissue regeneration is underpinned by the strategic distribution of stem cells across various tissues, where they lie in wait as resident repair mechanisms that can be promptly activated in response to injury. Lieber and Mahowald's (1996) pioneering study examining adult intestinal stem cells in Drosophila offers insight into these staples of life. The researchers extensively detailed the regeneration process in the gut epithelium triggered by injury. The resident, precursor-like intestinal stem cells (ISCs) of Drosophila could be clearly identified due to markers such as the expression of escargot gene. Following injury inflicted using cytotoxic agents, a rapid proliferation of these ISCs was noted under the influence of regeneration signals. By leveraging the power of lineage labeling, combined with the temporal control offered by targeted gene expression systems, the researchers traced the fate adopted by the progeny of these ISCs. These daughter cells, termed enteroblasts, were observed to follow a coordinated process of differentiation to functional gut cells, termed enterocytes and enteroendocrine cells, to repopulate the injured gut lining.

The authors also teased apart the role of multiple signaling pathways, including the evolutionarily conserved Wnt, JAK-STAT, Hippo and Notch pathways that orchestrate the remarkable regenerative process. Thus, the study highlights how stem cells, primed for action, stand ready to leap in and start the process of regeneration whenever necessary. This not only underlines the potential for regenerative medicine but also is exemplary of the intricate interactions inherent within cellular systems to maintain organismal homeostasis (see figure 76 for an example).

The process of tissue repair involves several key phases (Liu et al., 2022; Loconte et al., 2022; S. Lim et al., 2023). Initially, an inflammatory response is triggered, which clears the debris and pathogens from the damaged area. In this phase, immune cells and platelets infiltrate the site, setting the stage for subsequent repair mechanisms. Following inflammation, the proliferative phase ensues, during which stem cells, guided by specific signals, multiply and differentiate into the required cell types. The extracellular matrix, a scaffold of proteins and fibers, is crucial in guiding cell migration and tissue organization during this phase. Tissue repair, a fundamental biological process, encompasses several critical phases that coordinate to restore tissue integrity and homeostasis following an injury. One of the most comprehensive elucidations of this process can be found in a study by

Gurtner, Werner, Barrandon, and Longaker (2008), which delves into the biological and molecular processes governing wound healing.

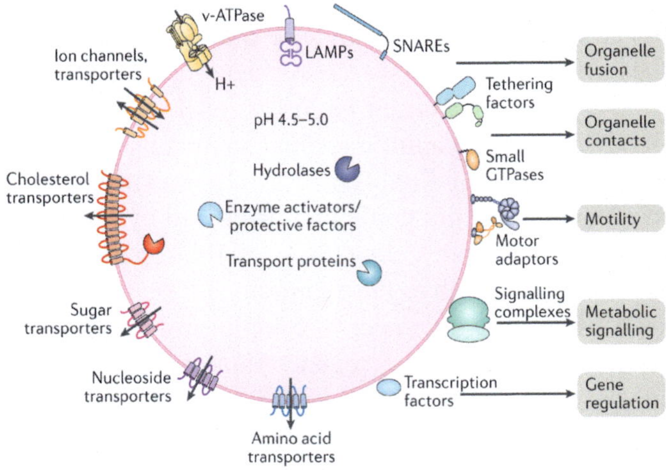

Figure 76. Lysosomes as dynamic regulators of cell and organismal homeostasis.

Source: Ballabio, A., Bonifacino, J.S. Lysosomes as dynamic regulators of cell and organismal homeostasis. Nat Rev Mol Cell Biol **21**, 101–118 (2020). https://doi.org/10.1038/s41580-019-0185-4

Initiated by tissue injury, the first phase, hemostasis, stems from the interplay between platelets, endothelial cells, and coagulation pathways to form a fibrin clot. The study, using in vivo models, gives particular emphasis to platelet-derived growth factor (PDGF) and transforming growth-factor-beta (TGF-β), which are secreted by platelets to plug the wound and initiate the subsequent phase, inflammation. The inflammatory phase, marked by vasodilation, infiltration of leukocytes, and removal of necrotic tissue and pathogens, sets the stage for the next phase: proliferation. Herein, the authors exploited gene knockout models of growth factors and cytokines, illustrating how factors like TGF-β, fibroblast growth factor (FGF), and vascular endothelial growth factor (VEGF) orchestrate the proliferation of fibroblasts and endothelial cells, leading to an elaborate process of neovascularization and formation of granulation tissue. Concomitantly, the migration and proliferation of epithelial cells across the wound, known as re-epithelialization, seals off the wound. The final phase, tissue remodeling, embodies the transformation of granulation tissue into a scar, maturing over time to restore tissue strength. This process, driven by matrix metalloproteinases (MMPs), forms a delicate balance between collagen deposition and degradation. Gurtner et al.'s study accentuates the profundity and complexity of the repair process, thereby acknowledging the key phases that ensure the maintenance of organismal integrity in the face of injury.

The concept of regeneration extends beyond simple tissue repair (S. Y. Li et al., 2021; Y. Li et al., 2021; Y. H. Li et al., 2021). Some organisms, particularly certain amphibians and zebrafish, exhibit a remarkable ability for tissue regeneration that far surpasses that of humans. These organisms can regrow entire limbs, spinal cords, and even parts of the heart. This exceptional regenerative capacity is often attributed to the presence of specialized stem cells that are primed for rapid division and differentiation, along with an environment that supports efficient regeneration. However, mammals, including humans, also possess regenerative capabilities to varying degrees. For instance, the liver can regenerate a substantial portion of its tissue, and the bone marrow constantly replenishes blood cells. The skin can repair itself through a process called wound healing, where epithelial cells migrate and proliferate to cover the wound. Furthermore, certain tissues, like skeletal muscle, possess a limited capacity for regeneration through the activation of satellite cells—specialized stem cells that contribute to muscle repair and growth. Harnessing the regenerative potential of stem cells has far-reaching implications for medicine. The field of regenerative medicine aims to exploit the ability of stem cells to replace damaged tissues or even generate entire organs. Techniques such as stem cell transplantation, tissue engineering, and induced pluripotent stem cells (iPSCs) hold promise in treating conditions ranging from degenerative diseases to injuries.

In conclusion, the regenerative abilities of stem cells stand as a testament to the remarkable intricacies of life's self-preserving mechanisms. From wound healing to tissue repair, stem cells orchestrate complex processes that enable organisms to recover from injuries and maintain functional integrity. Understanding and harnessing the potential of stem cells and regenerative processes not only sheds light on the fundamental principles of biology but also opens doors to revolutionary medical treatments that could transform the landscape of healthcare. The penultimate section in this final chapter focuses on aging and senescence.

8.4 AGING AND SENESCENCE: THE IMPACT ON CELLS AND ORGANISMS

The passage of time leaves its indelible mark on all living organisms, culminating in the intricate phenomenon of aging—a process that entails a gradual decline in cellular and physiological functions (Chen et al., 2021; Chen et al., 2022; Chen et al., 2023). This relentless march of time brings forth senescence, a state characterized by diminished cellular vitality and increased susceptibility to diseases. At the cellular level, aging entails a complex interplay of genetic, environmental, and molecular factors that collectively influence the fate of cells and, consequently, the overall health of organisms. The inexorable voyage through time demonstrably imprints an indelible mark on all organisms, manifesting experientially through the multifaceted process known as aging—denoted by a systematic whittling down of cellular and physiological functions. Groundbreaking work by López-Otín et al (2013) on the hallmarks of aging provides exceptional insights into this complex process which delicately straddles myriad mechanisms from the cellular level to the organismal realm. The researchers, in this seminal study, suggested nine tentative hallmarks that grouped common denominators of aging across diverse organisms. These included genomic instability, telomere attrition, epigenetic alterations, loss of proteostasis, deregulated nutrient sensing, mitochondrial dysfunction, cellular senescence, stem cell exhaustion, and altered intercellular communication. Utilizing a wide spectrum of molecular and biochemical techniques coupled with model organisms ranging from yeast to mammals, they elaborated on the interconnectedness of these hallmarks. For instance, the effects of telomere attrition in human fibroblasts, outlined by evaluating telomerase reverse transcriptase (TERT) overexpression, demarcated the impacts of this process in the context of cell replicative lifespan and cancer. Further, the team underscored the synergies and antagonisms between various hallmarks, pointing towards the delicate balance within this complex network of aging mechanisms, such as between mitochondrial dysfunction and cellular senescence or between stem cell exhaustion and tissue dysfunction. A comprehensive survey by López-Otín et al of aging thus captures an integrative view of the gradual decline in cellular and physiological functions characterizing the process, offering insights into the potential interventions to tackle age-related diseases.

Central to the aging process is the gradual deterioration of cellular functions (Cahan et al., 2021; Caldas et al., 2021; Burke et al., 2023). Over time, cells accumulate damage from various sources, including reactive oxygen species, radiation, and metabolic byproducts. This damage impairs cellular components such as DNA, proteins, and lipids, thereby compromising cellular integrity and function. DNA mutations accumulate, potentially leading to genomic instability, while protein misfolding and aggregation disrupt proper cellular machinery. Such cellular damage forms the basis for the "wear and tear" theory of aging, highlighting the accumulating toll of stressors on cells over a lifetime.

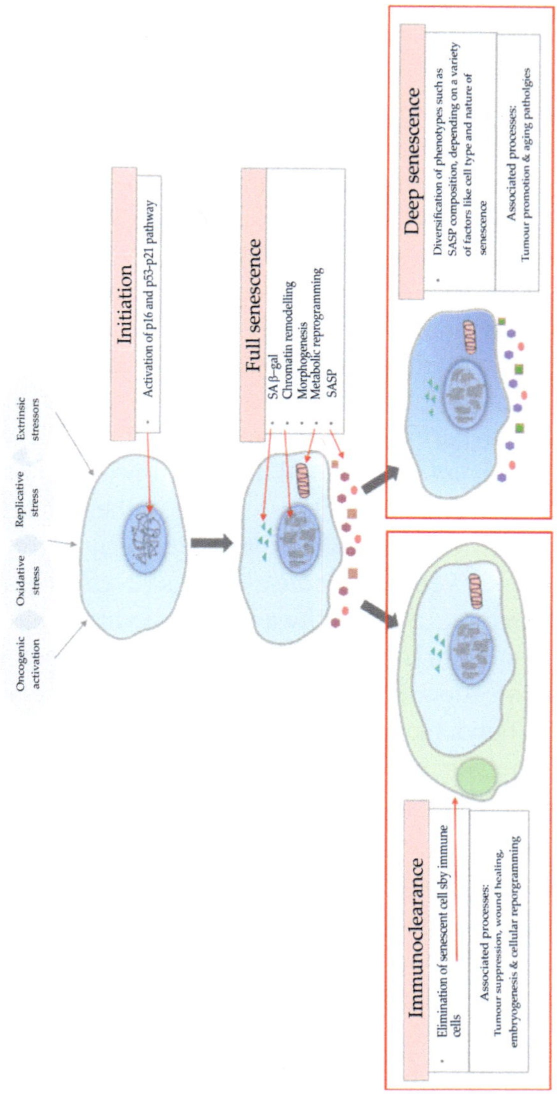

Figure 77. Various elements of cellular aging in senescence.

Source: Liao Z, Yeo HL, Wong SW, Zhao Y. Cellular Senescence: Mechanisms and Therapeutic Potential. Biomedicines. 2021; 9(12):1769. https://doi.org/10.3390/biomedicines9121769

Central to the aging process is the inexorable deterioration of cellular functions (see figure 77), a fundamental biological transformation that causes a cascade of changes at the level of tissues, organs, and ultimately, the organism. A model study delineating this at a microscopic level is the research conducted by López-Otín, Blasco, Partridge, Serrano, and Kroemer (2013). This extensive study persuasively demonstrated how multifarious changes at the cellular level recalibrate the overall physiology toward aging. Especially at the root of this gradual downswing in cellular operations was the acknowledgment of nine hallmarks of aging. One exemplar is genomic instability resulting from DNA damage accumulation over time, denoted by the progressive increment of double-strand breaks or replication errors. Another hallmark, telomere attrition, drawn from studies on cultured human cells, illuminated how iterative cell division erodes the chromosomal ends, thereby disrupting cellular functions and provoking senescence. Yet another cornerstone of cellular deterioration was the loss of proteostasis – as substantiated by experiments on C. elegans showing how aberrations in chaperone-mediated autophagy and ubiquitin-proteasome system led to proteome instability and precipitated aging. Other notable hallmarks include deregulated nutrient sensing and mitochondrial dysfunction, where corresponding studies on yeast and mouse models revealed how aberrations in Insulin/IGF-1 signaling, TOR pathway, and impaired mitophagy incite age-linked cellular deterioration. From cellular senescence to perturbed intercellular communication, López-Otín et al. craft a granular narrative of the aging process that underscores how integral the gradual decay of cellular activities is to the emergence of aging phenotypes.

Senescence, at both the cellular and organismal levels, is influenced by a myriad of factors (Altulea et al., 2021; Alvord et al., 2022; Aniento et al., 2022; Aprile et al., 2023). Genetic predisposition plays a role, with certain genes and alleles linked to longevity or susceptibility to age-related diseases. Epigenetic modifications, which regulate gene expression patterns without altering DNA sequences, also contribute to the aging process. Epigenetic changes can impact cellular functions and influence the timing of senescence. The phenomenon of senescence, perceptible at both cellular and organismal levels, is shaped by an array of interlaced factors. In order to decode these interdependencies, a landmark study by Campisi and d'Adda di Fagagna (2007) presents a comprehensive exploration of cellular senescence and its underpinning triggers and impacts. Documenting cellular senescence in diverse cell types and species, the researchers demarcate how this state of proliferative arrest in cells is induced by a wide variety of stressors such as telomere shortening, oncogene activation, oxidative stress, and DNA damage. Using techniques such as fluorescence-activated cell sorting (FACS) with senescence-associated β-galactosidase staining, alongside assessment of DNA damage response markers like γ-H2AX and 53BP1, the study plumbs the depths of cellular senescence manifesting in response to these stressors. At the systemic level, the authors highlight the critical role of senescent cells in shaping the organismal phenotype, acknowledging their identity as a source of pro-inflammatory cytokines, growth factors, and proteases, collectively termed the senescence-associated secretory phenotype (SASP). Further, the researchers probe into the complex interplay between genetic and environmental factors that modulate the onset and progression of senescence. For instance, genes regulating DNA repair and metabolic processes along with environmental stressors like lifestyle and chronic inflammation significantly modulate the "aging clock". The insights offered by Campisi and d'Adda di Fagagna's study into the precipitants and consequences of senescence aptly illustrate how both cellular and organismal levels of life are steered by a multitude of interdependent factors, cultivating a more profound understanding of the aging process.

One of the hallmarks of cellular aging is the decline in stem cell function (Pendharkar et al., 2021; Perera, 2021; Peng et al., 2022). As stem cells divide and differentiate over time, their regenerative potential wanes, leading to reduced tissue repair and replacement. This decline in stem cell function contributes to the age-related decline in tissue maintenance and repair mechanisms, leaving organisms more vulnerable to diseases and slower to recover from injuries. Cellular aging presents numerous diverse phenomena that are intimately intertwined, yet one of the most significant markers among this gossamer web of biological interactions is the consequential decline in stem cell function. A detailed investigation of this principle, and its underlying mechanisms, using a case study of hematopoietic stem cells (HSCs) based in the field of cellular microbiology may render a comprehensive understanding of this widely acknowledged phenomenon. In the mammalian bone marrow, HSCs play a pivotal role in continually reconstituting all blood lineages, thereby maintaining the homeostasis of blood component production. As aging encroaches, these HSCs undergo several significant morphological and functional changes, often referred to as "aging-related changes," which directly undermine their ability to differentiate and proliferate aptly. As a consequence of this, the organism experiences an aberrant reconstitution of blood components marked by an imbalance between lymphoid and myeloid cell lineages. Cellular and molecular studies of HSCs have attributed these alterations to key factors such as disruption of critical signaling pathways, genomic instability, stress responses, and epigenetic alterations, all increasing in manifestation intensity with age. For example, the TGF-beta signaling pathway, instrumental in maintaining the quiescence and self-renewal capacity of HSCs, has been found to be dysregulated significantly in aged HSCs, leading to their reduced regeneration potential. Another example involves epigenetic modifiers; with advancing age, the dynamic modifications in patterns of DNA methylation in HSCs lead to their impaired function, notably featuring a skewed differentiation potential. Collectively, these factors set foot a progressive decrement in HSCs' functionality, outlining their diminished capacity for self-renewal, homing, and differentiation, thereby underscoring the pivotal role of stem cell function decline in cellular aging. Hence, the observed correlation between cellular aging and stem cell function decline is a direct consequence of the mechanistic alterations that surge with aging, which if unraveled effectively, holds tremendous potential in age-associated clinical ramifications.

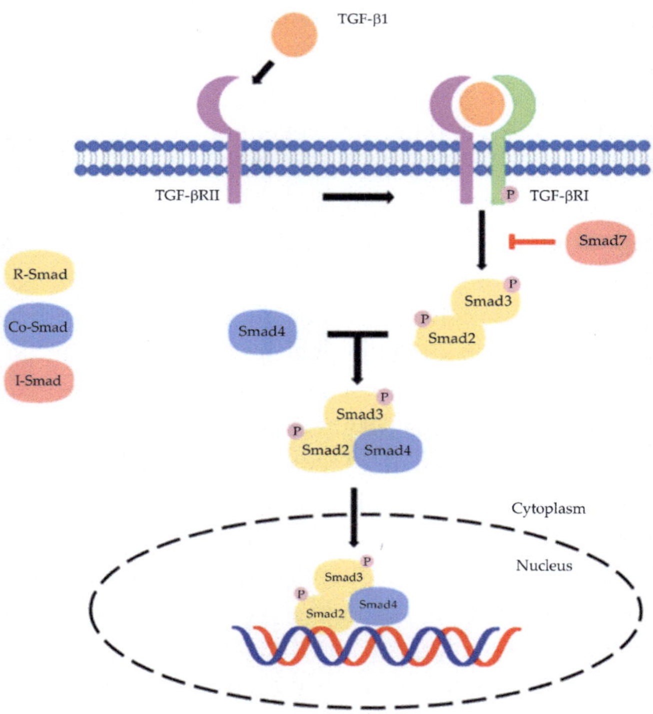

Figure 78. Mechanisms of TGF-β signaling from the cell membrane to the nucleus.

Source: Chung JY-F, Chan MK-K, Li JS-F, Chan AS-W, Tang PC-T, Leung K-T, To K-F, Lan H-Y, Tang PM-K. TGF-β Signaling: From Tissue Fibrosis to Tumor Microenvironment. International Journal of Molecular Sciences. 2021; 22(14):7575. https://doi.org/10.3390/ijms22147575Mechanisms of TGF-β signaling from the cell membrane to the nucleus

Cellular communication, an essential process for maintaining homeostasis, also falters with age (Lewis & Wells, 2021; D. Li et al., 2021; Lerit, 2022). Signaling pathways become dysregulated, impacting the transmission of information between cells and leading to impaired responses to growth factors and hormones. This disruption in communication can hinder the body's ability to mount effective responses to challenges, such as infections or tissue damage. The deterioration of cellular communication, a fundamental biological process crucial for maintaining homeostasis, is closely associated with the aging process. Considering the scenario of neurodegenerative ailments such as Alzheimer's disease, a specific manifestation of age-related decline in cellular communication, sheds light on the statement's ramifications. Primarily, Alzheimer's disease is characterized by compromised neuronal communication, which aggravates with the progression of age. Intricate neuronal networks in the central nervous system rely on synapses for intercellular communication. However, the aging process gives rise to synaptic dysfunction, which weakens this communication system. Amyloid-beta protein, intrinsic to Alzheimer's disease, plays a critical role here. Amyloid-beta oligomers federate to form plaques that disrupt synaptic function, primarily by impairing the integrity of neuronal membranes and triggering an aberrant immune response. Moreover, these oligomers resist degradation and clearance, owing to age-related modifications in the endolysosomal system responsible for the digestion and removal of extracellular debris. Consequently, the accumulation of amyloid-beta contributes to a self-perpetuating cycle of disrupted cellular communication leading to synaptic and neuronal loss. Additionally, the age-related deterioration in mitophagy regulation, a form of autophagy that is essential for the removal of dysfunctional mitochondria, negatively impacts neuronal function. Impaired mitophagy coupled with a high metabolic rate makes neurons particularly vulnerable to mitochondrial damage, further exacerbating the deterioration of cellular communication. Furthermore, impaired calcium homeostasis, an essential cellular communication element, has been observed in aged neurons. In fact, this disruption is thought to be a potential trigger for neurodegenerative pathways. Collectively, these factors contribute to the disruption of cellular communication, and hence, the overall functioning and survival of neurons, displaying a clear link between aging and compromised cellular communication. Understanding these underlying mechanisms might pave the way for interventions that can slow down or reverse the effects of aging on cellular communication, holding promise for the treatment of neurodegeneration and cognitive decline.

The immune system, responsible for defending the body against pathogens, also undergoes significant changes with age (Soon et al., 2021; Smole et al., 2022; Spangler et al., 2022). This phenomenon, known as immunosenescence, is characterized by a decline in immune cell function, reduced response to vaccinations, and an increased susceptibility to infections. The gradual failure of the immune system contributes to the higher incidence of infections and certain cancers in older individuals. The immune system, the instrumental biological framework tasked with defending the organism against a gamut of pathogens, is subject to significant alterations along with the progression of age, a phenomenon characterized as immunosenescence. This concept of an aging immune system can be perceptively elucidated through a case study revolving around T lymphocytes, a critical component of the adaptive immune response. These cells, sourced originally from the thymus, undergo age-consistent thymic involution resulting in a reduced output of naive T cells, the immune system's frontline warriors. With advancing age, this diminution of naive T cells hampers the body's ability to respond to new antigens and generate effective adaptive immune responses. Moreover, absent in the naive T cell repertoire, these gaps are filled in by memory T cells, contributing to the shrinkage of T cell diversity in the aged immune system. This skewing towards memory phenotype is complimented by a notable expansion in the population of senescent T cells, often characterized by shortened telomeres and loss of CD28 expression. Senescent T cells exhibit an impaired proliferative capacity and a paradoxical pro-inflammatory status, known as inflamm-aging, resulting in a net detrimental effect by amplifying systemic inflammation while failing to contribute towards productive immunity against pathogens. Concurrently, the aging immune system encounters a shift in T cell metabolism, wielding a profound impact on immune responses. Predominantly, a reduction in glycolysis and an increase in fatty acid oxidation reflect an energy conservation strategy in aged T cells; however, it also coincides with a compromised capacity to mount a swift and robust immune response. In this context, the aging immune system, particularly in the frame of T lymphocytes, exhibits debilitated adaptability to new infections and enhanced susceptibility to autoimmune disorders and chronic inflammatory diseases. Therefore, to preserve immune health with advancing age, understanding the intricate underlying molecular mechanisms driving immunosenescence holds significant potential.

Aging is not limited to the cellular level; it has profound effects on the physiology and function of entire organisms (Scott-Solomon & Hsu, 2022; Scott et al., 2022; Sezin et al., 2022). Organ systems, including the cardiovascular, respiratory, and musculoskeletal systems, undergo structural and functional changes with age. These changes manifest as decreased cardiac output, reduced lung capacity, and diminished muscle mass, contributing to the decline in overall physiological performance. Aging, a universal biological phenomenon, is often perceived as an unidirectional erosion of the functional and physiological homeostasis, conspicuously transcending beyond the cellular level to leave significant impacts on entire organisms. This proposition can be better understood by examining the inexorable trajectory of cardiovascular aging. On the whole, aging affects every facet of cardiovascular health, altering the structure and function of the heart and the vasculature it interacts with. Studies have demonstrated that the heart undergoes distinctive anatomical changes with age, including myocardial hypertrophy. On a physiological level, the decline in cardiac myocytes primarily due to apoptosis and senescence, coupled with the diminished compensatory proliferation, contributes to the loss of myocardial mass and left ventricular hypertrophy often witnessed with aging. In parallel, perturbations in the extracellular matrix lead to progressive fibrosis, thereby affecting cardiac stiffness and elasticity, consequently impairing diastolic function. Furthermore, aging brings an onslaught of oxidative stress and inflammation, instigating detrimental modifications within cardiac proteins and genetic machinery, which can culminate in functional impairments. Simultaneously, the vascular aging, characterized by arterial stiffening due to the loss of elastin and accumulation of collagen in the arterial walls, and endothelial dysfunction, has been noted to augment the age-dependent risk of developing cardiovascular diseases. The decline in nitric oxide bioavailability, a vital mediator of endothelial function, with age also impacts the vessel's ability to properly dilate in response to increased blood flow or constrict during rest or distress. Consequently, these changes contribute to elevated systemic vascular resistance, thereby imposing an increased load on the heart and augmenting the risk for hypertension and heart failure. Therefore, aging is a multifaceted process characterized by significant alterations at the cellular level, which cascade into substantial modifications in the physiology and function of entire organisms, blatantly exemplified by cardiovascular aging.

The study of aging, a complex and multifaceted phenomenon, has given rise to the field of gerontology (Katunin et al., 2021; Keenan et al., 2021; Khampang et al., 2023). Researchers seek to unravel the intricacies of aging, aiming to extend healthy lifespan and ameliorate the impact of age-related diseases. While aging is an inevitable process, understanding its underlying mechanisms allows for the identification of potential interventions to delay its onset and mitigate its effects. Aging, a complex and multifaceted phenomenon integral to all forms of life, is a matter of profound scientific interest due to its intricate dimensions and profound implications, thereby giving the momentum to the evolution of the discipline of gerontology, dedicated to the comprehensive study of aging processes. A case in point to highlight this progression is the exploration of aging in the cellular context through senescence, a pathway that underpins the functional and phenotypic changes that occur in cells as part of the normal aging process. Cellular senescence is characterized by a durable arrest of the cell cycle, preventing cells from undergoing mitosis, even under favorable circumstances. Over time, the accrual of senescent cells in tissues throughout the organism has been correlated with the natural aging process and related degenerative diseases. Mechanistically, the aging cell is a flurry of disruptions: It deals with DNA damage accumulated over time due to endogenous factors such as reactive oxygen species and exogenous factors such as we, ultraviolet radiation and chemicals from environmental exposures. The aging cell navigates these disruptions with the interplay of multiple complex pathways such as the TP53 and Rb tumor suppressor pathways, regulators of the cell cycle checkpoint functions. Over time, as the cells repeatedly rely on these pathways to manage the accrued DNA damage, they enter a state of permanent growth arrest termed as 'senescence.' This cellular degeneration with aging is complemented by a chronic inflammatory response, often dubbed 'senescence-associated secretory phenotype' (SASP), which leads to the release of numerous inflammatory cytokines, growth factors, and proteases, exacerbating tissue dysfunction and contributing to age-related pathology. This intricate portrayal of senescence is an emblem of how deeply aging pervades life processes, and that these complexities necessitated the amalgamation of genetics, molecular biology, physiology, and epidemiology into a new interdisciplinary field known as gerontology. We owe our understanding of aging, its underlying mechanistic pathways, implications on health, and potential therapeutics to this field. Gerontology stands as an embodiment of the scientific journey of decoding aging, a testament to the intricacy of biological phenomena.

In conclusion, aging and senescence represent the cumulative effects of time and environmental factors on cells and organisms. The decline in cellular functions, coupled with the disruption of molecular pathways, contributes to the gradual deterioration of health and the increased susceptibility to diseases that accompany aging. The pursuit of unraveling the complexities of aging holds promise not only for extending a healthy lifespan but also for enhancing our understanding of the fundamental principles that govern life and its limits. The final section of this chapter focuses on cancer biology.

8.5 CANCER BIOLOGY: CELL CYCLE DYSREGULATION AND TUMOR FORMATION

Cancer, an intricate and multifaceted class of diseases, arises from the disruption of normal cellular processes and the unchecked proliferation of cells (Tolkin & Hubbard, 2021; Tomoda & Kime, 2021; Tuieng et al., 2021). This phenomenon, underpinned by intricate genetic and molecular changes, stands as a striking example of the complex interplay between cellular biology, genetics, and disease progression. Cancer, a highly complex and multifaceted class of diseases, is the profound manifestation of disrupted normal cellular processes compounded by unchecked cell proliferation. Delving into the domain of non-small cell lung cancer (NSCLC), the most commonly diagnosed type of lung cancer, can provide a detailed perspective on this intricate relationship. Cancer initiation and progression in NSCLC, like other cancers, is a multistage process, with oncogenic transformation arising from the cumulative effects of genetic and epigenetic alterations that disrupt core cellular processes such as cell cycle regulation, apoptosis, and DNA repair. The aberrant activation of cell signaling pathways, known to govern cell proliferation, growth, and survival, is central to this process. For instance, alterations in the epidermal growth factor receptor (EGFR) have been commonly observed in NSCLC. Under normal conditions, EGFR, being a tyrosine kinase, helps regulate cell growth and proliferation through a faultlessly calibrated signaling cascade. However, mutations or amplifications in the EGFR gene lead to an overactive receptor, resulting in unchecked cell division and the formation of tumors. Additionally, mutations in tumor suppressor genes like TP53, which usually function to prevent inappropriate cell proliferation and promote apoptosis, are frequented in NSCLC. When impaired, this paves the way for the cancerous cells to evade critical checkpoints, thereby promoting unregulated cell division and growth. Furthermore, the tumor microenvironment, hypoxic and acidic due to rapid cell division, contributes to this uncontrolled proliferation by destabilizing the genome and increasing mutation rates, fueling the progression of the disease. Complimenting this, the tumor cells tend to manipulate the host immune system and establish angiogenesis, facilitating the supply of nutrients and oxygen for the proliferating cells, promoting metastasis and resisting cell death. Thus, cancer, specifically evidenced in NSCLC, involves the transformation from normal to malignant state through the accumulation of genetic and epigenetic adjustments, which disrupt normal cellular processes and lead to the excessive proliferation of cells, forming the crux of cancer pathogenesis.

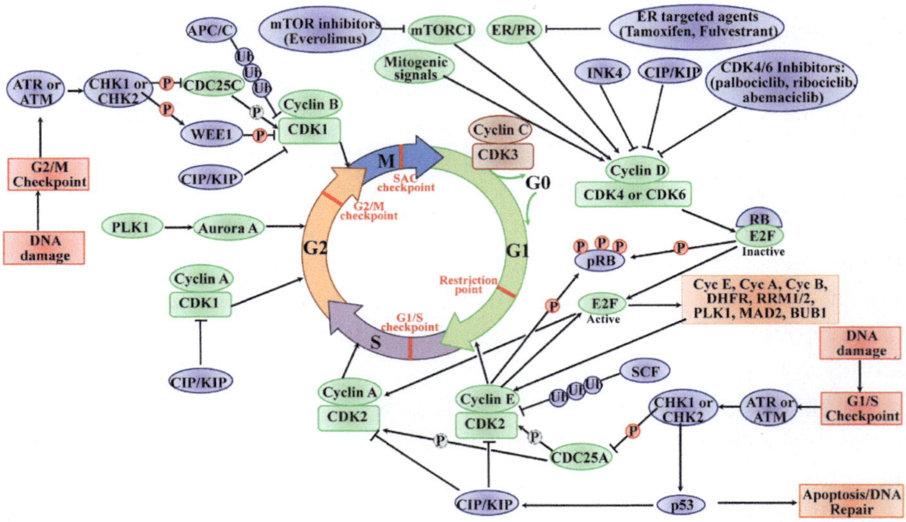

Figure 79. Cyclin D-Cdk4,6 drives cell cycle progression.

Source: Ding L, Cao J, Lin W, Chen H, Xiong X, Ao H, Yu M, Lin J, Cui Q. The Roles of Cyclin-Dependent Kinases in Cell-Cycle Progression and Therapeutic Strategies in Human Breast Cancer. International Journal of Molecular Sciences. 2020; 21(6):1960. https://doi.org/10.3390/ijms21061960

At the heart of cancer biology lies the cell cycle, a precisely orchestrated series of events that guide a cell's growth, replication, and division (Jaspersen, 2021; Jaufmann et al., 2021; Ji et al., 2021; Jeziorski et al., 2023). The cell cycle comprises distinct phases, including G1 (gap 1), S (synthesis), G2 (gap 2), and M (mitosis), each marked by specific molecular checkpoints that ensure the fidelity of DNA replication and cell division. A critical player in this process is the group of proteins known as cyclins and cyclin-dependent kinases (CDKs), which regulate the progression of the cell cycle by activating various downstream targets. At the very core of the sprawling scope of cancer biology lies the cell cycle, a meticulously orchestrated sequence of events tasked with precisely guiding a cell through phases of growth, replication, and eventual division. A specific illustration of this dynamic can be well comprehended in the context of breast cancer, which globally co-occurs as one of the leading types of cancer in women. Predominantly, the cell cycle is partitioned into four phases, each vital in its own right; the G1 phase, wherein the cell grows and prepares for DNA synthesis; the S phase, when the genetic information is duplicated; the G2 phase, when further growth and preparation for mitosis ensue; and finally, the M phase, when the cell's nucleus, and then its cytoplasm, divides resulting in two daughter cells. The transition between these phases is regulated by a series of proteins, more specifically, cyclins and cyclin-dependent kinases (CDKs).

When functioning correctly, these regulators ensure cell cycle progression and division only when conditions are suitable and upon the satisfaction of numerous molecular checkpoints. However, dysregulation of these regulators directly contributes to uncontrolled cell proliferation, a defining characteristic of cancer. Breast cancer offers a salient example, with a significant number of cases attributed to aberrations in cyclin D and CDK4/6. The cyclin D-CDK4/6 complex (see figure 79), active during the G1 phase, drives the transition from G1 to S phase, hence propelling the cell into the phase of active DNA synthesis. Amplifications or overexpression of cyclin D and CDK4/6 observed in breast cancer result in an expedited G1 to S phase transition, thus perturbing the regulation of the cell cycle and promoting unchecked proliferation. Furthermore, mutation or loss of CDK inhibitors, such as P16, known to counteract the action of cyclin D-CDK4/6 complex, also contribute to the unrestrained cell cycle progression. Therefore, the understanding of cell cycle dysregulation, as highlighted in breast cancer, not only fortifies its positioning as the heart of cancer biology but also proposes exploitability for potential therapeutic strategies in cancer treatment.

Cancer cells, in their quest for uncontrolled growth, often exploit vulnerabilities in the cell cycle machinery (Pandey et al., 2021; Panicker et al., 2021; Paliulis et al., 2023). Genetic mutations or epigenetic changes can lead to the overexpression or underexpression of specific cyclins and CDKs, disrupting the finely tuned balance that governs cell cycle progression. Dysregulation of these proteins can result in uncontrolled cell division, driving the formation of tumors. The interplay between cancer cells and the cell cycle machinery delineates a compelling narrative of sustained biological exploitation, whereby cancer cells, in their relentless pursuit of uncontrolled growth, often leverage existing vulnerabilities in the cell cycle system. This proposition can be soundly examined through the lens of hepatocellular carcinoma (HCC), the most common type of primary liver cancer. At the beating heart of eukaryotic cell division lies the cell cycle, an intricate sequence of bio-molecular events comprising phases of cell growth (G1), DNA replication (S), preparation for division (G2), and finally, mitosis (M). The fidelity of cell cycle progression is vigilantly monitored by a series of checkpoints and controlled by a cascade of cyclin-dependent kinases (CDKs) and their regulatory cyclins. In normal cells, these binary entities operate in harmony to regulate precise cell replication. However, in the hostile takeover by cancer cells, this harmonious regulation is subverted. In the context of HCC, mutations and overexpression of cyclin D1, CDK4, and CDK6, primarily involved in transitioning cells from the growth phase (G1) into the DNA synthesis phase (S), have been observed. This aberration effectively bulldozes the G1 checkpoint, breaking the control mechanism that checks DNA integrity before replication, manifesting in the unrestrained proliferation of hepatocytes. Concurrently, the machinery intended to halt this process, namely CDK inhibitors such as P16, P21, and P27, are often found to be downregulated or functionally inactive in HCC. Moreover, TP53, a key tumor suppressor protein that acts as a guardian of genomic integrity and enforcer of the G1/S and G2/M checkpoints, is frequently mutated or deleted in HCC, further enabling cancer cells to evade cycle arrest and apoptosis. Illustratively, these concerted manipulations of the cell cycle machinery pave the way for HCC cancer cells to proliferate rampantly and develop into aggressive tumors, reinforcing the notion that the abuse of cell cycle vulnerabilities forms a cornerstone of cancer cell survival and growth.

One of the key hallmarks of cancer is the evasion of growth suppressors (Gu & Rasmussen, 2022; Hammond et al., 2022; Gu et al., 2023). The tumor suppressor gene p53, often called the "guardian of the genome," plays a central role in monitoring DNA damage and initiating cell cycle arrest or apoptosis (cell death) when damage is detected. Mutations in the p53 gene can lead to its inactivation or loss of function, impairing the cell's ability to respond to DNA damage. This deficiency allows damaged cells to evade apoptosis and continue proliferating, promoting tumor development. Cancer, as a multifaceted disease, demonstrates several distinguishing characteristics, one of which is the remarkable capability to evade growth suppressors, essentially circumventing the checks and balances that keep cellular proliferation under control in healthy cells. An insightful example of this phenomenon could be explicated with chronic myeloid leukemia (CML), a bone marrow stem-cell malignancy marked by the deregulated growth of myeloid cells. The transformation to CML is primarily triggered by the 'Philadelphia chromosome,' a hallmark translocation t(9;22)(q34;q11), which results in the fusion of the Breakpoint Cluster Region (BCR) gene from chromosome 22 with the ABL1 gene from chromosome 9. This fusion gives rise to the BCR-ABL1 oncogene, which encodes a constitutively active BCR-ABL protein with tyrosine kinase activity. In normal cells, the ABL tyrosine kinase aids in regulating cell growth and division; however, when fused to BCR, it exists in an active state, promoting uncontrolled cell proliferation. Importantly, one may note that tumor suppressors such as TP53 and pRB work to inhibit cell cycle progression and promote apoptosis in the event of cellular abnormalities. However, these natural defenses are overwhelmed in CML, as the BCR-ABL1 fusion protein has been found to downregulate the function of these tumor suppressor genes, allowing cells to grow and divide without control. Additionally, the BCR-ABL1 also upregulates anti-apoptotic proteins and downregulates pro-apoptotic proteins, supporting cell survival. Hence, the ability of cancer cells to bypass the regulatory constraints imposed by growth suppressor mechanisms, as clearly illustrated in CML, underpins one of the key hallmarks of cancer biology. This deep understanding of the evasion of growth suppressors provides essential insights for developing targeted therapies, such as tyrosine kinase inhibitors, in combating cancer.

The concept of oncogenes further illuminates the intricate web of cancer biology (Han et al., 2023; Haraoka et al., 2023; Hartmann & Mayor, 2023). Oncogenes are mutated forms of normal genes, called proto-oncogenes, that drive excessive cell growth when activated. These mutations can result in the production of hyperactive proteins that continuously stimulate cell division, circumventing the regulatory mechanisms that normally restrict cell cycle progression. The oncogene Ras, for instance, is involved in transmitting signals for cell growth. Mutations in Ras can lead to its constant activation, promoting uncontrolled proliferation and tumor formation. Central to the complexities of cancer biology, the concept of oncogenes represents a critical element offering illuminating insights into the mechanistic labyrinth underlying the disease's initiation and progression. A vivid understanding of this concept may be attained by examining the case of colorectal cancer, a widely recognized malignancy where the implications of oncogene activation are particularly distinct. Oncogenes are fundamentally altered versions of normal genes, or proto-oncogenes, that when dysregulated or mutated encourage unlimited cellular growth and proliferation, contributing to cancer development.

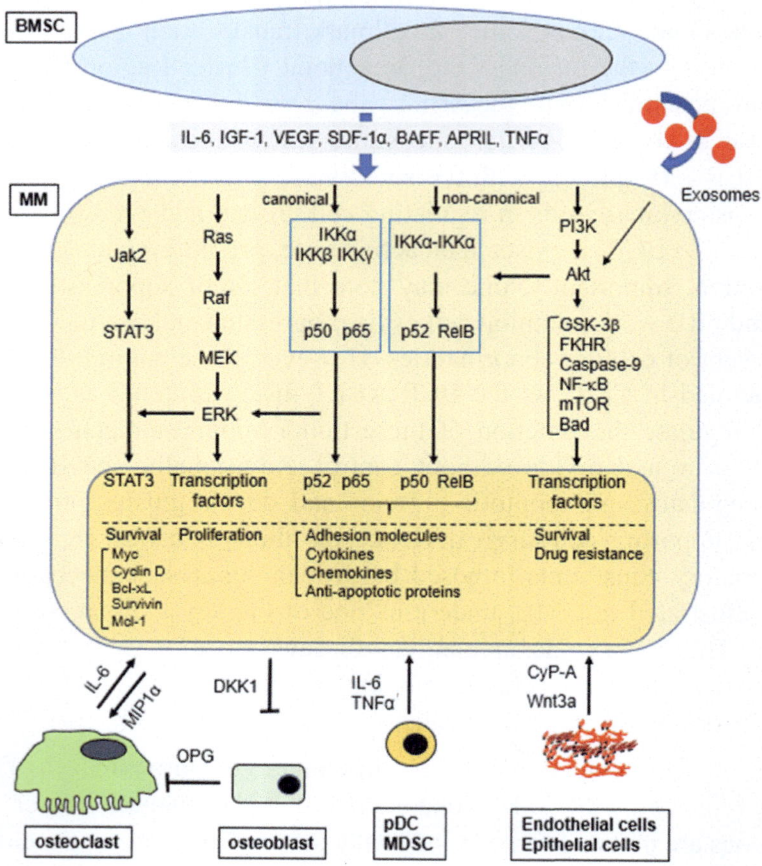

Figure 80. Mutations in the RAF/MEK/ERK and PI3K/AKT pathways.

Source: Hideshima T, Anderson KC. Signaling Pathway Mediating Myeloma Cell Growth and Survival. Cancers. 2021; 13(2):216. https://doi.org/10.3390/cancers13020216

Crucial to colorectal cancer genesis and progression is the oncogene KRAS, a GTPase involved in transmitting signals from growth factor receptors on the cell surface to the nucleus. Under normal circumstances,

KRAS oscillates between an active and inactive state, playing a carefully calibrated role in cell proliferation and differentiation. However, a commonly detected mutation at codons 12 or 13 in colorectal cancer stabilizes KRAS in a permanently active state, creating a continuous growth signal independent of external stimuli and driving unrestricted cellular proliferation and survival. This dysregulated signal transduction cascades down multiple pathways including the RAF/MEK/ERK and PI3K/AKT pathways (see figure 80) , both pivotal for regulating cellular growth, survival, and angiogenesis. Concurrently, the regular functioning of TP53, often deemed the guardian of the genome, and APC, a tumor suppressor essential for appropriate cell cycle regulation, is disrupted in many colorectal cancer instances, further facilitating uncontrolled cellular proliferation. The close interplay between oncogenes and tumor suppressor genes thus outlines the delicate balance that when disturbed leads to the cellular chaos characterizing cancer. Therefore, the decoding of oncogenes, as evidenced with KRAS in colorectal cancer, deepens our comprehension of cancer biology and caters to the development of innovative therapeutic strategies, solidifying the value of oncogene study within the cancer research domain.

Tumor suppressor genes and oncogenes do not operate in isolation (Jaspersen, 2021; Jaufmann et al., 2021; Ji et al., 2021; Jeziorski et al., 2023). Their interactions form a complex network that governs cellular fate. The "two-hit" hypothesis, proposed by Alfred Knudson, describes how the loss of function of a single allele of a tumor suppressor gene can be tolerated, but the loss of the second allele can lead to uncontrolled growth. This concept is exemplified in retinoblastoma, a rare cancer that can arise from mutations in the RB gene, which regulates the cell cycle. The seminal "two-hit" hypothesis, propounded by Alfred Knudson, elucidates a critical gambit in the genesis of many cancers, asserting that while the loss of function of a single allele of a tumor suppressor gene can be endured, the subsequent inactivation of the second allele induces conditions favorable for uncontrolled cell proliferation, and hence, tumorigenesis. A case that exemplifies this principle with vivid conceptual detail is retinoblastoma, a rare form of eye cancer predominantly affecting children. Central to the development of retinoblastoma is the mutation in the RB1 gene, a quintessential tumor suppressor gene that codes for the retinoblastoma (Rb) protein. This protein plays an indispensable role in regulating the cell cycle transition from the growth phase (G1) to the synthesis phase (S), thus exerting control on cell proliferation. According to the "two-hit" hypothesis, retinoblastoma will only manifest when both copies, or alleles, of the RB1 gene are mutated or lost; the first "hit" usually being a germline mutation inherited or occurring during early development, followed by a second "hit" in the susceptible retinal cells. The first "hit" indeed puts the individual at risk, but the cell can still function with the remaining normal allele. However, the occurrence of the second "hit" obliterates regulatory constraints on cell division, engendering unregulated cell proliferation characteristic of retinoblastoma development, as the safeguard provided by the tumor suppressor gene is completely lost. This dual allelic inactivation makes Knudson's "two-hit" hypothesis a crucial blueprint for understanding the mechanistic evolution of retinoblastoma, and generally, the molecular pathway leading to tumorigenesis in similar cancers. Thus, the Knudson "two-hit" hypothesis has contributed profoundly to the foundational understanding of cancer genetics, providing insight into the theoretical basis of tumorigenesis and guiding future directions in cancer biology research.

As cancer cells proliferate, they acquire additional mutations that enable them to invade neighboring tissues and metastasize to distant sites (MacArthur & Dawlaty, 2021; Ma et al., 2022; Ma & Wu, 2022). This process, known as metastasis, involves the breakdown of cellular adhesion, enabling cancer cells to detach from the primary tumor and invade surrounding tissues or enter the bloodstream. In the evolutionary landscape of cancer biology, a critical hallmark of advancing malignancies is the ability of cancer cells to proliferate rampantly, acquiring additional mutations that empower them to invade adjoining tissues and metastasize to remote sites. This transformative journey can be particularly dissected in the context of melanoma, an aggressive form of skin cancer infamous for its proclivity for metastasis. Over time, melanocytes, the pigment-producing cells in the skin, may accumulate genetic mutations due to factors such as chronic UV exposure, leading initially to non-invasive forms of skin cancer. However, the advent of further mutations enables these cells to invade the underlying skin layers (dermis) and the lymphatic and circulatory systems, a process significantly driven by the activation of specific genes controlling cell motility, extracellular matrix degradation and angiogenesis. A prime example is the BRAF gene which codes for a component of the mitogen-activated protein kinase (MAPK) pathway. Mutations in BRAF (especially BRAF V600E) result in constitutive activation of the MAPK pathway, leading to increased cell proliferation and survival. Simultaneously, the activation of other genes such as MMPs (matrix metalloproteinases), VEGF (vascular endothelial growth factor) and interleukins guide the degradation of the extracellular matrix, formation of new blood vessels and evasion of the immune system, all crucial steps in facilitating the invasive ethos of melanoma cells. In parallel, mutations that confer resistance to cell death, like alterations in the TP53 or the activation of anti-apoptotic mechanisms, aid the survival and growth of these nascent colonies at distant sites, eventually leading to the establishment of metastatic tumors. Therefore, the progression and aggressiveness of cancers, exemplified in melanoma, are intrinsically tied to the proliferation of cancer cells and their acquisition of genetic mutations that enable local invasion and distant dissemination, underscoring the mutable prowess of cancer biology.

Cancer is not a uniform disease; its diverse manifestations arise from a combination of genetic, environmental, and lifestyle factors (Gan et al., 2022; Garcia-Gutierrez & Garcia-Dominguez, 2022; Geers et al., 2022). The study of cancer biology delves into the intricate mechanisms that govern cell cycle dysregulation, oncogene activation, and tumor suppressor gene inactivation. Through this understanding, researchers aim to develop targeted therapies that exploit the vulnerabilities of cancer cells while minimizing harm to healthy tissues. Unraveling the science of cancer biology necessitates intensive exploration into the multifaceted mechanisms that govern cell cycle dysregulation, oncogene activation, and tumor suppressor gene inactivation. An embodiment of these principles can be portrayed using the case of glioblastoma multiforme (GBM), the most aggressive and deadliest form of brain cancer whose pathogenesis relies heavily upon these mechanistic entities. A typical eukaryotic cell cycle adheres to a specific order of stages — Growth 1 (G1), Synthesis (S), Growth 2 (G2), and Mitosis (M) — regulated by a series of "stop" and "go" signals referred to as "checkpoints." Overarching these checkpoints and underlying any malignancy like GBM are genomic aberrations that fuel this dysregulation leading to the unrestrained proliferation of cells. One such alteration involves the activation of oncogenes: EGFR amplification, the most frequent genomic alteration in GBM, produces a constitutively active receptor that drives cellular proliferation relentlessly. Similarly, mutations and/or deletions of tumor suppressor genes allow GBM cells to outmaneuver the restrictions that curtail abnormal cell growth. For instance, alterations in TP53, a tumor suppressor gene that halts the cell cycle to rectify DNA damage or stimulate apoptosis if damage is irreparable, are common in GBM and they foil this crucial checkpoint of genomic integrity. Moreover, mutations in PTEN, another tumor suppressor, disable its ability to control the PI3K/Akt/mTOR pathway — a regulator of cell proliferation and survival — thereby imparting survival advantage to GBM cells. Hence, the intricate interplay of cell cycle dysregulation, oncogene activation, and tumor suppressor gene inactivation, as delineated in the GBM model, forms the crux of cancer biology, guiding scientists in the quest to decipher the pathogenesis of cancer and strategize apt therapeutics.

In conclusion, cancer biology offers a window into the complexities of cell cycle regulation, genetic mutations, and the unrelenting drive for uncontrolled growth. The disruption of normal cellular processes, driven by genetic and molecular changes, culminates in the formation of tumors and the intricate web of cancer progression. Understanding the underlying mechanisms not only sheds light on the origins of cancer but also paves the way for innovative approaches to diagnosis, treatment, and prevention. Now, let us summarize the final chapter of this book.

CHAPTER 8 SUMMARY

The exploration of cell reproduction and development has traversed a landscape of intricacies, unveiling the profound mechanisms that underlie the birth, growth, and transformation of living organisms. The chapters within this section have unfurled a tapestry of cellular processes, from the intimate union of gametes to the complex journey of embryonic development, offering insights into the symphony of life's progression.

In the opening chapter, we delved into the realm of fertilization, witnessing the remarkable convergence of gametes to form the zygote—a union that symbolizes the inception of a new life. This foundational event marks the initiation of a cascade of cellular processes that pave the way for the embryonic journey ahead.

Embryonic development emerged as a captivating narrative within this exploration, unveiling the stages of cleavage, gastrulation, and organogenesis. These stages, driven by intricate genetic programs and signaling networks, delineate the transformation of a single cell into a complex multicellular organism. The orchestration of these phases underpins the blueprint of life, wherein distinct cell types and tissues arise to fulfill specific roles within the organism's architecture.

Amidst this intricate developmental landscape, the potential of stem cells and regeneration has shone as a beacon of hope. The chapters within this section unraveled the remarkable capacity of stem cells to heal and repair, offering insights into tissue regeneration and the restoration of function. These cellular architects have the power to rejuvenate, offering promise in the realm of regenerative medicine and advancing our understanding of cellular resilience.

As the chapter unfolded, the phenomenon of aging and senescence cast its shadow, reminding us of the passage of time and its indelible effects on cellular and organismal function. The intricate dance of genetics, environmental influences, and molecular changes came to the forefront, illustrating the complexities that define the later stages of life. This exploration illuminates the boundaries of human existence, while also offering avenues for comprehending age-related diseases and interventions.

Finally, the journey led us through the labyrinth of cancer biology, where the disruption of the cell cycle and unchecked proliferation culminate in the formation of tumors. The unraveling of cellular dysregulation underscored the fragility of cellular processes, revealing the intricate web of interactions that drive tumorigenesis. This journey into the realm of cancer not only illuminated the intricacies of cellular biology but also illuminated the path toward targeted therapies and innovative treatments.

In culmination, Chapter 8 has navigated the intricate landscape of cell reproduction and development, unveiling the beauty and complexity that define life's progression. From the inception of a zygote to the challenges posed by aging and disease, this exploration stands as a testament to the relentless pursuit of knowledge in our quest to unravel the mysteries of life itself. As we conclude this chapter, we carry with us a deeper understanding of the cellular underpinnings that shape our existence and inspire us to continue probing the frontiers of scientific discovery.

CONCLUDING REMARKS

"Introduction to Living Cells: The Building Blocks of Life" is not just a compilation of chapters; it is a voyage into the microscopic universe that underpins all life forms. By critically examining the themes, ideas, and concepts interwoven throughout the book, we can unearth the profound implications for learners, practitioners, policymakers, and the public, while also acknowledging the book's limitations and outlining pathways for deeper knowledge acquisition.

At its core, this book delves into the cellular realm where the secrets of existence are whispered. The thematic thread of unity and diversity pulsates through its pages, revealing the intricate harmony of life's basic units. The early pioneers—Hooke, Schleiden, Schwann, and Virchow—etched the foundation of cell theory, a cornerstone that unified our understanding of living entities. These foundational principles ripple through time, connecting us to contemporary explorations of cellular diversity and specialization. The book masterfully unveils how cells, though adhering to common principles, metamorphose into a bewildering array of forms and functions, from muscle cells that power motion to nerve cells that orchestrate communication. This theme echoes the symphonic blend of unity and diversity that characterizes life itself.

As we dissect the implications of these themes, a multi-faceted tapestry emerges, illuminating the path for different stakeholders. For learners, the book serves as a scaffold for comprehending the exquisite complexities that transpire within each cell. However, the true magnificence lies in how these understandings resonate in practical realms. Practitioners, whether clinicians, researchers, or technologists, can harness this knowledge to sculpt innovations in diagnosis, treatment, and beyond. By understanding cellular machinery, practitioners can decipher the intricacies of disease progression, design targeted interventions, and craft strategies that redefine the contours of medical practice.

Policymakers are not left untouched. The ethical implications of stem cell research, genetic manipulation, and scientific advancement reverberate throughout society. Armed with insights from this book, policymakers grapple with the balancing act between scientific progress and societal values. This balance is essential to ensuring that the potential of cellular research is harnessed for the greater good without compromising ethical integrity. Yet, like any literary work, this book bears limitations. The dynamic nature of science means that contemporary breakthroughs might not be fully encapsulated within its pages. The book paints with broad strokes, occasionally leaving readers craving in-depth explorations of certain topics. The complexity of cellular biology is itself a double-edged sword, often demanding more elaboration and detailed dissections to foster deeper understanding. Furthermore, the expansive topic of stem cells and regeneration, a tantalizing field fraught with potential, could have warranted a dedicated treatment to delve into cutting-edge research and future applications.

To transcend these limitations, proactive readers can embark on a journey of self-expansion. Beyond these pages lie endless avenues for learning and discovery. The world of scholarly journals awaits, offering the latest revelations and igniting curiosity. Online courses and workshops can provide interactive platforms to engage with experts, enabling in-depth discussions on intricate subjects. Collaborative dialogues with peers and mentors propel explorations into the uncharted territories of cellular intricacies. Venturing into interdisciplinary domains, where cellular biology meets genetics, epigenetics, and beyond, opens a gateway to holistic understanding.

In conclusion, this book represents an intellectual voyage through the tapestry of life, stitched together by cells. It is not just a mere guide but an expedition that unveils the microscopic foundations that define existence. As we bid adieu to this exploration, our gratitude extends to readers who embarked on this journey with us. Your engagement transforms this book into a beacon of knowledge and curiosity. We invite you to continue your quest for understanding, to peer deeper into the cellular abyss, to dissect the complex mechanisms that underlie life. In doing so, we kindle the fervent hope that this journey is but a prologue to a lifelong pursuit of knowledge, and that you will join us in continuing to unravel the mysteries of life's most fundamental constituents.

REFERENCES

1. Ajayi-Smith, A., van der Watt, P., Mkwanazi, N., Carden, S., Trent, J. O., & Leaner, V. D. (2021). Novel small molecule inhibitor of Kpn beta 1 induces cell cycle arrest and apoptosis in cancer cells. *Experimental Cell Research*, *404*(2), Article 112637. https://doi.org/10.1016/j.yexcr.2021.112637
2. Al-Marsoummi, S., Vomhof-DeKrey, E. E., &Basson, M. D. (2021). Schlafens: Emerging Proteins in Cancer Cell Biology. *Cells*, *10*(9), Article 2238. https://doi.org/10.3390/cells10092238
3. Alao, J. P., Legon, L., & Rallis, C. (2021). Crosstalk between the mTOR and DNA Damage Response Pathways in Fission Yeast. *Cells*, *10*(2), Article 305. https://doi.org/10.3390/cells10020305
4. Albrecht, L. V., Tejeda-Munoz, N., & De Robertis, E. M. (2021). Cell Biology of Canonical Wnt Signaling. In R. Lehmann (Ed.), *Annual Review of Cell and Developmental Biology, Vol 37* (Vol. 37, pp. 369-389). https://doi.org/10.1146/annurev-cellbio-120319-023657
5. Altulea, D., Maassen, S., Baranov, M. V., & van den Bogaart, G. (2021). What makes (hydroxy)chloroquine ineffective against COVID-19: insights from cell biology. *Journal of Molecular Cell Biology*, *13*(3), 175-184. https://doi.org/10.1093/jmcb/mjab016
6. Alvord, V. M., Kantra, E. J., &Pendergast, J. S. (2022). Estrogens and the circadian system. *Seminars in Cell & Developmental Biology*, *126*, 56-65. https://doi.org/10.1016/j.semcdb.2021.04.010
7. Aniento, F., Hernandez, V. S. D., Dagdas, Y., Rojas-Pierce, M., &Russinova, E. (2022). Molecular mechanisms of endomembrane trafficking in plants. *Plant Cell*, *34*(1), 146-173. https://doi.org/10.1093/plcell/koab235

8. Aprile, D., Alessio, N., Squillaro, T., Di Bernardo, G., Peluso, G., &Galderisi, U. (2023). Role of glycosphingolipid SSEA-3 and FGF2 in the stemness and lineage commitment of multilineage differentiating stress enduring (MUSE) cells. *Cell Proliferation*, *56*(1). https://doi.org/10.1111/cpr.13345
9. Arber, C., Lovejoy, C., Harris, L., Willumsen, N., Alatza, A., Casey, J. M., Lines, G., Kerins, C., Mueller, A. K., Zetterberg, H., Hardy, J., Ryan, N. S., Fox, N. C., Lashley, T., & Wray, S. (2021). Familial Alzheimer's Disease Mutations in PSEN1 Lead to Premature Human Stem Cell Neurogenesis. *Cell Reports*, *34*(2), Article 108615. https://doi.org/10.1016/j.celrep.2020.108615
10. Ardalan, M., Khatibi, S. M. H., Saadat, Y. R., Bastami, M., Nariman-Saleh-Fam, Z., Abediazar, S., Khalilov, R., &Vahed, S. Z. (2022). Migrasomes and exosomes; different types of messaging vesicles in podocytes. *Cell Biology International*, *46*(1), 52-62. https://doi.org/10.1002/cbin.11711
11. Arhontoulis, D. C., Kerr, C. M., Richards, D., Tjen, K., Hyams, N., Jones, J. A., Deleon-Pennell, K., Menick, D., Brauninger, H., Lindner, D., Westermann, D., & Mei, Y. (2022). Human cardiac organoids to model COVID-19 cytokine storm induced cardiac injuries. *Journal of Tissue Engineering and Regenerative Medicine*, *16*(9), 799-811. https://doi.org/10.1002/term.3327
12. Aridor, M. (2022). A tango for coats and membranes: New insights into ER-to-Golgi traffic. *Cell Reports*, *38*(3). https://doi.org/10.1016/j.celrep.2021.110258
13. Arjmand, B., Alavi-Moghadam, S., Aghayan, H. R., Rezaei-Tavirani, M., Goodarzi, P., Tayanloo-Beik, A., Biglar, M., Rajaeinejad, M., Shouroki, F. F., & Larijani, B. (2023). How to establish infrastructures to achieve more efficient regenerative medicine? *Cell and Tissue Banking*, *24*(1), 1-9. https://doi.org/10.1007/s10561-022-10028-2
14. Baluska, F., & Yokawa, K. (2021). Anaesthetics and plants: from sensory systems to cognition-based adaptive behaviour. *Protoplasma*, *258*(2), 449-454. https://doi.org/10.1007/s00709-020-01594-x
15. Bennison, S. A., Liu, X. A., & Toyo-oka, K. (2022). Nuak kinase signaling in development and disease of the central nervous system. *Cellular Signalling*, *100*, Article 110472. https://doi.org/10.1016/j.cellsig.2022.110472
16. Beppler, C., Eichorst, J., Marchuk, K., Cai, E., Castellanos, C. A.,

Sriram, V., Roybal, K. T., &Krummel, M. F. (2022). Hyperstabilization of T cell microvilli contacts by chimeric antigen receptors. *Journal of Cell Biology*, *222*(3), Article e202205118. https://doi.org/10.1083/jcb.202205118

17. Bettegazzi, B., Monasor, L. S., Bellani, S., Codazzi, F., Restelli, L. M., Colombo, A. V., Deigendesch, N., Frank, S., Saito, T., Saido, T. C., Lammich, S., Tahirovic, S., Grohovaz, F., &Zacchetti, D. (2021). Casein Kinase 2 dependent phosphorylation of eIF4B regulates BACE1 expression in Alzheimer's disease. *Cell Death & Disease*, *12*(8), Article 769. https://doi.org/10.1038/s41419-021-04062-3
18. Biermann, M., & Reya, T. (2022). Hematopoietic Stem Cells and Regeneration. *Cold Spring Harbor Perspectives in Biology*, *14*(8), Article a040774. https://doi.org/10.1101/cshperspect.a040774
19. Bohnert, K. A., & Johnson, A. E. (2022). Branching Off: New Insight Into Lysosomes as Tubular Organelles. *Frontiers in Cell and Developmental Biology*, *10*, Article 863922. https://doi.org/10.3389/fcell.2022.863922
20. Bomont, P. (2021). The dazzling rise of neurofilaments: Physiological functions and roles as biomarkers. *Current Opinion in Cell Biology*, *68*, 181-191. https://doi.org/10.1016/j.ceb.2020.10.011
21. Bonilha, C. S. (2022). BCyto: A shiny app for flow cytometry data analysis. *Molecular and Cellular Probes*, *65*, Article 101848. https://doi.org/10.1016/j.mcp.2022.101848
22. Boukhatmi, H. (2021). Drosophila, an Integrative Model to Study the Features of Muscle Stem Cells in Development and Regeneration. *Cells*, *10*(8), Article 2112. https://doi.org/10.3390/cells10082112
23. Boumard, B., & Bardin, A. J. (2021). An amuse-bouche of stem cell regulation: Underlying principles and mechanisms from adult Drosophila intestinal stem cells. *Current Opinion in Cell Biology*, *73*, 58-68. https://doi.org/10.1016/j.ceb.2021.05.007
24. Burke, D. F., Bryant, P., Barrio-Hernandez, I., Memon, D., Pozzati, G., Shenoy, A., Zhu, W. S., Dunham, A. S., Albanese, P., Keller, A., Scheltema, R. A., Bruce, J. E., Leitner, A., Kundrotas, P., Beltrao, P., &Elofsson, A. (2023). Towards a structurally resolved human protein interaction network. *Nature Structural & Molecular Biology*, *30*(2), 216-+. https://doi.org/10.1038/s41594-022-00910-8
25. Cahan, P., Cacchiarelli, D., Dunn, S. J., Hemberg, M., Lopes, S., Morris,

S. A., Rackham, O. J., del Sol, A., & Wells, C. A. (2021). Computational Stem Cell Biology: Open Questions and Guiding Principles. *Cell Stem Cell*, *28*(1), 20-32. https://doi.org/10.1016/j.stem.2020.12.012

26. Caldas, L. A., Carneiro, F. A., Monteiro, F. L., Augusto, I., Higa, L. M., Miranda, K., Tanuri, A., & de Souza, W. (2021). Intracellular host cell membrane remodelling induced by SARS-CoV-2 infection in vitro. *Biology of the Cell*, *113*(6), 281-293. https://doi.org/10.1111/boc.202000146
27. Carroll, E. C., & Marqusee, S. (2022). Site-specific ubiquitination: Deconstructing the degradation tag. *Current Opinion in Structural Biology*, *73*, Article 102345. https://doi.org/10.1016/j.sbi.2022.102345
28. Chang, D. H., Yang, X. T., Fan, S. Y., Fan, T. B., Zhang, M. K., & Ono, M. (2023). Engineering of MSCs sheet for the prevention of myocardial ischemia and for left ventricle remodeling. *Stem Cell Research & Therapy*, *14*(1), Article 102. https://doi.org/10.1186/s13287-023-03322-7
29. Charpentier, J. C., & King, P. D. (2021). Mechanisms and functions of endocytosis in T cells. *Cell Communication and Signaling*, *19*(1), Article 92. https://doi.org/10.1186/s12964-021-00766-3
30. Chen, J. L., Li, X., Jia, Y. F., Xia, Z. Y., & Ye, J. S. (2021). Publication Trends on Mitophagy in the World and China: A 16-Year Bibliometric Analysis. *Frontiers in Cell and Developmental Biology*, *9*, Article 793772. https://doi.org/10.3389/fcell.2021.793772
31. Chen, J. Y., Wang, J., Wu, X. J., Simon, N., Svensson, C. I., Yuan, J., Hart, D. A., Ahmed, A. S., & Ackermann, P. W. (2023). eEF2 improves dense connective tissue repair and healing outcome by regulating cellular death, autophagy, apoptosis, proliferation and migration. *Cellular and Molecular Life Sciences*, *80*(5), Article 128. https://doi.org/10.1007/s00018-023-04776-x
32. Chen, X. Y., Huang, J., Wu, J., Hao, J., Fu, B. Q., Wang, Y., Zhou, B., Na, T., Wei, J., Zhang, Y., Li, Q. Y., Hu, S. J., Zhou, J. X., Yu, J. Y., Wu, Z. H., Zhu, H. X., Cao, J. N., Wang, L., Peng, Y. J., . . . Xiang, A. P. (2022). Human mesenchymal stem cells. *Cell Proliferation*, *55*(4), Article e13141. https://doi.org/10.1111/cpr.13141
33. Cheung, A. Y., Cosgrove, D. J., Hara-Nishimura, I., Jurgens, G., Lloyd, C., Robinson, D. G., Staehelin, L. A., &Weijers, D. (2022). A rich and bountiful harvest: Key discoveries in plant cell biology. *Plant Cell*, *34*(1), 53-71. https://doi.org/10.1093/plcell/koab234

34. Chhabra, Y., Seiffert, P., Gormal, R. S., Vullings, M., Lee, C. M. M., Wallis, T. P., Dehkhoda, F., Indrakumar, S., Jacobsen, N. L., Lindorff-Larsen, K., Durisic, N., Waters, M. J., Meunier, F. A., Kragelund, B. B., & Brooks, A. J. (2023). Tyrosine kinases compete for growth hormone receptor binding and regulate receptor mobility and degradation. *Cell Reports*, *42*(5), Article 112490. https://doi.org/10.1016/j.celrep.2023.112490

35. Choezom, D., & Gross, J. C. (2022). Neutral sphingomyelinase 2 controls exosome secretion by counteracting V-ATPase-mediated endosome acidification. *Journal of Cell Science*, *135*(5), Article jcs259324. https://doi.org/10.1242/jcs.259324

36. Chohra, I., Chung, K. S., Giri, S., &Malgrange, B. (2023). ATP-Dependent Chromatin Remodellers in Inner Ear Development. *Cells*, *12*(4), Article 532. https://doi.org/10.3390/cells12040532

37. Choo, J., Glisovic, N., &Vignjevic, D. M. (2022). Gut homeostasis at a glance. *Journal of Cell Science*, *135*(21), Article jcs. https://doi.org/10.1242/jcs.260248

38. Cibelli, A., Scemes, E., & Spray, D. C. (2022). Activity and Stability of Panx1 Channels in Astrocytes and Neuroblastoma Cells Are Enhanced by Cholesterol Depletion. *Cells*, *11*(20), Article 3219. https://doi.org/10.3390/cells11203219

39. Clarke, R., Terry, A. R., Pennington, H., Hasty, C., MacDougall, M. S., Regan, M., & Merrill, B. J. (2021). Sequential Activation of Guide RNAs to Enable Successive CRISPR-Cas9 Activities. *Molecular Cell*, *81*(2), 226-+. https://doi.org/10.1016/j.molcel.2020.12.003

40. Claude-Taupin, A., Dupont, N., &Codogno, P. (2022). Autophagy and the primary cilium in cell metabolism: What's upstream? *Frontiers in Cell and Developmental Biology*, *10*, Article 1046248. https://doi.org/10.3389/fcell.2022.1046248

41. Colin, L., Martin-Arevalillo, R., Bovio, S., Bauer, A., Vernoux, T., Caillaud, M. C., Landrein, B., &Jaillais, Y. (2022). Imaging the living plant cell: From probes to quantification. *Plant Cell*, *34*(1), 247-272. https://doi.org/10.1093/plcell/koab237

42. Colin, L., Ruhnow, F., Zhu, J. K., Zhao, C. Z., Zhao, Y., & Persson, S. (2023). The cell biology of primary cell walls during salt stress. *Plant Cell*, *35*(1), 201-217. https://doi.org/10.1093/plcell/koac292

43. Collins, M. A., Coon, L. A., Thomas, R., Mandigo, T. R., Wynn, E.,

&Folker, E. S. (2021). Ensconsin-dependent changes in microtubule organization and LINC complex-dependent changes in nucleus-nucleus interactions result in quantitatively distinct myonuclear positioning defects. *Molecular Biology of the Cell, 32*(21), Article ar27. https://doi.org/10.1091/mbc.E21-06-0324

44. Costa, A. C., & Sousa, M. M. (2022). The Role of Spastin in Axon Biology. *Frontiers in Cell and Developmental Biology, 10*, Article 934522. https://doi.org/10.3389/fcell.2022.934522
45. Cristiani, A., Dutta, A., Poveda-Cuevas, S. A., Kern, A., &Bhaskara, R. M. (2023). Identification of potential selective autophagy receptors from protein-content profiling of autophagosomes. *Journal of Cellular Biochemistry*. https://doi.org/10.1002/jcb.30405
46. Critcher, M., & Huang, M. L. (2022). Excavating proteoglycan structure-function relationships: modern approaches to capture the interactions of ancient biomolecules. *American Journal of Physiology-Cell Physiology, 323*(2), C415-C422. https://doi.org/10.1152/ajpcell.00222.2022
47. Cui, L., Ma, R., Cai, J. L. Y., Guo, C. Y., Chen, Z. Y., Yao, L. L., Wang, Y. Y., Fan, R., Wang, X., & Shi, Y. L. (2022). RNA modifications: importance in immune cell biology and related diseases. *Signal Transduction and Targeted Therapy, 7*(1), Article 334. https://doi.org/10.1038/s41392-022-01175-9
48. Das, M. K., Lunavat, T. R., Miletic, H., & Hossain, J. A. (2021). The Potentials and Pitfalls of Using Adult Stem Cells in Cancer Treatment. In K. Turksen (Ed.), *Cell Biology and Translational Medicine, Vol 12: Stem Cells in Development and Disease* (Vol. 1326, pp. 139-157). https://doi.org/10.1007/5584_2021_619
49. de Albuquerque, J. B., Mueller, C., &Gungor, B. (2021). Tissue-Resident T Cells in Chronic Relapsing-Remitting Intestinal Disorders. *Cells, 10*(8), Article 1882. https://doi.org/10.3390/cells10081882
50. De Camilli, P. (2021). How a first research experience had an impact on my scientific journey. *Molecular Biology of the Cell, 32*(22), Article ae1. https://doi.org/10.1091/mbc.E21-08-0397
51. de Torres-Jurado, A., Manzanero-Ortiz, S., &Carmena, A. (2022). Glial-secreted Netrins regulate Robo1/Rac1-Cdc42 signaling threshold levels during Drosophila asymmetric neural stem/progenitor cell division. *Current Biology, 32*(10), 2174-+. https://doi.org/10.1016/j.cub.2022.04.001

52. Domcke, S., & Shendure, J. (2023). A reference cell tree will serve science better than a reference cell atlas. *Cell, 186*(6), 1103-1114. https://doi.org/10.1016/j.cell.2023.02.016

53. Don, F. K., & Kalebic, N. (2022). Forebrain Organoids to Model the Cell Biology of Basal Radial Glia in Neurodevelopmental Disorders and Brain Evolution. *Frontiers in Cell and Developmental Biology*, *10*, Article 917166. https://doi.org/10.3389/fcell.2022.917166

54. Donato, M., Bin, E. P., D'Annunzio, V., &Gelpi, R. J. (2021). Myocardial remote ischemic preconditioning: from cell biology to clinical application. *Molecular and Cellular Biochemistry*, *476*(10), 3857-3867. https://doi.org/10.1007/s11010-021-04192-4

55. Doostmohammadi, A., & Ladoux, B. (2022). Physics of liquid crystals in cell biology. *Trends in Cell Biology*, *32*(2), 140-150. https://doi.org/10.1016/j.tcb.2021.09.012

56. Drozdz, M. M., Doane, A. S., Alkallas, R., Desman, G., Bareja, R., Reilly, M., Bang, J., Yusupova, M., You, J., Eraslan, Z., Wang, J. Z., Verma, A., Aguirre, K., Kane, E., Watson, I. R., Elemento, O., Piskounova, E., Merghoub, T., &Zippin, J. H. (2022). A nuclear cAMP microdomain suppresses tumor growth by Hippo pathway inactivation. *Cell Reports*, *40*(13). https://doi.org/10.1016/j.celrep.2022.111412

57. Du, X. F., Amponsah, A. E., Kong, D. S., He, J. J., Ma, Z. H., Ma, J., & Cui, H. X. (2023). hiPSC-Neural Stem/Progenitor Cell Transplantation Therapy for Spinal Cord Injury. *Current Stem Cell Research & Therapy*, *18*(4), 487-498. https://doi.org/10.2174/1574888x17666220509222520

58. Duncan, M. C. (2022). New directions for the clathrin adaptor AP-1 in cell biology and human disease. *Current Opinion in Cell Biology*, *76*, Article 102079. https://doi.org/10.1016/j.ceb.2022.102079

59. Duncan, M. C. (2023). A practical guide to expert learner skills in the research environment. *Molecular Biology of the Cell*, *34*(5), 1-4. https://doi.org/10.1091/mbc.E22-11-0517

60. Dunkley, S., Scheffler, K., &Mogessie, B. (2022). Cytoskeletal form and function in mammalian oocytes and zygotes. *Current Opinion in Cell Biology*, *75*, Article 102073. https://doi.org/10.1016/j.ceb.2022.02.007

61. Efstathiou, C., & Draviam, V. M. (2021). Electrically tunable lenses - eliminating mechanical axial movements during high-speed 3D live imaging. *Journal of Cell Science*, *134*(16), Article jcs258650. https://

doi.org/10.1242/jcs.258650

62. Erickson, A. G., Kameneva, P., &Adameyko, I. (2023). The transcriptional portraits of the neural crest at the individual cell level. *Seminars in Cell & Developmental Biology*, *138*, 68-80. https://doi.org/10.1016/j.semcdb.2022.02.017
63. Eskandari, E., & Eaves, C. J. (2022). Paradoxical roles of caspase-3 in regulating cell survival, proliferation, and tumorigenesis. *Journal of Cell Biology*, *221*(6), Article e202201159. https://doi.org/10.1083/jcb.202201159
64. Farese, R. V., & Walther, T. C. (2023). Glycerolipid Synthesis and Lipid Droplet Formation in the Endoplasmic Reticulum. *Cold Spring Harbor Perspectives in Biology*, *15*(5), Article a041246. https://doi.org/10.1101/cshperspect.a041246
65. Farzadfard, F., Gharaei, N., Citorik, R. J., & Lu, T. K. (2021). Efficient retroelement-mediated DNA writing in bacteria. *Cell Systems*, *12*(9), 860-+. https://doi.org/10.1016/j.cels.2021.07.001
66. Fauser, J., Brennan, M., Tsygankov, D., &Karginov, A. V. (2021). Methods for assessment of membrane protrusion dynamics. In M. A. Model & I. Levitan (Eds.), *New Methods and Sensors for Membrane and Cell Volume Research* (Vol. 88, pp. 205-234). https://doi.org/10.1016/bs.ctm.2021.09.005
67. Favre-Bulle, I. A., & Scott, E. K. (2022). Optical tweezers across scales in cell biology. *Trends in Cell Biology*, *32*(11), 932-946. https://doi.org/10.1016/j.tcb.2022.05.001
68. Featherston, E. R., & Cotruvo, J. A. (2021). The biochemistry of lanthanide acquisition, trafficking, and utilization. *Biochimica Et Biophysica Acta-Molecular Cell Research*, *1868*(1), Article 118864. https://doi.org/10.1016/j.bbamcr.2020.118864
69. Ferrari, M., Cirisano, F., & Moran, M. C. (2022). Super Liquid-repellent Surfaces and 3D Spheroids Growth. *Frontiers in Bioscience-Landmark*, *27*(5), Article 144. https://doi.org/10.31083/j.fbl2705144
70. Ferrari, S., Valeri, E., Conti, A., Scala, S., Aprile, A., Di Micco, R., Kajaste-Rudnitski, A., Montini, E., Ferrari, G., Aiuti, A., &Naldini, L. (2023). Genetic engineering meets hematopoietic stem cell biology for next-generation gene therapy. *Cell Stem Cell*, *30*(5), 549-570. https://doi.org/10.1016/j.stem.2023.04.014
71. Ford, M. J., Harwalkar, K., Pacis, A. S., Maunsell, H., Wang, Y. C.,

Badescu, D., Teng, K. T., Yamanaka, N., Bouchard, M., Ragoussis, J., & Yamanaka, Y. (2021). Oviduct epithelial cells constitute two developmentally distinct lineages that are spatially separated along the distal-proximal axis. *Cell Reports*, *36*(10), Article 109677. https://doi.org/10.1016/j.celrep.2021.109677

72. Forveille, S., Leduc, M., Sauvat, A., Cerrato, G., Kroemer, G., &Kepp, O. (2021). High throughput screening for autophagy. In O. Kepp& L. Galluzzi (Eds.), *Monitoring Vesicular Trafficking in Cellular Responses to Stress - Pt B* (Vol. 165, pp. 89-101). https://doi.org/10.1016/bs.mcb.2020.12.011
73. Gallusser, B., Maltese, G., Di Caprio, G., Vadakkan, T. J., Sanyal, A., Somerville, E., Sahasrabudhe, M., O'Connor, J., Weigert, M., &Kirchhausen, T. (2022). Deep neural network automated segmentation of cellular structures in volume electron microscopy. *Journal of Cell Biology*, *222*(2), Article e202208005. https://doi.org/10.1083/jcb.202208005
74. Gan, B. Y. (2021). Mitochondrial regulation of ferroptosis. *Journal of Cell Biology*, *220*(9), Article e202105043. https://doi.org/10.1083/jcb.202105043
75. Gan, Y. J., Sha, H. H., Zou, R. R., Xu, M., Zhang, Y., Feng, J. F., & Wu, J. Z. (2022). Research Progress on Mono-ADP-Ribosyltransferases in Human Cell Biology. *Frontiers in Cell and Developmental Biology*, *10*, Article 864101. https://doi.org/10.3389/fcell.2022.864101
76. Garcia-Gutierrez, P., & Garcia-Dominguez, M. (2022). SUMO control of nervous system development. *Seminars in Cell & Developmental Biology*, *132*, 203-212. https://doi.org/10.1016/j.semcdb.2021.11.022
77. Geers, B., Hagenstein, J., Endig, J., Ulrich, H., Fleig, L., Sprezyna, P., Mikulec, J., Heukamp, L., Tiegs, G., & Diehl, L. (2022). The ADP-Ribosylation Factor 4d Restricts Regulatory T-Cell Induction via Control of IL-2 Availability. *Cells*, *11*(17), Article 2639. https://doi.org/10.3390/cells11172639
78. Gelles, J. D., Mohammed, J. N., Chen, Y. Y., Sebastian, T. M., &Chipuk, J. E. (2022). A kinetic fluorescence polarization ligand assay for monitoring BAX early activation. *Cell Reports Methods*, *2*(3). https://doi.org/10.1016/j.crmeth.2022.100174
79. Giannuzzi, F., Maiullari, S., Gesualdo, L., &Sallustio, F. (2023). The Mission of Long Non-Coding RNAs in Human Adult Renal Stem/Progenitor Cells and Renal Diseases. *Cells*, *12*(8), Article 1115. https://

doi.org/10.3390/cells12081115

80. Ginther, D. K. (2022). Reflections on race, ethnicity, and NIH research awards. *Molecular Biology of the Cell*, *33*(1), Article ae1. https://doi.org/10.1091/mbc.E21-08-0403
81. Giolito, M. V., & Plateroti, M. (2022). Thyroid hormone signaling in the intestinal stem cells and their niche. *Cellular and Molecular Life Sciences*, *79*(9), Article 476. https://doi.org/10.1007/s00018-022-04503-y
82. Gordeeva, O., Gordeev, A., &Erokhov, P. (2022). Archetypal Architecture Construction, Patterning, and Scaling Invariance in a 3D Embryoid Body Differentiation Model. *Frontiers in Cell and Developmental Biology*, *10*, Article 852071. https://doi.org/10.3389/fcell.2022.852071
83. Gowrishankar, S., Lyons, L., Rafiq, N. M., Roczniak-Ferguson, A., De Camilli, P., & Ferguson, S. M. (2021). Overlapping roles of JIP3 and JIP4 in promoting axonal transport of lysosomes in human iPSC-derived neurons. *Molecular Biology of the Cell*, *32*(11), 1094-1103. https://doi.org/10.1091/mbc.E20-06-0382
84. Granados-Riveron, J. T., & Aquino-Jarquin, G. (2021). CRISPR/Cas13-Based Approaches for Ultrasensitive and Specific Detection of microRNAs. *Cells*, *10*(7), Article 1655. https://doi.org/10.3390/cells10071655
85. Graziadei, A., & Rappsilber, J. (2022). Leveraging crosslinking mass spectrometry in structural and cell biology. *Structure*, *30*(1), 37-54. https://doi.org/10.1016/j.str.2021.11.007
86. Gu, Y., & Rasmussen, C. G. (2022). Cell biology of primary cell wall synthesis in plants. *Plant Cell*, *34*(1), 103-128. https://doi.org/10.1093/plcell/koab249
87. Gu, Y. L., Zhang, X., Li, Y., Shi, J. K., Cui, H. R., Ren, Y. X., Liu, S. N., Qiao, Y. C., Cheng, Y., & Liu, Y. W. (2023). MiR-204-5p-targeted AP1S2 is necessary for papillary thyroid carcinoma. *Molecular and Cellular Endocrinology*, *574*, Article 111993. https://doi.org/10.1016/j.mce.2023.111993
88. Hammond, G. R. V., Ricci, M. M. C., Weckerly, C. C., & Wills, R. C. (2022). An update on genetically encoded lipid biosensors. *Molecular Biology of the Cell*, *33*(5), Article tp2. https://doi.org/10.1091/mbc.E21-07-0363
89. Han, X., Hu, Z. X., Surya, W., Ma, Q. Q., Zhou, F., Nordenskiold, L.,

Torres, J., Lu, L. Y., & Miao, Y. S. (2023). The intrinsically disordered region of coronins fine-tunes oligomerization and actin polymerization. *Cell Reports*, *42*(6), Article 112594. https://doi.org/10.1016/j.celrep.2023.112594

90. Haraoka, Y., Miyake, M., &Ishitani, T. (2023). Zebrafish imaging reveals hidden oncogenic-normal cell communication during primary tumorigenesis. *Cell Structure and Function*, *48*(1), 113-121. https://doi.org/10.1247/csf.23026
91. Hartmann, J., & Mayor, R. (2023). Self-organized collective cell behaviors as design principles for synthetic developmental biology. *Seminars in Cell & Developmental Biology*, *141*, 63-73. https://doi.org/10.1016/j.semcdb.2022.04.009
92. He, L. G., Qian, X. H., & Cui, Y. X. (2021). Advances in ER-Phagy and Its Diseases Relevance. *Cells*, *10*(9), Article 2328. https://doi.org/10.3390/cells10092328
93. Henderson, J., Havranek, O., Ma, M. C. J., Herman, V., Kupcova, K., Chrbolkova, T., Pacheco-Blanco, M., Wang, Z. Q., Comer, J. M., Zal, T., & Davis, R. E. (2022). Detecting Forster resonance energy transfer in living cells by conventional and spectral flow cytometry. *Cytometry Part A*, *101*(10), 818-834. https://doi.org/10.1002/cyto.a.24472
94. Herdman, M., von Kugelgen, A., Kureisaite-Ciziene, D., Duman, R., El Omari, K., Garman, E. F., Kjaer, A., Kolokouris, D., Lowe, J., Wagner, A., Stansfeld, P. J., & Bharat, T. A. M. (2022). High-resolution mapping of metal ions reveals principles of surface layer assembly in Caulobacter crescentus cells. *Structure*, *30*(2), 215-+. https://doi.org/10.1016/j.str.2021.10.012
95. Hernandez-Gonzalez, M., Larocque, G., & Way, M. (2021). Viral use and subversion of membrane organization and trafficking. *Journal of Cell Science*, *134*(5), Article jcs252676. https://doi.org/10.1242/jcs.252676
96. Herranz-Iturbide, M., Penuelas-Haro, I., Espinosa-Sotelo, R., Bertran, E., &Fabregat, I. (2021). The TGF-beta/NADPH Oxidases Axis in the Regulation of Liver Cell Biology in Health and Disease. *Cells*, *10*(9), Article 2312. https://doi.org/10.3390/cells10092312
97. Hiatt, J., Cavero, D. A., McGregor, M. J., Zheng, W. H., Budzik, J. M., Roth, T. L., Haas, K. M., Wu, D., Rathore, U., Meyer-Franke, A., Bouzidi, M. S., Shifrut, E., Lee, Y., Kumar, V. E., Dang, E. V., Gordon, D. E., Wojcechowskyj, J. A., Hultquist, J. F., Fontaine, K. A.,

. . . Marson, A. (2021). Efficient generation of isogenic primary human myeloid cells using CRISPR-Cas9 ribonucleoproteins. *Cell Reports*, *35*(6), Article 109105. https://doi.org/10.1016/j.celrep.2021.109105

98. Hickey, S. M., Ung, B., Bader, C., Brooks, R., Lazniewska, J., Johnson, I. R. D., Sorvina, A., Logan, J., Martini, C., Moore, C. R., Karageorgos, L., Sweetman, M. J., & Brooks, D. A. (2022). Fluorescence Microscopy-An Outline of Hardware, Biological Handling, and Fluorophore Considerations. *Cells*, *11*(1), Article 35. https://doi.org/10.3390/cells11010035
99. Ho, B., Loll-Krippleber, R., & Brown, G. W. (2022). Yeast goes viral: probing SARS-CoV-2 biology using S. cerevisiae. *Microbial Cell*, *9*(4), 80-83. https://doi.org/10.15698/mic2021.12.766
100. Hu, C., Wang, B. C., Liu, Z. G., Chen, Q. L., Ishikawa, M., Lin, H., Lian, Q. Q., Li, J., Li, J. V., Ma, D. Q., & Gr, E.-I. O.-A. R. (2022). Sevoflurane but not propofol enhances ovarian cancer cell biology through regulating cellular metabolic and signaling mechanisms. *Cell Biology and Toxicology*. https://doi.org/10.1007/s10565-022-09766-6
101. Hu, Q. X., Wu, D., Walker, M., Wang, P., Tian, R., & Wang, W. (2021). Genetically encoded biosensors for evaluating NAD(+)/NADH ratio in cytosolic and mitochondrial compartments. *Cell Reports Methods*, *1*(7), Article 100116. https://doi.org/10.1016/j.crmeth.2021.100116
102. Hudson, W. H., & Wieland, A. (2023). Perspective Technology meets TILs: Deciphering T cell function in the -omics era. *Cancer Cell*, *41*(1), 41-57. https://doi.org/10.1016/j.ccell.2022.09.011
103. Hummert, J., Tashev, S. A., & Herten, D. P. (2021). An update on molecular counting in fluorescence microscopy. *International Journal of Biochemistry & Cell Biology*, *135*, Article 105978. https://doi.org/10.1016/j.biocel.2021.105978
104. Husnik, F., Tashyreva, D., Boscaro, V., George, E. E., Lukes, J., & Keeling, P. J. (2021). Bacterial and archaeal symbioses with protists. *Current Biology*, *31*(13), R862-R877. https://doi.org/10.1016/j.cub.2021.05.049
105. Hynds, R. E., Zacharias, W. J., Nikolic, M. Z., Konigshoff, M., Eickelberg, O., Gosens, R., de Coppi, P., Janes, S. M., Morrisey, E., Clevers, H., Ryan, A. L., Stripp, B. R., Sun, X., Kim, C. F., & Lin, Q. S. (2021). National Heart, Lung, and Blood Institute and Building Respiratory Epithelium and Tissue for Health (BREATH) Consortium Workshop Report Moving Forward in Lung Regeneration. *American*

Journal of Respiratory Cell and Molecular Biology, *65*(1), 22-29. https://doi.org/10.1165/rcmb.2020-0397WS

106. Iachettini, S., Ciccarone, F., Maresca, C., D' Angelo, C., Petti, E., Di Vito, S., Ciriolo, M. R., Zizza, P., &Biroccio, A. (2023). The telomeric protein TERF2/TRF2 impairs HMGB1-driven autophagy. *Autophagy*, *19*(5), 1479-1490. https://doi.org/10.1080/15548627.2022.2138687

107. Jardin, I., Lopez, J. J., Sanchez-Collado, J., Gomez, L. J., Salido, G. M., & Rosado, J. A. (2022). Store-Operated Calcium Entry and Its Implications in Cancer Stem Cells. *Cells*, *11*(8), Article 1332. https://doi.org/10.3390/cells11081332

108. Jaspersen, S. L. (2021). Anatomy of the fungal microtubule organizing center, the spindle pole body. *Current Opinion in Structural Biology*, *66*, 22-31. https://doi.org/10.1016/j.sbi.2020.09.008

109. Jaufmann, J., Franke, F. C., Sperlich, A., Blumendeller, C., Kloos, I., Schneider, B., Sasaki, D., Janssen, K. P., & Beer-Hammer, S. (2021). The emerging and diverse roles of the SLy/SASH1-protein family in health and disease Overview of three multifunctional proteins. *Faseb Journal*, *35*(4), Article e21470. https://doi.org/10.1096/fj.202002495R

110. Jeziorski, J., Brandt, R., Evans, J. H., Campana, W., Kalichman, M., Thompson, E., Goldstein, L., Koch, C., &Muotri, A. R. (2023). Brain organoids, consciousness, ethics and moral status. *Seminars in Cell & Developmental Biology*, *144*, 97-102. https://doi.org/10.1016/j.semcdb.2022.03.020

111. Ji, Y. G., Lotfollahi, M., Wolf, F. A., &Theis, F. J. (2021). Machine learning for perturbational single-cell omics. *Cell Systems*, *12*(6), 522-537. https://doi.org/10.1016/j.cels.2021.05.016

112. Jiang, B., Zhang, J., Zhao, G. H., Liu, M. J., Hu, J. L., Lin, F. R., Wang, J. Y., Zhao, W. T., Ma, H. H., Zhang, C. X., Wu, C. M., Yao, L. M., Liu, Q. F., Chen, X., Cao, Y. T., Zheng, Y., Zhang, C. S., Han, A. D., Lin, D. H., & Li, Q. X. (2022). Filamentous GLS1 promotes ROS-induced apoptosis upon glutamine deprivation via insufficient asparagine synthesis. *Molecular Cell*, *82*(10), 1821-+. https://doi.org/10.1016/j.molcel.2022.03.016

113. Jiang, J., Cheong, K. Y., Falkowski, P. G., & Dai, W. (2021). Integrating on-grid immunogold labeling and cryo-electron tomography to reveal photosystem II structure and spatial distribution in thylakoid membranes. *Journal of Structural Biology*, *213*(3), Article 107746. https://doi.org/10.1016/j.jsb.2021.107746

114. Jinkerson, R. E., Russo, J. A., Newkirk, C. R., Kirk, A. L., Chi, R. J., Martindale, M. Q., Grossman, A. R., Hatta, M., & Xiang, T. T. (2022). Cnidarian-Symbiodiniaceae symbiosis establishment is independent of photosynthesis. *Current Biology*, *32*(11), 2402-+. https://doi.org/10.1016/j.cub.2022.04.021

115. Johnson, M. E., Chen, A., Faeder, J. R., Henning, P., Moraru, II, Meier-Schellersheim, M., Murphy, R. F., Prustel, T., Theriot, J. A., &Uhrmacher, A. M. (2021). Quantifying the roles of space and stochasticity in computer simulations for cell biology and cellular biochemistry. *Molecular Biology of the Cell*, *32*(2), 186-210. https://doi.org/10.1091/mbc.E20-08-0530

116. Jones, C. L., Inguva, A., & Jordan, C. T. (2021). Targeting Energy Metabolism in Cancer Stem Cells: Progress and Challenges in Leukemia and Solid Tumors. *Cell Stem Cell*, *28*(3), 378-393. https://doi.org/10.1016/j.stem.2021.02.013

117. Ju, J., Lee, H. N., Ning, L., Ryu, H., Zhou, X. X., Chun, H., Lee, Y. W., Lee-Richerson, A. I., Jeong, C., Lin, M. Z., &Seong, J. (2022). Optical regulation of endogenous RhoA reveals selection of cellular responses by signal amplitude. *Cell Reports*, *40*(2), Article 111080. https://doi.org/10.1016/j.celrep.2022.111080

118. Karlsson, R., & Draber, P. (2021). Profilin-A master coordinator of actin and microtubule organization in mammalian cells. *Journal of Cellular Physiology*, *236*(10), 7256-7265. https://doi.org/10.1002/jcp.30379

119. Katunin, P., Zhou, J. B., Shehata, O. M., Peden, A. A., Cadby, A., & Nikolaev, A. (2021). An Open-Source Framework for Automated High-Throughput Cell Biology Experiments. *Frontiers in Cell and Developmental Biology*, *9*, Article 697584. https://doi.org/10.3389/fcell.2021.697584

120. Keenan, K. E., Zachman, D. K., &Hirschey, M. D. (2021). Discovering the landscape of protein modifications. *Molecular Cell*, *81*(9), 1868-1878. https://doi.org/10.1016/j.molcel.2021.03.015

121. Khampang, S., Jiamvoraphong, N., Laowtammathron, C., Lorthongpanich, C., Klaihmon, P., Srisook, P., Kang, X., Luanpitpong, S., U-pratya, Y., &Issaragrisil, S. (2023). CRISPR/Cas9 mediated approach to generate YAP-depleted human embryonic stem cell line (MUSIe002-A-1). *Stem Cell Research*, *66*, Article 102990. https://doi.org/10.1016/j.scr.2022.102990

122. Khanna, K., & Villa, E. (2022). Revealing bacterial cell biology using

cryo-electron tomography. *Current Opinion in Structural Biology*, *75*, Article 102419. https://doi.org/10.1016/j.sbi.2022.102419

123. Kikuchi, N., Matsusaka, H., Bai, L. L., Sano, H., Eitsuka, T., Nakagawa, K., Sugano, E., Ozaki, T., Tomita, H., Kiyono, T., & Fukuda, T. (2023). Sheep-derived cell immortalization through the expression of cell cycle regulators and biological characterization using transcriptomes. *Cell Biology International*. https://doi.org/10.1002/cbin.12034
124. Kim, G. R., & Choi, J. M. (2022). Current Understanding of Cytotoxic T Lymphocyte Antigen-4 (CTLA-4) Signaling in T-Cell Biology and Disease Therapy. *Molecules and Cells*, *45*(8), 513-521. <Go to ISI>://WOS:000888494800001
125. Kim, T. M., Lee, R. H., Kim, M. S., Lewis, C. A., & Park, C. (2023). ETV2/ER71, the key factor leading the paths to vascular regeneration and angiogenic reprogramming. *Stem Cell Research & Therapy*, *14*(1), Article 41. https://doi.org/10.1186/s13287-023-03267-x
126. Klingemann, H. (2023). The NK-92 cell line-30 years later: its impact on natural killer cell research and treatment of cancer. *Cytotherapy*, *25*(5), 451-457. https://doi.org/10.1016/j.jcyt.2022.12.003
127. Kroll, K. W., Shah, S. V., Lucar, O. A., Premeaux, T. A., Shikuma, C. M., Corley, M. J., Mosher, M., Woolley, G., Bowler, S., Ndhlovu, L. C., & Reeves, R. K. (2022). Mucosal-homing natural killer cells are associated with aging in persons living with HIV. *Cell Reports Medicine*, *3*(10), Article 100773. https://doi.org/10.1016/j.xcrm.2022.100773
128. Laiman, J., Lin, S. S., & Liu, Y. W. (2023). Dynamins in human diseases: differential requirement of dynamin activity in distinct tissues. *Current Opinion in Cell Biology*, *81*, Article 102174. https://doi.org/10.1016/j.ceb.2023.102174
129. Lamb, M. G., Rangarajan, H. G., Tullius, B. P., & Lee, D. A. (2021). Natural killer cell therapy for hematologic malignancies: successes, challenges, and the future. *Stem Cell Research & Therapy*, *12*(1), Article 211. https://doi.org/10.1186/s13287-021-02277-x
130. Lamptey, J., Czika, A., Aremu, J. O., Pervaz, S., Adu-Gyamfi, E. A., Otoo, A., Li, F. F., Wang, Y. X., & Ding, Y. B. (2021). The role of fascin in carcinogenesis and embryo implantation. *Experimental Cell Research*, *409*(1), Article 112885. https://doi.org/10.1016/j.yexcr.2021.112885
131. Langford, G. M. (2022). Inclusivity in cell biology: cultural identity and the power of authenticity. *Molecular Biology of the Cell*, *33*(14),

Article ae5. https://doi.org/10.1091/mbc.E22-08-0378

132. Lannagan, T. R. M., Jackstadt, R., Leedham, S. J., & Sansom, O. J. (2021). Advances in colon cancer research: *in vitro* and animal models. *Current Opinion in Genetics & Development*, *66*, 50-56. https://doi.org/10.1016/j.gde.2020.12.003
133. Lanz, M. C., Yugandhar, K., Gupta, S., Sanford, E. J., Faca, V. M., Vega, S., Joiner, A. M. N., Fromme, J. C., Yu, H. Y., &Smolka, M. B. (2021). In-depth and 3-dimensional exploration of the budding yeast phosphoproteome. *EMBO reports*, *22*(2), Article e51121. https://doi.org/10.15252/embr.202051121
134. Larson, R. T., & McFarlane, H. E. (2021). Small but Mighty: An Update on Small Molecule Plant Cellulose Biosynthesis Inhibitors. *Plant and Cell Physiology*, *62*(12), 1828-1838. https://doi.org/10.1093/pcp/pcab108
135. Le Goff, E., Martinand-Mari, C., Belkhir, K., Vacelet, J., Nidelet, S., Godefroy, N., &Baghdiguian, S. (2022). Molecular complexity and gene expression controlling cell turnover during a digestive cycle of carnivorous sponge Lycopodina hypogea. *Cell and Tissue Research*, *388*(2), 399-416. https://doi.org/10.1007/s00441-022-03610-3
136. Leao, M. H. C., Costa, M. L., &Mermelstein, C. (2023). Epithelial-to-mesenchymal transition as a learning paradigm of cell biology. *Cell Biology International*, *47*(2), 352-366. https://doi.org/10.1002/cbin.11967
137. Lee, V., & Rompolas, P. (2022). Corneal regeneration: insights in epithelial stem cell heterogeneity and dynamics. *Current Opinion in Genetics & Development*, *77*, Article 101981. https://doi.org/10.1016/j.gde.2022.101981
138. Lefranc, F. (2021). Transient Receptor Potential (TRP) Ion Channels Involved in Malignant Glioma Cell Death and Therapeutic Perspectives. *Frontiers in Cell and Developmental Biology*, *9*, Article 618961. https://doi.org/10.3389/fcell.2021.618961
139. LeGuennec, M., Klena, N., Aeschlimann, G., Hamel, V., &Guichard, P. (2021). Overview of the centriole architecture. *Current Opinion in Structural Biology*, *66*, 58-65. https://doi.org/10.1016/j.sbi.2020.09.015
140. Leidal, A. M., & Debnath, J. (2021). Emerging roles for the autophagy machinery in extracellular vesicle biogenesis and secretion. *FasebBioadvances*, *3*(5), 377-386. https://doi.org/10.1096/fba.2020-00138

141. Lendahl, U. (2022). 100 plus years of stem cell research-20 years of ISSCR. *Stem Cell Reports*, *17*(6), 1248-1267. https://doi.org/10.1016/j.stemcr.2022.04.004

142. Lerit, D. A. (2022). Reflections on mentorship as an early career researcher. *Molecular Biology of the Cell*, *33*(14), Article ae3. https://doi.org/10.1091/mbc.E22-08-0359

143. Lewis, P. L., & Wells, J. M. (2021). Engineering-inspired approaches to study beta-cell function and diabetes. *Stem Cells*, *39*(5), 522-535. https://doi.org/10.1002/stem.3340

144. Li, D., Taylor, D. H., & van Wolfswinkel, J. C. (2021). PIWI-mediated control of tissue-specific transposons is essential for somatic cell differentiation. *Cell Reports*, *37*(1), Article 109776. https://doi.org/10.1016/j.celrep.2021.109776

145. Li, S. Y., Li, Y., Yao, J. N., Chen, B., Song, J. Y., Xue, Q., & Yang, X. N. (2021). Label-free classification of dead and live colonic adenocarcinoma cells based on 2D light scattering and deep learning analysis. *Cytometry Part A*, *99*(11), 1134-1142. https://doi.org/10.1002/cyto.a.24475

146. Li, Y., Yuan, R. Y., Ren, T., Yang, B., Miao, H. J., Liu, L. G., Li, Y. S., Cai, C., Yang, Y., Hu, Y. P., Jiang, C. K., Xu, Q. D., Zhang, Y. J., & Liu, Y. B. (2021). Role of Sciellin in gallbladder cancer proliferation and formation of neutrophil extracellular traps. *Cell Death & Disease*, *12*(1), Article 30. https://doi.org/10.1038/s41419-020-03286-z

147. Li, Y. H., Li, J. B., Zhou, T. C., Pan, G. J., & Huang, K. (2021). Generation of PARP1 gene knockout human embryonic stem cell line using CRISPR/Cas9. *Stem Cell Research*, *53*, Article 102288. https://doi.org/10.1016/j.scr.2021.102288

148. Li, Z. X., Gao, E., Zhou, J. X., Han, W. K., Xu, X. P., & Gao, X. (2023). Applications of deep learning in understanding gene regulation. *Cell Reports Methods*, *3*(1), Article 100384. https://doi.org/10.1016/j.crmeth.2022.100384

149. Liang, R. G., Lin, Y. H., & Zhu, H. (2022). Genetic and Cellular Contributions to Liver Regeneration. *Cold Spring Harbor Perspectives in Biology*, *14*(9), Article a040832. https://doi.org/10.1101/cshperspect.a040832

150. Liang, W. Q., Wu, X. D., Dong, Y. Q., Chen, X. R., Zhou, P., & Xu, F. M. (2021). Mechanical stimuli-mediated modulation of bone cell

function-implications for bone remodeling and angiogenesis. *Cell and Tissue Research*, *386*(3), 445-454. https://doi.org/10.1007/s00441-021-03532-6

151. Liao, X. M., Wu, C. X., Shao, Z. M., Zhang, S. Y., Zou, Y., Wang, K. K., Ha, Y. P., Xing, J. C., Zheng, A. X., Shen, Z. H., Zheng, S. J., Guo, J. L., &Jie, W. (2021). SETD4 in the Proliferation, Migration, Angiogenesis, Myogenic Differentiation and Genomic Methylation of Bone Marrow Mesenchymal Stem Cells. *Stem Cell Reviews and Reports*, *17*(4), 1374-1389. https://doi.org/10.1007/s12015-021-10121-1
152. Liedtke, J., Depelteau, J. S., &Briegel, A. (2022). How advances in cryo-electron tomography have contributed to our current view of bacterial cell biology. *Journal of Structural Biology-X*, *6*, Article 100065. https://doi.org/10.1016/j.yjsbx.2022.100065
153. Lim, K., Donovan, A. P. A., Tang, W., Sun, D. W., He, P., Pett, J. P., Teichmann, S. A., Marioni, J. C., Meyer, K. B., Brand, A. H., & Rawlins, E. L. (2023). Organoid modeling of human fetal lung alveolar development reveals mechanisms of cell fate patterning and neonatal respiratory disease. *Cell Stem Cell*, *30*(1), 20-+. https://doi.org/10.1016/j.stem.2022.11.013
154. Lim, S., Lee, D. E., da Silva, F. M., Koopmans, P. J., Vechetti, I. J., von Walden, F., Greene, N. P., &Murach, K. A. (2023). MicroRNA control of the myogenic cell transcriptome and proteome: the role of miR-16. *American Journal of Physiology-Cell Physiology*, *324*(5), C1101-C1109. https://doi.org/10.1152/ajpcell.00071.2023
155. Liu, Y., Cui, D. X., Pan, Y., Yu, S. H., Zheng, L. W., & Wan, M. (2022). Metabolic-epigenetic nexus in regulation of stem cell fate. *World Journal of Stem Cells*, *14*(7), 490-502. https://doi.org/10.4252/wjsc.v14.i7.490
156. Loconte, V., Singla, J., Li, A. D., Chen, J. H., Ekman, A., McDermott, G., Sali, A., Gros, M. L., White, K. L., &Larabell, C. A. (2022). Soft X-ray tomography to map and quantify organelle interactions at the mesoscale. *Structure*, *30*(4), 510-+. https://doi.org/10.1016/j.str.2022.01.006
157. Lu, H. J., Li, J., Yang, G. D., Yi, C. J., Zhang, D. P., Yu, F. G., & Ma, Z. W. (2022). Circular RNAs in stem cells: from basic research to clinical implications. *Bioscience Reports*, *42*(1), Article Bsr20212510. https://doi.org/10.1042/bsr20212510
158. Lynes, M. D., Carlone, D. L., Townsend, K. L., Breault, D. T., & Tseng,

Y. H. (2022). Telomerase Reverse Transcriptase Expression Marks a Population of Rare Adipose Tissue Stem Cells. *Stem Cells*, *40*(1), 102-111. https://doi.org/10.1093/stmcls/sxab005

159. Ma, C., & Hoffmann, P. R. (2021). Selenoproteins as regulators of T cell proliferation, differentiation, and metabolism. *Seminars in Cell & Developmental Biology*, *115*, 54-61. https://doi.org/10.1016/j.semcdb.2020.11.006
160. Ma, Q. G., Song, C. H., Yin, B., Shi, Y., & Ye, L. (2022). The role of Trithorax family regulating osteogenic and Chondrogenic differentiation in mesenchymal stem cells. *Cell Proliferation*, *55*(5), Article e13233. https://doi.org/10.1111/cpr.13233
161. Ma, W., & Wu, T. (2022). RNA m6A modification in liver biology and its implication in hepatic diseases and carcinogenesis. *American Journal of Physiology-Cell Physiology*, *323*(4), C1190-C1205. https://doi.org/10.1152/ajpcell.00214.2022
162. MacArthur, I. C., & Dawlaty, M. M. (2021). TET Enzymes and 5-Hydroxymethylcytosine in Neural Progenitor Cell Biology and Neurodevelopment. *Frontiers in Cell and Developmental Biology*, *9*, Article 645335. https://doi.org/10.3389/fcell.2021.645335
163. Magalhaes, J., Eira, J., & Liz, M. A. (2021). The role of transthyretin in cell biology: impact on human pathophysiology. *Cellular and Molecular Life Sciences*, *78*(17-18), 6105-6117. https://doi.org/10.1007/s00018-021-03899-3
164. Mahaldashtian, M., Khalili, M. A., Vatanparast, M., Anbari, F., Nabi, A., &Mangoli, E. (2023). The effect of pentoxifylline and calcium ionophore treatment on sperm cell biology in oligoasthenoteratozoospermia samples. *Zygote*, *31*(1), 85-90, Article Pii s0967199422000582. https://doi.org/10.1017/s0967199422000582
165. Maib, H., & Murray, D. H. (2022). A mechanism for exocyst-mediated tethering via Arf6 and PIP5K1C-driven phosphoinositide conversion. *Current Biology*, *32*(13), 2821-+. https://doi.org/10.1016/j.cub.2022.04.089
166. Mangeat, T., Labouesse, S., Allain, M., Negash, A., Martin, E., Guenole, A., Poincloux, R., Estibal, C., Bouissou, A., Cantaloube, S., Vega, E., Li, T., Rouviere, C., Allart, S., Keller, D., Debarnot, V., Wang, X. B., Michaux, G., Pinot, M., . . . Sentenac, A. (2021). Super-resolved live-cell imaging using random illumination microscopy. *Cell Reports Methods*, *1*(1). https://doi.org/10.1016/j.crmeth.2021.100009

167. Manils, J., Marruecos, L., & Soler, C. (2022). Exonucleases: Degrading DNA to Deal with Genome Damage, Cell Death, Inflammation and Cancer. *Cells*, *11*(14), Article 2157. https://doi.org/10.3390/cells11142157

168. Maricchiolo, E., Panfili, E., Pompa, A., De Marchis, F., Bellucci, M., & Pallotta, M. T. (2022). Unconventional Pathways of Protein Secretion: Mammals vs. Plants. *Frontiers in Cell and Developmental Biology*, *10*, Article 895853. https://doi.org/10.3389/fcell.2022.895853

169. Marvin, J. C., Mochida, A., Paredes, J., Vaughn, B., &Andarawis-Puri, N. (2022). Detergent-Free Decellularization Preserves the Mechanical and Biological Integrity of Murine Tendon. *Tissue Engineering Part C-Methods*, *28*(12), 646-655. https://doi.org/10.1089/ten.tec.2022.0135

170. Masur, S. K., Goodenough, U., Kane, C. M., Marincola, E., Zavala, M. E., &Omotade, J. (2021). 50 Years of Women in Cell Biology: Where have we been? Where are we going? *Molecular Biology of the Cell*, *32*(22), Article pe2. https://doi.org/10.1091/mbc.E21-04-0186

171. McCutcheon, J. P. (2021). The Genomics and Cell Biology of Host-Beneficial Intracellular Infections. In R. Lehmann (Ed.), *Annual Review of Cell and Developmental Biology, Vol 37* (Vol. 37, pp. 115-142). https://doi.org/10.1146/annurev-cellbio-120219-024122

172. McFall, T., & Stites, E. C. (2021). Identification of RAS mutant biomarkers for EGFR inhibitor sensitivity using a systems biochemical approach. *Cell Reports*, *37*(11), Article 110096. https://doi.org/10.1016/j.celrep.2021.110096

173. Melo, G. A., Caloba, C., Brum, G., Passos, T. O., Martinez, G. J., & Pereira, R. M. (2022). Epigenetic regulation of T cells by Polycomb group proteins. *Journal of Leukocyte Biology*, *111*(6), 1253-1267. https://doi.org/10.1002/jlb.2ri0122-039r

174. Mendes, P. (2023). Reproducibility and FAIR principles: the case of a segment polarity network model. *Frontiers in Cell and Developmental Biology*, *11*, Article 1201673. https://doi.org/10.3389/fcell.2023.1201673

175. Mierke, C. T. (2023). Physical and biological advances in endothelial cell-based engineered co-culture model systems. *Seminars in Cell & Developmental Biology*, *147*, 58-69. https://doi.org/10.1016/j.semcdb.2023.01.012

176. Migliano, S. M., Wenzel, E. M., &Stenmark, H. (2022). Biophysical

and molecular mechanisms of ESCRT functions, and their implications for disease. *Current Opinion in Cell Biology*, *75*, Article 102062. https://doi.org/10.1016/j.ceb.2022.01.007

177. Montagnoli, T. L., da Silva, J. S., Sudo, S. Z., Santos, A. D., Gomide, G. F., de Sa, M. P. L., & Zapata-Sudo, G. (2021). ROCK Inhibition as Potential Target for Treatment of Pulmonary Hypertension. *Cells*, *10*(7), Article 1648. https://doi.org/10.3390/cells10071648
178. Montgomery, A. B., Chen, S. Y., Wang, Y. D., Gadhvi, G., Mayr, M. G., Cuda, C. M., Dominguez, S., Makinde, H. K. M., Gurra, M. G., Misharin, A. V., Mandelin, A. M., Ruderman, E. M., Thakrar, A., Brar, S., Carns, M., Aren, K., Akbarpour, M., Filer, A., Nayar, S., . . . Perlman, H. (2023). Tissue-resident, extravascular Ly6c-monocytes are critical for inflammation in the synovium. *Cell Reports*, *42*(5), Article 112513. https://doi.org/10.1016/j.celrep.2023.112513
179. Morris, E. V., & Edwards, C. M. (2021). Morphogens and growth factor signalling in the myeloma bone-lining niche. *Cellular and Molecular Life Sciences*, *78*(9), 4085-4093. https://doi.org/10.1007/s00018-021-03767-0
180. Murphy, K. J., Reed, D. A., Trpceski, M., Herrmann, D., & Timpson, P. (2021). Quantifying and visualising the nuances of cellular dynamics *in vivo* using intravital imaging. *Current Opinion in Cell Biology*, *72*, 41-53. https://doi.org/10.1016/j.ceb.2021.04.007
181. Nagle, M. P., Tam, G. S., Maltz, E., Hemminger, Z., & Wollman, R. (2021). Bridging scales: From cell biology to physiology using in situ single-cell technologies. *Cell Systems*, *12*(5), 388-400. https://doi.org/10.1016/j.cels.2021.03.002
182. Nieto, F., Garrido, F., Dinamarca, S., Cebrian, I., & Mayorga, L. S. (2022). Kinetics of antigen cross-presentation assessed experimentally and by a model of the complete endomembrane system. *Cellular Immunology*, *382*, Article 104636. https://doi.org/10.1016/j.cellimm.2022.104636
183. Nyberg, W. A., Ark, J., To, A., Clouden, S., Reeder, G., Muldoon, J. J., Chung, J. Y., Xie, W. H., Allain, V., Steinhart, Z., Chang, C., Talbot, A., Kim, S., Rosales, A., Havlik, L. P., Pimentel, H., Asokan, A., & Eyquem, J. (2023). An evolved AAV variant enables efficient genetic engineering of murine T cells. *Cell*, *186*(2), 446-+. https://doi.org/10.1016/j.cell.2022.12.022
184. Nygaard, R., Yu, J., Kim, J., Ross, D. R., Parisi, G., Clarke, O. B.,

Virshup, D. M., &Mancia, F. (2021). Structural Basis of WLS/Evi-Mediated Wnt Transport and Secretion. *Cell*, *184*(1), 194-+. https://doi.org/10.1016/j.cell.2020.11.038

185. Ohi, M. D., & Kenworthy, A. K. (2022). Emerging Insights into the Molecular Architecture of Caveolin-1. *Journal of Membrane Biology*, *255*(4-5), 375-383. https://doi.org/10.1007/s00232-022-00259-5
186. Okano, Y., Kase, Y., & Okano, H. (2023). A set-theoretic definition of cell types with an algebraic structure on gene regulatory networks and application in annotation of RNA-seq data. *Stem Cell Reports*, *18*(1), 113-130. https://doi.org/10.1016/j.stemcr.2022.10.015
187. Orr, J. C., & Hynds, R. E. (2021). Stem Cell-derived Respiratory Epithelial Cell Cultures as Human Disease Models. *American Journal of Respiratory Cell and Molecular Biology*, *64*(6), 657-668. https://doi.org/10.1165/rcmb.2020-0440TR
188. Ostrop, J., Zwiggelaar, R. T., Pedersen, M. T., Gerbe, F., Bosl, K., Lindholm, H. T., Diez-Sanchez, A., Parmar, N., Radetzki, S., von Kries, J. P., Jay, P., Jensen, K. B., Arrowsmith, C., &Oudhoff, M. J. (2021). A Semi-automated Organoid Screening Method Demonstrates Epigenetic Control of Intestinal Epithelial Differentiation. *Frontiers in Cell and Developmental Biology*, *8*, Article 618552. https://doi.org/10.3389/fcell.2020.618552
189. Ouyang, J. F., Chothani, S., & Rackham, O. J. L. (2023). Deep learning models will shape the future of stem cell research. *Stem Cell Reports*, *18*(1), 6-12. https://doi.org/10.1016/j.stemcr.2022.11.007
190. Ovciarikova, J., Souza, R. O. O., Arrizabalaga, G., &Sheiner, L. (2022). Protein control of membrane and organelle dynamics: Insights from the divergent eukaryote Toxoplasma gondii. *Current Opinion in Cell Biology*, *76*, Article 102085. https://doi.org/10.1016/j.ceb.2022.102085
191. Paliulis, L., Fabig, G., & Muller-Reichert, T. (2023). The X chromosome still has a lot to reveal-revisiting Hermann Henking?s work on firebugs. *Journal of Cell Science*, *136*(4), Article jcs260998. https://doi.org/10.1242/jcs.260998
192. Pandey, A., Shen, C., Feng, S. Y., & Man, S. M. (2021). Cell biology of inflammasome activation. *Trends in Cell Biology*, *31*(11), 924-939. https://doi.org/10.1016/j.tcb.2021.06.010
193. Panicker, N., Ge, P., Dawson, V. L., & Dawson, T. M. (2021). The cell biology of Parkinson's disease. *Journal of Cell Biology*, *220*(4), Article

e202012095. https://doi.org/10.1083/jcb.202012095

194. Panieri, E., Pinho, S. A., Afonso, G. J. M., Oliveira, P. J., Cunha-Oliveira, T., &Saso, L. (2022). NRF2 and Mitochondrial Function in Cancer and Cancer Stem Cells. *Cells*, *11*(15), Article 2401. https://doi.org/10.3390/cells11152401

195. Panstruga, R., Antonin, W., &Lichius, A. (2023). Looking outside the box: a comparative cross-kingdom view on the cell biology of the three major lineages of eukaryotic multicellular life. *Cellular and Molecular Life Sciences*, *80*(8), Article 198. https://doi.org/10.1007/s00018-023-04843-3

196. Park, J. Y., Won, H. Y., DiPalma, D. T., Kim, H. K., Kim, T. H., Li, C., Sato, N., Hong, C. W., Abraham, N., Gress, R. E., & Park, J. H. (2022). *In vivo* availability of the cytokine IL-7 constrains the survival and homeostasis of peripheral iNKT cells. *Cell Reports*, *38*(2), Article 110219. https://doi.org/10.1016/j.celrep.2021.110219

197. Passanha, F. R., Divinagracia, M. L., & LaPointe, V. L. S. (2022). Cadherin-11 Regulates Cell Proliferation via the PDGFR beta-ERK1/2 Signaling Pathway in Human Mesenchymal Stem Cells. *Stem Cells*, *40*(2), 165-174. https://doi.org/10.1093/stmcls/sxab019

198. Patriarca, E. J., Cermola, F., D'Aniello, C., Fico, A., Guardiola, O., De Cesare, D., &Minchiotti, G. (2021). The Multifaceted Roles of Proline in Cell Behavior. *Frontiers in Cell and Developmental Biology*, *9*, Article 728576. https://doi.org/10.3389/fcell.2021.728576

199. Pauletto, E., Eickhoff, N., Padrao, N. A., Blattner, C., & Zwart, W. (2021). TRIMming Down Hormone-Driven Cancers: The Biological Impact of TRIM Proteins on Tumor Development, Progression and Prognostication. *Cells*, *10*(6), Article 1517. https://doi.org/10.3390/cells10061517

200. Pendharkar, G., Lu, Y. T., Chang, C. M., Lu, M. P., Lu, C. H., Chen, C. C., & Liu, C. H. (2021). A Microfluidic Flip-Chip Combining Hydrodynamic Trapping and Gravitational Sedimentation for Cell Pairing and Fusion. *Cells*, *10*(11), Article 2855. https://doi.org/10.3390/cells10112855

201. Peng, Z. L., Wu, J. Y., Hu, S. J., Ma, A. J., Wang, L., Cao, N., Zhang, Y., Li, Q. Y., Yu, J. Y., Meng, S. F., Na, T., Shi, X. L., Li, M., Liu, H. D., Qian, L. G., Tian, E., Lin, F., Cao, J. N., Peng, Y. J., . . . Pan, G. Y. (2022). Requirments for primary human hepatocyte. *Cell Proliferation*, *55*(4), Article e13147. https://doi.org/10.1111/cpr.13147

202. Perera, R. M. (2021). Zooming in on the cell biology of disease. *Molecular Biology of the Cell*, *32*(22), Article ae4. https://doi.org/10.1091/mbc.E21-09-0459

203. Pinson, A., & Huttner, W. B. (2021). Neocortex expansion in development and evolution - from genes to progenitor cell biology. *Current Opinion in Cell Biology*, *73*, 9-18. https://doi.org/10.1016/j.ceb.2021.04.008

204. Postler, T. S. (2021). A most versatile kinase: The catalytic subunit of PKA in T-cell biology. In T. S. Postler & L. Galluzzi (Eds.), *Signal Transduction in Cancer and Immunity* (Vol. 361, pp. 301-318). https://doi.org/10.1016/bs.ircmb.2021.01.005

205. Pratt, E. P. S., Damon, L. J., Anson, K. J., & Palmer, A. E. (2021). Tools and techniques for illuminating the cell biology of zinc. *Biochimica Et Biophysica Acta-Molecular Cell Research*, *1868*(1), Article 118865. https://doi.org/10.1016/j.bbamcr.2020.118865

206. Prazak, V., Grunewald, K., & Kaufmann, R. (2021). Correlative super-resolution fluorescence and electron cryo-microscopy based on cryo-SOFI. In T. MullerReichert& P. Verkade (Eds.), *Correlative Light and Electron Microscopy Iv* (Vol. 162, pp. 253-271). https://doi.org/10.1016/bs.mcb.2020.10.021

207. Pressly, J. D., Gurumani, M. Z., Santos, J. T. V., Fornoni, A., Merscher, S., & Al-Ali, H. (2022). Adaptive and maladaptive roles of lipid droplets in health and disease. *American Journal of Physiology-Cell Physiology*, *322*(3), C468-C481. https://doi.org/10.1152/ajpcell.00239.2021

208. Qazi, T. H., Blatchley, M. R., Davidson, M. D., Yavitt, F. M., Cooke, M. E., Anseth, K. S., & Burdick, J. A. (2022). Programming hydrogels to probe spatiotemporal cell biology. *Cell Stem Cell*, *29*(5), 678-691. https://doi.org/10.1016/j.stem.2022.03.013

209. Quintero-Carmona, O. A. (2021). Inclusion: an individual pursuit requiring organization-level investment to sustain it. *Molecular Biology of the Cell*, *32*(22), Article ae2. https://doi.org/10.1091/mbc.E21-08-0402

210. Radulovic, S., Sunkara, S., Rachel, R., &Leitinger, G. (2022). Three-dimensional SEM, TEM, and STEM for analysis of large-scale biological systems. *Histochemistry and Cell Biology*, *158*(3), 203-211. https://doi.org/10.1007/s00418-022-02117-w

211. Rao, H., Che, J., Yin, C. H., Zhou, Y. F., Ma, Y. F., & Tian, R. L. (2022).

How many authors does it take to publish a high profile or classic paper? *Molecular Biology of the Cell, 33*(12), Article 33:pe6. https://doi.org/10.1091/mbc.E22-04-0140

212. Reiche, M. A., Aaron, J. S., Boehm, U., DeSantis, M. C., Hobson, C. M., Khuon, S., Lee, R. M., & Chew, T. L. (2022). When light meets biology - how the specimen affects quantitative microscopy. *Journal of Cell Science, 135*(6), Article jcs259656. https://doi.org/10.1242/jcs.259656
213. Reuschl, A. K., Mesner, D., Shivkumar, M., Whelan, M. V. X., Pallett, L. J., Guerra-Assuncao, J. A., Madansein, R., Dullabh, K. J., Sigal, A., Thornhill, J. P., Herrera, C., Fidler, S., Noursadeghi, M., Maini, M. K., & Jolly, C. (2022). HIV-1 Vpr drives a tissue residency-like phenotype during selective infection of resting memory T cells. *Cell Reports, 39*(2), Article 110650. https://doi.org/10.1016/j.celrep.2022.110650
214. Ricci, L., & Srivastava, M. (2021). Transgenesis in the acoel worm Hofsteniamiamia. *Developmental Cell, 56*(22), 3160-+. https://doi.org/10.1016/j.devcel.2021.10.012
215. Riley, E. P., Schwarz, C., Derman, A. I., & Lopez-Garrido, J. (2021). Milestones in *Bacillus subtilis* sporulation research. *Microbial Cell, 8*(1), 1-16. https://doi.org/10.15698/mic2021.01.739
216. Robaszkiewicz, K., Jurewicz, E., Moraczewska, J., &Filipek, A. (2021). Ca2+-dependentbinding of S100A6 to cofilin-1 regulates actin filament polymerization-depolymerization dynamics. *Cell Calcium, 99*, Article 102457. https://doi.org/10.1016/j.ceca.2021.102457
217. Roeder, A. H. K., Otegui, M. S., Dixit, R., Anderson, C. T., Faulkner, C., Zhang, Y., Harrison, M. J., Kirchhelle, C., Goshima, G., Coate, J. E., Doyle, J. J., Hamant, O., Sugimoto, K., Dolan, L., Meyer, H., Ehrhardt, D. W., Boudaoud, A., & Messina, C. (2022). Fifteen compelling open questions in plant cell biology. *Plant Cell, 34*(1), 72-102. https://doi.org/10.1093/plcell/koab225
218. Roth, J., & Taatjes, D. J. (2023). Histochemistry and Cell Biology-a glance into the past and a look ahead. *Histochemistry and Cell Biology, 159*(6), 465-475. https://doi.org/10.1007/s00418-023-02195-4
219. Roy, A., Padhi, S. S., Khyriem, I., Nikose, S., Sankar, S. H. H., &Bharathavikru, R. S. (2022). Resetting the epigenome: Methylation dynamics in cancer stem cells. *Frontiers in Cell and Developmental Biology, 10*, Article 909424. https://doi.org/10.3389/fcell.2022.909424

220. Russo, E., Runge, P., Jahromi, N. H., Naboth, H., Landtwing, A., Montecchi, R., Leicht, N., Hunter, M. C., Takai, Y., &Halin, C. (2021). CD112 Regulates Angiogenesis and T Cell Entry into the Spleen. *Cells*, *10*(1), Article 169. https://doi.org/10.3390/cells10010169

221. Ryder, L. S., Cruz-Mireles, N., Molinari, C., Eisermann, I., Eseola, A. B., & Talbot, N. J. (2022). The appressorium at a glance. *Journal of Cell Science*, *135*(14), Article jcs259857. https://doi.org/10.1242/jcs.259857

222. Sagwal, S. K., & Bekeschus, S. (2021). ROS Pleiotropy in Melanoma and Local Therapy with Physical Modalities. *Oxidative Medicine and Cellular Longevity*, *2021*, Article 6816214. https://doi.org/10.1155/2021/6816214

223. Sakamaki, J. I., & Mizushima, N. (2023). Cell biology of protein-lipid conjugation. *Cell Structure and Function*, *48*(1), 99-112. https://doi.org/10.1247/csf.23016

224. Sales-Gil, R., Kommer, D. C., de Castro, I. J., Amin, H. A., Vinciotti, V., Sisu, C., &Vagnarelli, P. (2021). Non-redundant functions of H2AZ1 and H2AZ2 in chromosome segregation and cell cycle progression. *EMBO reports*, *22*(11), Article e52061. https://doi.org/10.15252/embr.202052061

225. Salinas-Luypaert, C., Allu, P. K., Logsdon, G. A., Dawicki-McKenna, J. M., Gambogi, C. W., Fachinetti, D., & Black, B. (2021). Gene replacement strategies validate the use of functional tags on centromeric chromatin and invalidate an essential role for CENP-A(K124ub). *Cell Reports*, *37*(5), Article 109924. https://doi.org/10.1016/j.celrep.2021.109924

226. Schornig, M., & Taverna, E. (2021). A Closer Look to the Evolution of Neurons in Humans and Apes Using Stem-Cell-Derived Model Systems. *Frontiers in Cell and Developmental Biology*, *9*, Article 661113. https://doi.org/10.3389/fcell.2021.661113

227. Scott-Solomon, E., & Hsu, Y. C. (2022). Neurobiology, Stem Cell Biology, and Immunology: An Emerging Triad for Understanding Tissue Homeostasis and Repair. *Annual Review of Cell and Developmental Biology*, *38*, 419-446. https://doi.org/10.1146/annurev-cellbio-120320-032429

228. Scott, J. S., Nassar, Z. D., Swinnen, J. V., & Butler, L. M. (2022). Monounsaturated Fatty Acids: Key Regulators of Cell Viability and Intracellular Signaling in Cancer. *Molecular Cancer Research*, *20*(9),

1354-1364. https://doi.org/10.1158/1541-7786.Mcr-21-1069

229. Sezin, T., Selvakumar, B., &Scheffold, A. (2022). The role of A Disintegrin and Metalloproteinase (ADAM)-10 in T helper cell biology. *Biochimica Et Biophysica Acta-Molecular Cell Research*, *1869*(4), Article 119192. https://doi.org/10.1016/j.bbamcr.2021.119192
230. Shen, L. Y., Sharma, D., Yu, Y. L., Long, F. X., &Karner, C. M. (2021). Biphasic regulation of glutamine consumption by WNT during osteoblast differentiation. *Journal of Cell Science*, *134*(1), Article jcs251645. https://doi.org/10.1242/jcs.251645
231. Shi, L. L., Jian, Y. L., Li, M. J., Hao, T. C., Yang, C. L., & Wang, X. C. (2022). Filamin FLN-2 promotes MVB biogenesis by mediating vesicle docking on the actin cytoskeleton. *Journal of Cell Biology*, *221*(7), Article e202201020. https://doi.org/10.1083/jcb.202201020
232. Shoup, D., & Priola, S. A. (2023). Cell biology of prion strains *in vivo* and in vitro. *Cell and Tissue Research*, *392*(1), 269-283. https://doi.org/10.1007/s00441-021-03572-y
233. Siu, K. K., Serrao, V. H. B., Ziyyat, A., & Lee, J. F. E. (2021). The cell biology of fertilization: Gamete attachment and fusion. *Journal of Cell Biology*, *220*(10), Article e202102146. https://doi.org/10.1083/jcb.202102146
234. Skoniecka, A., Cichorek, M., Tyminska, A., Pelikant-Malecka, I., &Dziewiatkowski, J. (2021). Melanization as unfavorable factor in amelanotic melanoma cell biology. *Protoplasma*, *258*(5), 935-948. https://doi.org/10.1007/s00709-021-01613-5
235. Smith, P., King, O. N. F., Pennington, A., Tun, W., Basham, M., Jones, M. L., Collinson, L. M., Darrow, M. C., & Spiers, H. (2023). Online citizen science with the Zooniverse for analysis of biological volumetric data. *Histochemistry and Cell Biology*. https://doi.org/10.1007/s00418-023-02204-6
236. Smith, S. J., Guillon, E., & Holley, S. A. (2022). The roles of inter-tissue adhesion in development and morphological evolution. *Journal of Cell Science*, *135*(9), Article jcs259579. https://doi.org/10.1242/jcs.259579
237. Smole, A., Benton, A., Poussin, M. A., Eiva, M. A., Mezzanotte, C., Camisa, B., Greco, B., Sharma, P., Minutolo, N. G., Gray, F., Bear, A. S., Baroja, M. L., Cummins, C., Xu, C., Sanvito, F., Goldgewicht, A. L., Blanchard, T., Rodriguez-Garcia, A., Klichinsky, M., . . . Powell, D.

J. P. (2022). Expression of inducible factors reprograms CAR-T cells for enhanced function and safety. *Cancer Cell*, *40*(12), 1470-+. https://doi.org/10.1016/j.ccell.2022.11.006

238. Soon, K., Mourad, O., & Nunes, S. S. (2021). Engineered human cardiac microtissues: The state-of-the-(he)art. *Stem Cells*, *39*(8), 1008-1016. https://doi.org/10.1002/stem.3376
239. Spangler, C. J., Yadav, S. P., Li, D. X., Geil, C. N., Smith, C. B., Wang, G. G., Lee, T. H., & McGinty, R. K. (2022). DOT1L activity in leukemia cells requires interaction with ubiquitylated H2B that promotes productive nucleosome binding. *Cell Reports*, *38*(7), Article 110369. https://doi.org/10.1016/j.celrep.2022.110369
240. Spiers, H., Songhurst, H., Nightingale, L., de Folter, J., Hutchings, R., Peddie, C. J., Weston, A., Strange, A., Hindmarsh, S., Lintott, C., Collinson, L. M., & Jones, M. L. (2021). Deep learning for automatic segmentation of the nuclear envelope in electron microscopy data, trained with volunteer segmentations. *Traffic*, *22*(7), 240-253. https://doi.org/10.1111/tra.12789
241. Stacey, G. N., & Healy, L. (2021). The International Stem Cell Banking Initiative (ISCBI). *Stem Cell Research*, *53*, Article 102265. https://doi.org/10.1016/j.scr.2021.102265
242. Stallaert, W., Kedziora, K. M., Taylor, C. D., Zikry, T. M., Ranek, J. S., Sobon, H. K., Taylor, S. R., Young, C. L., Cook, J. G., & Purvis, J. E. (2022). The structure of the human cell cycle. *Cell Systems*, *13*(3), 230-+. https://doi.org/10.1016/j.cels.2021.10.007
243. Stancliffe, E., & Patti, G. J. (2022). Quick tips for re-using metabolomics data. *Nature Cell Biology*, *24*(11), 1560-1562. https://doi.org/10.1038/s41556-022-01019-2
244. Stowers, R. S. (2023). Advances in Extracellular Matrix-Mimetic Hydrogels to Guide Stem Cell Fate. *Cells Tissues Organs*, *211*(6), 703-720. https://doi.org/10.1159/000514851
245. Stratman, A. N., Crewe, C., & Stahl, P. D. (2022). The microenvironment-a general hypothesis on the homeostatic function of extracellular vesicles. *FasebBioadvances*, *4*(5), 284-297. https://doi.org/10.1096/fba.2021-00155
246. Sudhof, T. C. (2021). The cell biology of synapse formation. *Journal of Cell Biology*, *220*(7), Article e202103052. https://doi.org/10.1083/jcb.202103052

247. Summers, H. D., Wills, J. W., & Rees, P. (2022). Spatial statistics is a comprehensive tool for quantifying cell neighbor relationships and biological processes via tissue image analysis. *Cell Reports Methods*, *2*(11), Article 100348. https://doi.org/10.1016/j.crmeth.2022.100348

248. Sundaram, S., & Chen, C. S. (2022). Next-generation engineered microsystems for cell biology: a systems-level roadmap. *Trends in Cell Biology*, *32*(6), 490-500. https://doi.org/10.1016/j.tcb.2022.01.003

249. Tanaka, T., Warner, B. M., Michael, D. G., Nakamura, H., Odani, T., Yin, H. G., Atsumi, T., Noguchi, M., &Chiorini, J. A. (2022). LAMP3 inhibits autophagy and contributes to cell death by lysosomal membrane permeabilization. *Autophagy*, *18*(7), 1629-1647. https://doi.org/10.1080/15548627.2021.1995150

250. Tang, T., Han, Y. L., Wang, Y. R., Huang, H., & Qian, P. X. (2021). Programmable System of Cas13-Mediated RNA Modification and Its Biological and Biomedical Applications. *Frontiers in Cell and Developmental Biology*, *9*, Article 677587. https://doi.org/10.3389/fcell.2021.677587

251. Thakur, A., Ke, X. S., Chen, Y. W., Motallebnejad, P., Zhang, K., Lian, Q. Z., & Chen, H. J. (2022). The mini player with diverse functions: extracellular vesicles in cell biology, disease, and therapeutics. *Protein & Cell*, *13*(9), 631-654. https://doi.org/10.1007/s13238-021-00863-6

252. Tolic, I. M., & Pavin, N. (2021). Mitotic spindle: lessons from theoretical modeling. *Molecular Biology of the Cell*, *32*(3), 218-222. https://doi.org/10.1091/mbc.E20-05-0335

253. Tolkin, T., & Hubbard, E. J. A. (2021). Germline Stem and Progenitor Cell Aging in *C. elegans*. *Frontiers in Cell and Developmental Biology*, *9*, Article 699671. https://doi.org/10.3389/fcell.2021.699671

254. Tomoda, K., & Kime, C. (2021). Synthetic embryology: Early mammalian embryo modeling systems from cell cultures. *Development Growth & Differentiation*, *63*(2), 116-126. https://doi.org/10.1111/dgd.12713

255. Tuieng, R. J., Cartmell, S. H., Kirwan, C. C., &Sherratt, M. J. (2021). The Effects of Ionising and Non-Ionising Electromagnetic Radiation on Extracellular Matrix Proteins. *Cells*, *10*(11), Article 3041. https://doi.org/10.3390/cells10113041

256. Turner, N. L., Macrina, T., Bae, J. A., Yang, R. Z., Wilson, A. M., Schneider-Mizell, C., Lee, K., Lu, R., Wu, J. P., Bodor, A. L., Bleckert,

A. A., Brittain, D., Froudarakis, E., Dorkenwald, S., Collman, F., Kemnitz, N., Ih, D., Silversmith, W. M., Zung, J., . . . Seung, H. S. (2022). Reconstruction of neocortex: Organelles, compartments, cells, circuits, and activity. *Cell, 185*(6), 1082-+. https://doi.org/10.1016/j.cell.2022.01.023

257. Urmi, A. S., Inaba, H., Nomura, R., Yoshida, S., Ohara, N., Asai, F., Nakano, K., & Matsumoto-Nakano, M. (2021). Roles of Porphyromonasgulae proteases in bacterial and host cell biology. *Cellular Microbiology, 23*(8), Article e13312. https://doi.org/10.1111/cmi.13312

258. Usmani, A., Shavarebi, F., &Hiniker, A. (2021). The Cell Biology of LRRK2 in Parkinson's Disease. *Molecular and Cellular Biology, 41*(5), Article e00660-20. https://doi.org/10.1128/mcb.00660-20

259. Vakonakis, I. (2021). The centriolar cartwheel structure: symmetric, stacked, and polarized. *Current Opinion in Structural Biology, 66*, 1-7. https://doi.org/10.1016/j.sbi.2020.08.007

260. van den Dries, K., Fransen, J., &Cambi, A. (2022). Fluorescence CLEM in biology: historic developments and current super-resolution applications. *Febs Letters, 596*(19), 2486-2496. https://doi.org/10.1002/1873-3468.14421

261. Veerasamy, T., Simon, S. E., & Tan, K. O. (2021). Emerging strategies for sensitization of therapy resistant tumors toward cancer therapeutics by targeting the Bcl-2 family, TGF-ll, Wnt/ll-Catenin, RASSF and miRNA regulated signaling pathways. *International Journal of Biochemistry & Cell Biology, 137*, Article 106016. https://doi.org/10.1016/j.biocel.2021.106016

262. Vega-Lugo, J., da Rocha-Azevedo, B., Dasgupta, A., &Jaqaman, K. (2022). Analysis of conditional colocalization relationships and hierarchies in three-color microscopy images. *Journal of Cell Biology, 221*(7), Article e202106129. https://doi.org/10.1083/jcb.202106129

263. Venkatesan, M., Mohan, H., Ryan, J. R., Schurch, C. M., Nolan, G. P., Frakes, D. H., & Coskun, A. F. (2021). Virtual and augmented reality for biomedical applications. *Cell Reports Medicine, 2*(7), Article 100348. https://doi.org/10.1016/j.xcrm.2021.100348

264. Voisin, A., & Grinberg-Bleyer, Y. (2021). The many-sided contributions of NF-kappa B to T-cell biology in health and disease. In T. S. Postler & L. Galluzzi (Eds.), *Signal Transduction in Cancer and Immunity* (Vol. 361, pp. 245-300). https://doi.org/10.1016/bs.ircmb.2020.10.006

265. Wang, D., Liu, C., Liu, H., Meng, Y. L., Lin, F., Gu, Y. Q., Wang, H. R., Shang, M. Y., Tong, C., Sachinidis, A., Ying, Q. L., Li, L., & Peng, L. Y. (2021). ERG1 plays an essential role in rat cardiomyocyte fate decision by mediating AKT signaling. *Stem Cells*, *39*(4), 443-457. https://doi.org/10.1002/stem.3328

266. Wang, J. J., Sun, H. Y., Jiang, M. M., Li, J. Q., Zhang, P. J., Chen, H. D., Mei, Y. Q., Fei, L. J., Lai, S. J., Han, X. P., Song, X. H., Xu, S. H., Chen, M., Ouyang, H. W., Zhang, D., Yuan, G. C., & Guo, G. J. (2021). Tracing cell-type evolution by cross-species comparison of cell atlases. *Cell Reports*, *34*(9), Article 108803. https://doi.org/10.1016/j.celrep.2021.108803

267. Wang, L., Cao, J. N., Peng, Y. J., Fu, B. Q., Jin, Z. B., Hu, Y. T., Wu, W., Xiang, A. P., Hu, S. J., Yu, J. Y., Zhang, Y., Wei, J., Zhang, Y., Li, Q. Y., Zhou, J. X., Zhai, P. J., Zhu, H. X., Liang, L. M., Ma, A. J., . . . Hao, J. (2022). Human retinal pigment epithelial cells. *Cell Proliferation*, *55*(4), Article e13153. https://doi.org/10.1111/cpr.13153

268. Wang, X., & Chen, X. W. (2023). Cargo Receptor-Mediated ER Export in Lipoprotein Secretion and Lipid Homeostasis. *Cold Spring Harbor Perspectives in Biology*, *15*(6), Article a041260. https://doi.org/10.1101/cshperspect.a041260

269. Wang, X. C., Zhang, D. D., Zheng, C. N., Wu, S. A., Glotzer, M., &Tse, Y. C. (2021). Cortical recruitment of centralspindlin and RhoA effectors during meiosis I of Caenorhabditis elegans primary spermatocytes. *Journal of Cell Science*, *134*(3), Article jcs238543. https://doi.org/10.1242/jcs.238543

270. Wang, Y. F., & Chen, H. R. (2023). Protein glycosylation alterations in hepatocellular carcinoma: function and clinical implications. *Oncogene*, *42*(24), 1970-1979. https://doi.org/10.1038/s41388-023-02702-w

271. Warecki, B., & Sullivan, W. (2022). The Cell Biology of Heterochromatin. *Cells*, *11*(7), Article 1247. https://doi.org/10.3390/cells11071247

272. Watson, J. L., Aich, S., Oller-Salvia, B., Drabek, A. A., Blacklow, S. C., Chin, J., &Derivery, E. (2021). High-efficacy subcellular micropatterning of proteins using fibrinogen anchors. *Journal of Cell Biology*, *220*(2), Article e202009063. https://doi.org/10.1083/jcb.202009063

273. Weigert, A., & Herhaus, L. (2023). Immune modulation through

secretory autophagy. *Journal of Cellular Biochemistry*. https://doi.org/10.1002/jcb.30427

274. Westman, J., & Grinstein, S. (2021). Determinants of Phagosomal pH During Host-Pathogen Interactions. *Frontiers in Cell and Developmental Biology*, *8*, Article 624958. https://doi.org/10.3389/fcell.2020.624958
275. Whiteley, J. T., Fernandes, S., Sharma, A., Mendes, A. P. D., Racha, V., Benassi, S. K., &Marchetto, M. C. (2022). Reaching into the toolbox: Stem cell models to study neuropsychiatric disorders. *Stem Cell Reports*, *17*(2), 187-210. https://doi.org/10.1016/j.stemcr.2021.12.015
276. Wiedner, H. J., & Giudice, J. (2021). It's not just a phase: function and characteristics of RNA-binding proteins in phase separation. *Nature Structural & Molecular Biology*, *28*(6), 465-473. https://doi.org/10.1038/s41594-021-00601-w
277. Wolin, S. L. (2021). Support for a career in science. *Molecular Biology of the Cell*, *32*(22), Article ae6. https://doi.org/10.1091/mbc.E21-08-0399
278. Wong, M. N., Tharaux, P. L., Grahammer, F., &Puelles, V. G. (2021). Parietal epithelial cell dysfunction in crescentic glomerulonephritis. *Cell and Tissue Research*, *385*(2), 345-354. https://doi.org/10.1007/s00441-021-03513-9
279. Wood, B. M., Baena, V., Huang, H., Jorgens, D. M., Terasaki, M., & Kornberg, T. B. (2021). Cytonemes with complex geometries and composition extend into invaginations of target cells. *Journal of Cell Biology*, *220*(5), Article e202101116. https://doi.org/10.1083/jcb.202101116
280. Woods, B. L., & Gladfelter, A. S. (2021). The state of the septin cytoskeleton from assembly to function. *Current Opinion in Cell Biology*, *68*, 105-112. https://doi.org/10.1016/j.ceb.2020.10.007
281. Wu, M., & Liu, J. (2021). Mechanobiology in cortical waves and oscillations. *Current Opinion in Cell Biology*, *68*, 45-54. https://doi.org/10.1016/j.ceb.2020.08.017
282. Xie, S. C., Swaffer, M., &Skotheim, J. M. (2022). Eukaryotic Cell Size Control and Its Relation to Biosynthesis and Senescence. *Annual Review of Cell and Developmental Biology*, *38*, 291-319. https://doi.org/10.1146/annurev-cellbio-120219-040142
283. Xu, D. Z., & Li, C. (2021). Gene 33/Mig6/ERRFI1, an Adapter Protein

with Complex Functions in Cell Biology and Human Diseases. *Cells, 10*(7), Article 1574. https://doi.org/10.3390/cells10071574

284. Xue, Q., Yan, R., Ji, S. T., & Yu, S. (2022). Regulation of mitochondrial network homeostasis by O-GlcNAcylation. *Mitochondrion, 65*, 45-55. https://doi.org/10.1016/j.mito.2022.04.007

285. Yang, X. J., Wang, M. H., Wu, J. Q., Shi, D. C., Zhang, Y. F., Zeng, H. K., Zhu, Y., Lan, C. H., Deng, Y., Guo, S. X., Xu, L. J., Ma, C. Y., Zhang, Y. X., Ou, J. X., Liu, C. J., Chen, Y., Wang, Q. L., Xie, W. X., Guan, J. J., . . . Zhang, Z. H. (2021). Large-scale analysis of 2,152 Ig-seq datasets reveals key features of B cell biology and the antibody repertoire. *Cell Reports, 35*(6), Article 109110. https://doi.org/10.1016/j.celrep.2021.109110

286. Yasuhiko, O., Takeuchi, K., Yamada, H., & Ueda, Y. (2021). Single-shot quantitative phase imaging as an extension of differential interference contrast microscopy. *Genes to Cells, 26*(8), 596-610. https://doi.org/10.1111/gtc.12876

287. Ye, Y. S., Wang, M. W., & Huang, H. (2021). Follicular regulatory T cell biology and its role in immune-mediated diseases. *Journal of Leukocyte Biology, 110*(2), 239-255. https://doi.org/10.1002/jlb.1mr0321-601rr

288. Yla-Anttila, P. (2021). Autophagy receptors as viral targets. *Cellular & Molecular Biology Letters, 26*(1), Article 29. https://doi.org/10.1186/s11658-021-00272-x

289. Yoshimatsu, S., Yamazaki, A., Edamura, K., Koushige, Y., Shibuya, H., Qian, E. M., Sato, T., Okahara, J., Kishi, N., Noce, T., Yamaguchi, Y., & Okano, H. (2022). Step-by-step protocols for non-viral derivation of transgene-free induced pluripotent stem cells from somatic fibroblasts of multiple mammalian species. *Development Growth & Differentiation, 64*(6), 325-341. https://doi.org/10.1111/dgd.12798

290. Yu, M., Lei, W., Cao, J. N., Wang, L., Ma, A. J., Zhao, Z. A., Yang, H. T., Shen, Z. Y., Lan, F., Cao, F., Liang, P., Pei, X. T., Xiang, A. P., Yu, J. Y., Zhang, Y., Zhang, Y., Li, Q. Y., Zhou, J. X., Wei, J., . . . Hu, S. J. (2022). Requirements for human cardiomyocytes. *Cell Proliferation, 55*(4), Article e13150. https://doi.org/10.1111/cpr.13150

291. Yuan, B., Shen, C. Y., Luna, A., Korkut, A., Marks, D. S., Ingraham, J., & Sander, C. (2021). CellBox: Interpretable Machine Learning for Perturbation Biology with Application to the Design of Cancer Combination Therapy. *Cell Systems, 12*(2), 128-+. https://doi.org/10.1016/j.cels.2020.11.013

292. Zhang, X. J., Sridhran, S., Zagoriy, L., Oegema, C. E., Ching, C., Pflaesterer, T., Fung, H. K. H., Becher, I., Poser, I., Muller, C. W., Hyman, A. A., Savilski, M. M., &Mahamid, J. (2023). Molecular mechanisms of stress-induced reactivation in mumps virus condensates. *Cell*, *186*(9), 1877-+. https://doi.org/10.1016/j.cell.2023.03.015

293. Zhang, X. Y., Ma, D., Xuan, B. Q., Shi, D. B., He, J., Yu, M. H., Xiong, H., Ma, Y. R., Shen, C. Q., Guo, F. F., Cao, Y. Y., Yan, Y. Q., Gao, Z. Y., Tong, T. Y., Zhu, X. Q., Fang, J. Y., Chen, H. Y., & Hong, J. (2023). LncRNA CACClnc promotes chemoresistance of colorectal cancer by modulating alternative splicing of RAD51. *Oncogene*, *42*(17), 1374-1391. https://doi.org/10.1038/s41388-023-02657-y

294. Zhang, Y., Wei, J., Cao, J. N., Zhang, K. H., Peng, Y. J., Deng, H. K., Kang, J. H., Pan, G. J., Zhang, Y., Fu, B. Q., Hu, S. J., Na, J., Liu, Y., Wang, L., Liang, L. M., Zhu, H. X., Zhang, Y., Jin, Z. B., Hao, J., . . . Yu, J. Y. (2022). Requirements for human-induced pluripotent stem cells. *Cell Proliferation*, *55*(4), Article e13182. https://doi.org/10.1111/cpr.13182

295. Zhang, Y. F., Chen, Z. Y., Wang, F., Sun, H. H., Zhu, X. L., Ding, J. P., & Zhang, T. L. (2022). Nde1 is a Rab9 effector for loading late endosomes to cytoplasmic dynein motor complex. *Structure*, *30*(3), 386-+. https://doi.org/10.1016/j.str.2021.10.013

296. Zhao, H., Huang, X. Z., Liu, Z. X., Lai, L., Sun, R. L., Shen, R. L., Li, Y., He, L. J., Pu, W. J., Lv, Z., Li, Y., Han, X. M., Liu, X. X., & Zhou, B. (2023). Use of a dual genetic system to decipher exocrine cell fate conversions in the adult pancreas. *Cell Discovery*, *9*(1), Article 1. https://doi.org/10.1038/s41421-022-00485-0

297. Zhao, Y., Panzer, U., Bonn, S., & Krebs, C. F. (2021). Single-cell biology to decode the immune cellular composition of kidney inflammation. *Cell and Tissue Research*, *385*(2), 435-443. https://doi.org/10.1007/s00441-021-03483-y

298. Ziaja, M., Urbanek, K. A., Kowalska, K., &Piastowska-Ciesielska, A. W. (2021). Angiotensin II and Angiotensin Receptors 1 and 2-Multifunctional System in Cells Biology, What Do We Know? *Cells*, *10*(2), Article 381. https://doi.org/10.3390/cells10020381

299. Zidan, A. A., Perkins, G. B., Al-Hawwas, M., Elhossiny, A., Yang, J. Y., Bobrovskaya, L., Mourad, G. M., Zhou, X. F., & Hurtado, P. R. (2021). Urine stem cells are equipped to provide B cell survival signals. *Stem Cells*, *39*(6), 803-818. https://doi.org/10.1002/stem.3351

300. Zuch, D. T., Doyle, S. M., Majda, M., Smith, R. S., Robert, S., & Torii, K. U. (2022). Cell biology of the leaf epidermis: Fate specification, morphogenesis, and coordination. *Plant Cell*, *34*(1), 209-227. https://doi.org/10.1093/plcell/koab250

INDEX

C

R

S

T